KV-372-282

Contents

Structure of Synthetic and Natural Fibers and its Relationship to Properties

Chemical Modifications and Finishing of Fibers

Flame Retardancy

Medical Applications of Fibers

Fibers as Membranes

STRENGTH AND DURABILITY CHARACTERISTICS OF KEVLAR ARAMID FIBER

R. E. WILFONG and J. ZIMMERMAN

*Industrial Fibers Division, Textile Fibers Department,
E.I. du Pont de Nemours & Co., Inc., Wilmington, Delaware 19898*

SYNOPSIS

The characteristics of Kevlar aramid fiber, such as very high strength, modulus, and melting point, make it especially useful in end uses such as rubber reinforcement (e.g., in radial tires), ropes and cables, and protective clothing. Tensile strength in fiber assemblies depends not only on filament strength, but also on a number of configurational factors which affect stress transfer between filaments and angle between the filaments and the yarn axis. The dependence of strength on temperature and time under load is rationalized by the Absolute Rate Theory. Tensile fatigue behavior is predictable in terms of these considerations except where minimum load decreases to zero, in which case there is a modest reduction (about 7%) in effective strength. The moderate compressive strength of Kevlar prevents its use in many structural composites, but hybridization with graphite or glass can offer a desirable balance of properties. Fatigue in tires occurs when cords are put into compression. Under realistic conditions, abrasion is the predominant cause of strength loss which is negligible in radial tire belts or carcasses. In bias tires, use of proper twist multipliers and cord dip-stretching conditions which avoid additional fiber orientation minimizes fatigue strength loss.

INTRODUCTION

The commercial introduction of Kevlar aramid fibers, based on p-linked amide forming intermediates, has made available a highly desirable set of fiber characteristics which have opened new end use opportunities for organic fibers. These characteristics can be summarized as follows:

High melting point (zero strength temperature ∼640°C).
High glass transition temperature ($\geq$375°C).
Relatively high thermal stability and low combustibility.
Low density (1.44–1.45) vs. glass (2.55) or steel (7.86).
Nonconductive and good dielectric properties.
High specific strength (e.g., 5× steel and 2× nylon) and modulus (e.g., 2–4× steel and 10–20× nylon) along with low creep; strength and modulus per unit cross section approaching or equaling steel.
High damping vs. inorganic fibers.
High tensile fatigue and cut resistance.
Good flex resistance and textile processibility.

Journal of Applied Polymer Science: Applied Polymer Symposium 31, 1–21 (1977)
© 1977 by John Wiley & Sons, Inc.

Good stability in many environments such as sea water, oils, solvents; moderate stability in saturated steam in the pH range of 4–8.

Moderate abrasion resistance which is improvable by several approaches.

Good resistance to high-energy radiation but mediocre with respect to ultraviolet radiation, thus requiring some kind of shielding for some critical end uses. This is feasible. Indeed, Kevlar is a strong self-screener and ultraviolet resistance is better for larger assemblies.

Only moderate compressive strength which limits use in structural composites unless hybridized with inorganic fibers, a useful approach in certain applications.

The balance of properties thus offered has resulted in commercial or active development programs in a large variety of end uses, [1] some of which are listed below.

In *tires,* durability, soft ride, and good hazard resistance added to high specific strength and modulus result in an attractive cost/performance balance in radial tire belts and in carcasses for truck and other heavy-duty tires.

For *hose and conveyor belts,* thermal stability vs. rayon and polyester, and fatigue and impact resistance vs. glass and steel has prompted commercial acceptance in such uses as automotive radiator and heater hoses with a useful temperature range up to 150°C and in critical drive belts, such as for snowmobiles.

With respect to *ropes and cables,* the high specific strength of Kevlar can offer greatly increased payloads in long cables used in oceanographic and aerospace applications. The additional features of corrosion resistance, nonconductivity, and improved handleability point to applications in oil rig and buoy mooring, helicopter hoist and anchor pendant lines, and antenna guys.

Protective clothing such as ballistic vests and protective gloves, aprons, helmets, etc., capitalize on the high strength, cut resistance, and melting point of Kevlar.

Filament-wound pressure vessels based on Kevlar 49 composites can take advantage of its high specific strength and creep rupture and tensile fatigue resistance and its high modulus and low creep.

In *coated fabrics,* a favorable balance of strength, weight savings, and flexibility vs. incumbents such as glass, nylon, or polyester have stimulated developments in air support shelters, skirts for air-cushioned vehicles, and specialized conveyor belting. In these uses, where outdoor exposure is encountered, the coating should be formulated to screen most or all of the incident light.

Marine applications include improved stiffness and durability with reduced weight for hulls of boats where speed and durability are of prime importance; Kevlar 49 or combinations of Kevlar 49 with glass are being used in a number of such applications.

For *sporting goods,* stiffness, impact strength, and weight savings have been the driving forces for use of Kevlar, sometimes in combination with glass or graphite fibers. The scarcity of good wood has motivated development of combinations of Kevlar 49 in fabric, roving, or tape form with lower-grade woods to obtain superior performance (e.g., hockey sticks).

Aerospace composites (in addition to filament wound pressure vessels) in which the high specific strength, modulus, and creep rupture performance of Kevlar are not limited by its relatively low compressive strength, are commercial or under development in non- or semistructural end uses in interiors of commercial aircraft, fairings, and cargo compartment liners. The low radar profile of Kevlar, because of its relatively low dielectric constant and dielectric loss, has also been a driving force in some of these uses. Where compressive strength is a problem, hybrids with graphite show an attractive balance of fracture toughness, stiffness and reduced cost with adequate compressive strength [2,3].

These end uses have been described in some detail elsewhere; therefore, in this paper, we will focus on only those strength and durability characteristics of Kevlar which bear importantly on end-use performance.

TENSILE STRENGTH

It is already well known that Kevlar is higher by a factor of at least two than the incumbent commercial industrial fibers in specific tensile strength with a typical value of 21 gpd* for yarn versus 9–10 gpd for nylon, polyester, or glass, and 3.2 gpd for steel. In developing an understanding of various aspects of strength and strength conversion in fiber assemblies, we will consider such aspects as (1) single filament strength; (2) effects of twist (stress transfer between filaments and helix angle); (3) conversion in impregnated strands (importance of stress transfer, effect of resin modulus and strength, and contributions of small amounts of twist for high fiber volume content); (4) the effect of bundle length on strength and how this is related to filament uniformity and affected by twist; (5) effect of temperature and time under load, and the predicted long lifetime for safety factors ≥2 with Kevlar; (6) impact loading, and finally; (7) how the relatively low density of Kevlar versus steel affects their relative "free lengths" in air and in water (length at which cable will break of its own weight).

Typical single filament (2.5-cm length) average tensile strengths for Kevlar fibers are in the range of 24–28 gpd. When a yarn is examined we see a significant dependence of strength on twist multiplier (Fig. 1). At zero twist, the yarn strength (nominal 25-cm length) is well below that of the filaments, reflecting primarily the effects of interfilament strength and elongation uniformity and the stress concentration which occurs when the filaments with lowest break elongation fail. As twist is increased, lateral stress transfer results in a relatively sharp break so that only filaments which break in the ultimate failure region contribute effectively to stress concentration. Counteracting this are the effects of increasing helix angle and nonuniform loading with increasing twist. It turns out that the TM for maximum strength with Kevlar is generally about 1.

When epoxy impregnated strands based on zero twist yarns and relatively low fiber volume content (e.g., 40%) are tested (standard ASTM procedure, D-2343, Method A), strengths comparable to those of average 2.5-cm filament strength (or better) are usually obtained because stress transfer is provided

---

* In the text of this paper, we will use units to which we have been accustomed for many years, some metric, some English, etc. The graphs will generally show the currently accepted metric units where our customary units differ. For convenience, some of the needed conversion factors are listed here: for denier to tex, multiply by 0.111; for g/denier to N/tex, 0.0883; for lb./in.^{2} to Pa (N/m^2), 6.895 × 10^3; for g/denier to lb./in^2, 12,800 × density.

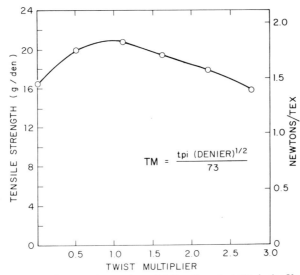

FIG. 1. Tensile strength vs. twist multiplier. Sample: 1500 denier Kevlar.

$$TM = \frac{tpi \; (DENIER)^{1/2}}{73}$$

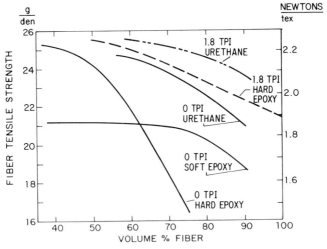

FIG. 2. Fiber tensile strength vs. vol % fiber. Sample: resin-impregnated Kevlar 49, 1420 denier (158 tex).

without the adverse consequences of twist. On the other hand, practical considerations of composite weight push us in the direction of the highest possible fiber volume content; in this case, the fiber tensile strength depends strongly on the strength, modulus, and break elongation of the resin (Fig. 2). With hard epoxy resins having moduli in the range of 600 Mpsi and elongations of about 4%, fiber strength of 1420 denier Kevlar 49 strands is relatively high, about 4 gpd greater than for 1.1 TM yarn at 40 vol % fiber, but falls off rapidly at higher fiber content because of loss of strand integrity under tensile loads. Soft epoxies having 340 psi modulus and 70% elongation are less effective in transferring stress at 40% fiber content but don't show significant losses until beyond 70%

fiber content. A urethane resin with a much lower modulus of 3300 psi and 300% elongation shows best retention of strength at high fiber contents. The use of about 1 TM yarn increases strength considerably for high fiber volume content and reduces sensitivity to resin type. If one calculates the tensile strength of the strand based on its total cross section (including resin) (Fig. 3), then strength for zero twist yarns reaches a maximum at about 60% fiber for the hard epoxy versus 85–90% for the soft epoxy or the urethane and 90–100% for the twisted yarns. On the assumption that 80% is largest fiber content which will permit reasonable strand integrity, there is a strength improvement opportunity for the strand of as much as 50% by optimizing resin and twist or of 40% through resin selection alone.

Similar considerations apply when considering the effect of yarn or cable length on strength (Fig. 4). With zero twist yarn, tensile strength drops fairly

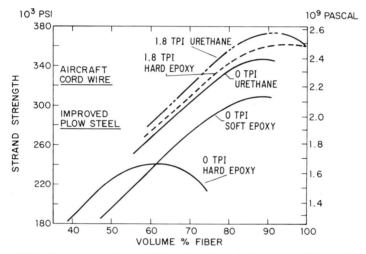

FIG. 3. Strand strength vs. vol % fiber. Sample: resin-impregnated Kevlar 49.

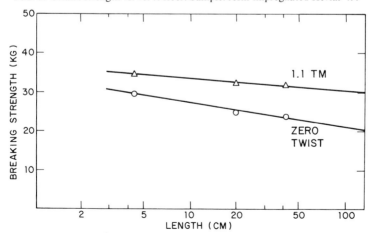

FIG. 4. Breaking strength vs. sample length. Sample: Kevlar 49 1420 denier (158 tex) yarn.

rapidly with length (about 20% per decade over the range shown). On the other hand, at 1.1 TM, the loss in strength is only 3.5–5% per decade. Similar results are obtained with impregnated strands, even at zero twist.

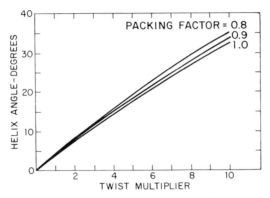

FIG. 5. Theoretical helix angle vs. twist multiplier. Sample: balanced 2-ply cord.

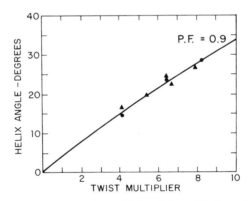

FIG. 6. Experimental helix angle vs. twist multiplier. Balanced 2-ply cord measured at 1.2 gpd cord extension: (●) hot-stretched (measured P.F. = 0.88); (▲) nonstretched.

- DENIER

  $$- \; D_c = \frac{\Sigma D_y}{\mathrm{Cos}\ Q}$$

- TENSILE STRENGTH

  $$- \; T_c = T_y \ \mathrm{Cos}^2\ Q$$

- MODULUS

  $$- \; M_c = M_y \ \mathrm{Cos}^4\ Q$$

- HELIX ANGLE

  $$- \; \mathrm{TAN}\ Q = \frac{0.063\ \mathrm{TM}}{(\mathrm{P.F.})^{1/2}}$$

FIG. 7. Yarn-to-cord conversion (theoretical).

The effects of helix angle become dominating in tire cords where twist multipliers of 6 or greater are frequently used to achieve desired fatigue resistance. Helix angle depends on twist, fiber density, and fiber packing factor. Calculated helix angles for Kevlar tire yarn versus TM for various packing factors are shown in Figure 5. Experimentally determined helix angles, measured under 1.2 gpd tension for 2-ply cords before and after hot-stretching at 1 gpd tension (245°C), are shown in Figure 6. The results are consistent with a packing factor of about 0.90, which is about what is measured. Theoretical equations for the effects of helix angle on cord denier, tensile strength, and modulus [4] are shown in Figure 7. These predict, for example, conversions of 84% for strength and 71% for modulus at 6.5 TM, in reasonably good agreement with results obtained for stretched cords (85–90% strength and 69–73% modulus conversion).

Many end uses involve operation at elevated temperatures; therefore, and for theoretical reasons, the temperature dependence of tensile strength is of interest. The data (Fig. 8) for Kevlar in the range of room temperature to 200°C show about 70% retention at 200°C (about twice that of nylon 66 or polyester) and extrapolate to zero strength at about 640°C. This compares with nylon whose *extrapolated* zero strength temperature determined below the triclinic/hexagonal transition temperature (of about 150°C) is about 390°C; the actual zero strength temperature is its melting point (about 265°C). These relationships provide one of the reasons for our belief that tensile fatigue is a creep process involving slippage of chains in crystallites.

The well-known exponential dependence of time of break on applied stress is also consistent with a creep failure process. This, together with the linear decrease of tensile strength with increasing temperature, can be correlated by the Absolute Rate Theory of Creep Failure [5,6] (Fig. 9). One consequence of these considerations is that one would expect the slope of the log time to break versus stress curve and the extrapolated log time to failure at zero stress (a) to be governed by the extrapolated zero strength temperature (T_0). One of the

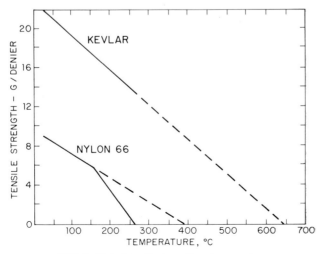

FIG. 8. Plot of tensile strength vs. temperature.

$$\text{LOG } t_b = a - bf$$

$$= \text{LOG}\left(\frac{h}{kT}\right) + \text{LOG}\left(\frac{d_b}{d_1}\right) + \frac{\Delta F^*}{2.3RT} - \frac{fAd_1}{4.6kT\ W_l}$$

LOG $(h/kT) = -12.8$ (AT 25 °C)

d_b/d_1 IS NUMBER OF JUMPS TO FAILURE (e.g. 1–10)

A IS CROSS-SECTIONAL AREA PER POLYMER CHAIN

d_1 IS THE DISTANCE PER JUMP

ΔF^* IS THE FREE ENERGY OF ACTIVATION FOR THE CREEP PROCESS

W_l IS THE FRACTION OF CHAINS BEARING THE APPLIED LOAD

FOR 1 sec. TESTING TIME (LOG $t_b = 0$), AT THE EXTRAPOLATED
ZERO STRENGTH TEMPERATURE (°K), WHERE THE LAST TERM IS 0:

$$\frac{\Delta F^*}{2.3RT_0} = 12.8 \quad (\text{OR } 11.8) \qquad a \sim (12-13)\left(\frac{T_0}{T} - 1\right)$$

FIG. 9. Absolute rate theory of creep failure for a fixed load.

uncertainties in the theory is the number of jumps required to achieve the degree of stress concentration needed for catastrophic acceleration of the failure process. Within a reasonable range of uncertainty (i.e., 1–10 jumps), we can derive the relationship between a and extrapolated zero strength temperature, as shown in Figure 9. Estimates of a are shown in Table I for Kevlar ($T_0 \sim 640°C$), nylon ($T_0 \sim 400°C$), and polyethylene ($T_0 \sim 150°C$).

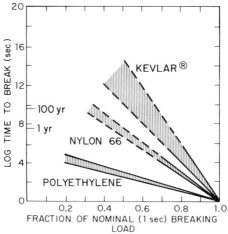

FIG. 10. Plot of tensile strength vs. time under fixed load.

TABLE I

Estimated Zero Stress Intercepts versus Extrapolated Zero Strength Temperature

	a	
T_0(°C)	25°C	120°C
640	25–27	16–17
400	15–16	7–8
150	5–5.5	~1

These estimated values are reasonably good guides to actual behavior as shown in Figure 10, where typical experimental plots of log time to break at 25°C versus load expressed as fraction of the 1-sec breaking load are shown (with some extrapolation) within the range of variability normally encountered. From this graph, we can see that high-melting fibers will, in general, support a larger fraction of their nominal breaking load for long periods of time. Thus, safety factors of 2 should provide very long life-times (e.g., ≥ 100 years) under fixed load for Kevlar as opposed to 3 for nylon (in the absence of other degradation processes such as abrasion, ultraviolet, oxidation, etc.), while polyethylene is evidently incapable of sustaining significant loads for extended periods of time even if its nominal strength were as high as Kevlar.

Impact or shock loading performance of Kevlar can be characterized as follows: (1) breaking strengths at high rates of loading are comparable to or slightly higher than at normal testing rates; (2) energy absorption at failure is about half that of nylon, as one would expect from their relative tensile strengths and break elongations; and (3) repeated shock loading (100,000 times) at 40% of the ultimate tensile strength of a Kevlar "Uniline" cable (data from Wall Rope Works, Beverly, N.J., July 1975) resulted in only 13% strength loss versus failure after 50 megacycles for a steel cable similarly loaded (i.e., to 40% of its ultimate strength).

In long cable applications, the density of the cable has a strong influence on the size of the payload which it can carry. The ultimate measure of this is the "free length," beyond which the cable breaks of its own weight. In water, buoyancy is very helpful for fibers whose density is not too high. Thus, Figure 11 shows calculated values for "free length" for 70 v/o Kevlar 49/urethane strands (density ~1.34; and steel density ~7.80), revealing a 5.8/1 advantage for the Kevlar strand in air for equal psi tensile strength and a 21.5/1 advantage in sea water.

FREE LENGTH = TENSILE STRENGTH / DENSITY

DENSITY: KEVLAR® STRAND ~1.34, STEEL ~7.80, SEA WATER ~1.025

FOR EQUAL TENSILE STRENGTH (UNIT AREA)

$$\frac{\text{F.L. OF KEVLAR}^{®}}{\text{F.L. OF STEEL}} = \begin{array}{l} 5.8 \text{ IN AIR} \\ 21.5 \text{ IN SEA WATER} \end{array}$$

FOR 280 M PSI TENSILE STRENGH

F.L. KEVLAR®	~480,000 FT.	AIR
F.L. STEEL	~83,000 FT.	
F.L. KEVLAR®	~1.9 MM FT.	SEA WATER
F.L. STEEL	95,000 FT.	

FIG. 11. Calculated values for free length for 70 vol % Kevlar 49 strand vs. steel.

TENSILE FATIGUE

The results of cyclic tension fatigue experiments follow the same general trends as do simple creep failure, but with some additional complications (Fig. 12). Experiments at 30 Hz in which the minimum load was kept at 0.1 of the maximum ($R = 0.1$) show the expected exponential dependence on load of cycles (or time) to failure. At zero minimum stress, it appears that the lifetime may drop by about 1 order of magnitude so that in one test at 80% of the nominal breaking stress, the sample still survived at 10^7 cycles but 80% of the filaments were broken. At 60% of the nominal stress there were no broken filaments at 10^7 cycles and no strength loss, so that we would expect very long lifetimes. In these tests, we saw little difference between Kevlar 29 and 49.

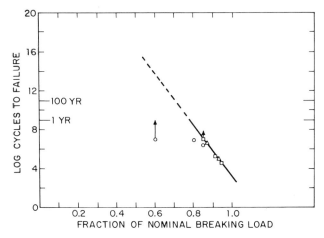

FIG. 12. Cyclic tensile fatigue of Kevlar at 30 Hz: (□) Kevlar 29, R = 0.1; (△) Kevlar 49, R = 0.1; (O) Kevlar 29, R = 0; arrow indicates "runout" at 10^7 cycles.

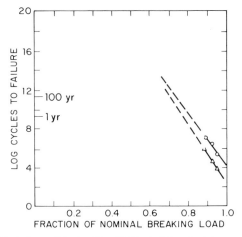

FIG. 13. Cyclic tensile fatigue of Kevlar 49 at 50 Hz: (O) 8 gpd minimum; (△) 0 gpd minimum. Data of Bunsell [9].

TABLE II
Tensile Fatigue

- TENSION–TENSION   (R = 0.1)
    - LITTLE STRENGTH LOSS BEFORE FAILURE

- TENSION–ZERO TENSION
    - REDUCED LIFETIME (1–2 DECADES)
    - EQUIVALENT TO 7% LOWER STRENGTH

- PROBABLY NOT A SIGNIFICANT FACTOR IN DURABILITY

These same kinds of results were reported earlier by Bunsell and co-workers for both Kevlar and other fibers [7–10]. In Figure 13, we have replotted the data of Bunsell [9] for single filaments of Kevlar 49 for 8 gpd and zero minimum load (50 Hz). The results again show a 1–2 decade reduction in cycles to failure at zero minimum load. This amounts to a reduction in effective working strength of at most 7% versus the higher minimum load. In summary (Table II), we conclude that tensile fatigue will not be a significant factor in durability at the kinds of safety factors normally used in actual end use conditions (e.g., 2–10).

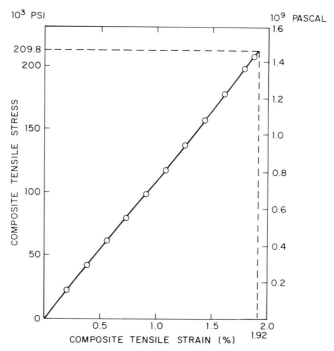

FIG. 14. Tensile stress-strain curve for Kevlar 49/epoxy unidirectional composite. Tensile modulus = 10.8×10^6 psi.

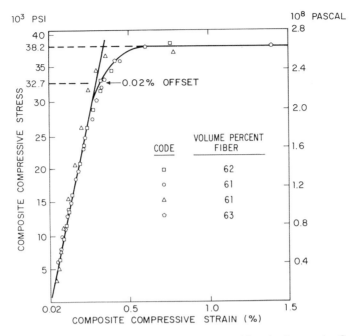

FIG. 15. Compressive stress-strain curve for Kevlar 49/epoxy unidirectional composite. Compressive modulus = 10.5×10^6 psi = 72.4×10^9 Pa.

COMPRESSIVE PROPERTIES

The response of Kevlar to compressive stresses differs significantly from that described above. Initial compressive modulus is the same as that in tension, as shown in the stress–strain curves (Figs. 14 and 15) for unidirectional 60 v/o Kevlar 49/epoxy composites in tension and compression [11]. However, at about 0.3% compressive strain, yielding is initiated with an ultimate compressive failure stress in the composite of 38.2 Mpsi in this example, only 18% of the tensile failure stress. (Values up to 40 Mpsi are frequently seen.) This ultimate stress is reached at about 0.6% compressive strain. The frequently quoted 0.02% offset yield stress is about 33 Mpsi (about 85% of the ultimate). Normalized to 100% Kevlar 49, the ultimate compressive stress is about 62–67 Mpsi (3.3–3.6 gpd). The ultimate compressive stress for the lower-modulus Kevlar 29 is about the same as for Kevlar 49 but is reached at about 0.8% strain. For comparison, unoriented nylon 66 plastic has an ultimate compressive strength of about 15 Mpsi or about 40% of that of a unidirectional Kevlar 49/epoxy composite.

The nature of the failure process in compression probably involves a different structural element from that in tensile failure as indicated (Fig. 16) by an extrapolated zero strength temperature of about 400°C (vs. 640°C for tensile failure). The similarity of this temperature to the estimated T_g suggests that compressive failure may involve buckling of the amorphous regions between Kevlar crystallites. However, much remains to be learned about the mechanism of compressive failure.

The observations are summarized in Table III. Beyond these we should rec-

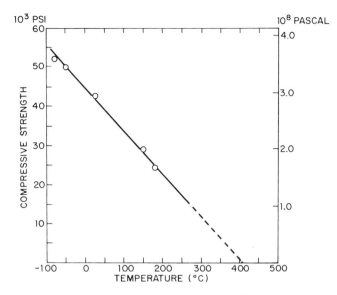

FIG. 16. Ultimate compressive strength vs. temperature for Kevlar 49/epoxy composites.

TABLE III
Compressive Properties of Kevlar 49

- MODULUS
    - EQUAL TO TENSILE MODULUS (BEFORE YIELD)
- STRENGTH
    - 60-68 M PSI (ULTIMATE)
    - ~18% OF TENSILE
    - 38-40 M PSI IN 60 VOL. % COMPOSITES
    - 0.02% OFFSET YIELD STRESS ~85% OF ULTIMATE
    - ZERO STRENGTH TEMPERATURE (EXTRAPOLATED):
        ~400 °C (vs TENSILE TO ~640 °C)
- CRITICAL STRAIN
    - ULTIMATE ~0.6% (0.8% FOR KEVLAR)
    - YIELD   ~0.3%
- KINK BANDS
    - CRITICAL STRAIN FUNCTION OF ORIENTATION
        0.7% FOR KEVLAR® 49 TO 2% FOR KEVLAR 29.
    - PRODUCED BY SNAPBACK FROM 67 M PSI LOAD
        BUT NOT FROM 37 M PSI.

ognize that one obvious manifestation of compressive failure in fibers is formation of kink bands at angles of about 60° from the fiber axis which reflect buckling of fibrils. The critical compressive strain for seeing obvious kink bands is about 0.7% for Kevlar 49, only slightly higher than was indicated in Figure 16 (0.6%), but is 2% for Kevlar 29, surprisingly higher than indicated from its compressive stress–strain curve (0.8%). A possible interpretation of this difference in behavior is that, in the lower-modulus fibers, kink bands are initiated at a higher frequency so that more fiber compressive strain is needed to develop them to the point where they are readily visible.

Another interesting way of generating kink bands is to loop a finer (e.g., 400 denier) yarn bundle around a heavier bundle (e.g., 1500 denier) and increase stress until the finer yarn breaks. It is noteworthy that snap back of the unbroken yarn from a 67 Mpsi (3.6 gpd) load results in kink bands but none are observed when the snap back is from 37 Mpsi (2 gpd). The rapid snap back evidently produces a compressive wave.

This compressive behavior prompted research to determine whether hybrids of Kevlar with inorganic fibers having high-compressive failure stresses would lead to an improved balance of properties. We believe that this is indeed the case (Table IV). In composites with graphite (e.g., Thornel 300), hybrids with Kevlar offer, as compared to all-graphite, lower cost, better impact resistance, noncatastrophic failure and higher damping, and versus all-Kevlar, improved compressive and flex strength and stiffness. In composites with glass, hybrids with Kevlar offer as compared to all-glass, higher stiffness and impact resistance and lower weight, although at higher cost. Further improvements in balance of properties can be achieved in some cases by strategic placement of the fibers so that the Kevlar is placed where stiffness is needed at a reasonable cost, but where compressive stresses are low to moderate, while the graphite or glass is

TABLE IV
Hybrids with Graphite or Glass

KEVLAR 49 HYBRIDS:

- vs GRAPHITE
    - LOWER COST
    - BETTER IMPACT RESISTANCE
    - NON-CATASTROPHIC FAILURE
    - HIGHER DAMPING

- vs KEVLAR 49
    - HIGHER COMPRESSIVE YIELD AND ULTIMATE STRENGTHS
    - HIGHER STIFFNESS

- vs GLASS
    - HIGHER STIFFNESS
    - LOWER WEIGHT
    - BETTER IMPACT RESISTANCE

TABLE V
Applications of Hybrids

KEVLAR 49/GRAPHITE
- GOLF CLUB SHAFTS
- BICYCLE FRAMES
- FISHING RODS
- JET ENGINE FAN BLADES
- HELICOPTER FUSELAGE PANEL

KEVLAR 49/GLASS
- OCEAN RACING YACHTS
- CANOES AND KAYAKS
- PADDLES
- PULTRUDED SHAPES

in the region of higher compressive stress. Some current applications of hybrids are shown in Table V.

Examples of the effects of hybridization of Kevlar 49 with graphite on tensile strength, modulus, impact energy, and flexural strength of 8-harness satin balanced fabric epoxy composites (~70 v/o) are shown in Figures 17 and 18 [2]. Note (1) the almost linear modulus relationship, (2) the small effect on tensile strength as Kevlar 49 content is increased from 0–50%, (3) the rapid increase in impact strength over this same range, and (4) that dynamic flexural

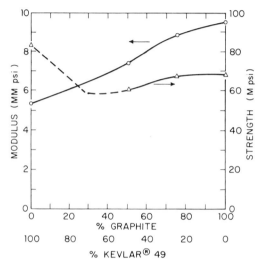

FIG. 17. Effect of Kevlar 49/"Thornel" 300 hybrid composites on tensile strength and modulus of balanced fabrics. Arrows indicate warp direction.

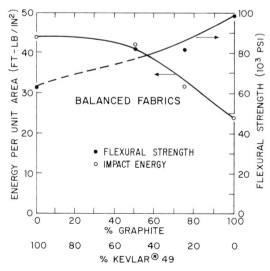

FIG. 18. Absorbed energy and flexural strength vs. graphite content from instrumented Charpy tests of balanced Kevlar 49/"Thornel" 300 hybrid fabric laminates.

strength was decreased by only 20–25%, still 30% higher than all-Kevlar composites at the 50/50 ratio. Based on recent price quotes, a 50/50 (by volume) Kevlar 49/graphite composite appears to offer the best overall balance of cost and properties, especially where impact performance is critical, with a fabric cost about 40% lower than that of an all-graphite fabric (although about four times higher than an all-Kevlar fabric of equal weight) with the mechanical properties as noted above.

FLEX LIFE

One common type of fiber fatigue test which involves compressive fatigue is the single-fiber flex test, which commonly involves flexing a filament around a 0.076-mm wire at some standard applied load (normally 0.6 gpd). As might be expected, flex life depends exponentially on the applied load [6] and on filament thickness. As can be seen from Figure 19, Kevlar at 1.5 denier per filament has similar performance to nylon 66 tire yarn at 6 denier per filament with somewhat lesser stress dependence. An increase in Kevlar filament thickness to 3 denier per filament reduces flex life by a factor of 3–5. Thus, while the intrinsic flex resistance of Kevlar appears to be less than that of nylon 66, proper choice of dpf can provide fully acceptable flex resistance for textile processibility and durability of fiber assemblies to flexing. In fact, we do not believe that sin-

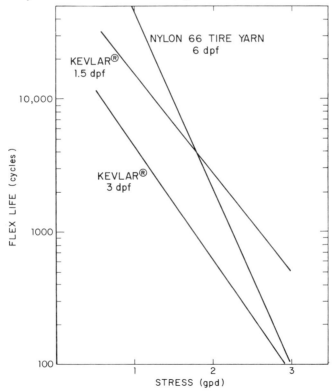

FIG. 19. Single-fiber flex life vs. stress for Kevlar and nylon 66 tire yarn (0.076-mm wire used).

gle-fiber flex is an important mechanism of fatigue in any of the end uses for which Kevlar is a candidate when appropriate constructions are used. Generally, abrasion or some kind of degradative process which decreases polymer molecular weight is involved.

FATIGUE IN TIRES

Our analysis of fatigue strength loss in tires in road tests or in laboratory cord tests which simulate intire stresses has shown the following [12]. (1) There is no strength loss unless the cord goes into compression (in the absence of thermal degradation of molecular weight). Thus, in a 4-ply bias nylon passenger tire, the inner ply, where compression occurs, loses strength while the outer ply does not. (2) At low twist, fiber buckling and true flex failure does occur with nylon, Kevlar polyester, or any tire yarn. However, choice of a proper cord twist multiplier or helix angle provides sufficient compressional capacity to avoid filament buckling and then, at realistic extents of compressions, abrasion becomes the dominant mechanism. This has been shown convincingly for nylon as well as Kevlar tire cords.

The points can be exemplified by the results of investigations involving the use of the so-called "Disk Fatigue Tester" in which cords are embedded in rubber blocks and these are mounted in the periphery of two canted circular disks which, as they rotate, alternately extend and compress the blocks at frequencies appropriate to tire operation at high speeds. Stress is transmitted to the cords by shear through the rubber with maximum cord stress at the center of the block. This stress depends mainly on the degree of extension and the modulus of the

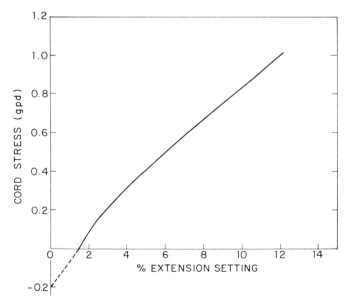

FIG. 20. Values for cord stress vs. percent extension setting for 1500/$\frac{1}{2}$ Kevlar cords (disk fatigue).

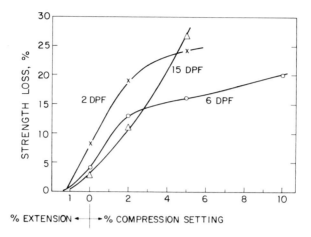

FIG. 21. Disk fatigue strength loss (24 hr) of nylon of various dpf at 110°C ambient temperature (maximum extension 7.5% constant).

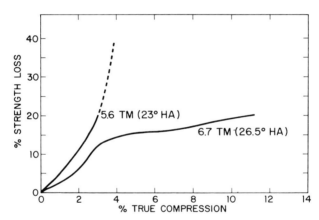

FIG. 22. Disk fatigue strength loss of nylon (6 dpf) vs. percent compression at 130°C.

rubber but only very mildly on cord modulus. In Figure 20, we show typical values for cord stress with Kevlar 6.5 TM cords as a function of percent of extension. Note that the 1.5% extension setting gives zero stress in our apparatus. In Figure 21, we show, for nylon 66 cords at 6.7 TM, the effect of percent compression on disk fatigue strength loss for cords with 2, 6 and 15 denier filaments. These results are consistent with an abrasion strength loss mechanism up to about 3.5% true compression, with buckling and flex fatigue coming in strongly for the 15 denier per filament sample beyond this point. We believe that realistic compressions in the flex zones of bias tires are in the range of 2%. From Figure 22, we can see that a modest reduction in TM from 6.7 to 5.6 (23 vs. 26.5° helix angle) causes a rapid increase in the sensitivity to percent compression, probably because buckling and flex fatigue become more prominent.

Since Kevlar has a higher density than nylon (1.44 vs. 1.14), it requires (Fig. 23) a slightly higher TM for a given helix angle (e.g., 7.5 vs. 6.7 TM for 26.5°). At these constructions it can be seen that, at the lower compressions and at equal

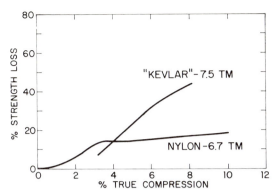

FIG. 23. Disk fatigue strength loss of "Kevlar" (1.5 dpf) vs. nylon (6 dpf) (26.5° helical angle).

tensile stress, Kevlar and nylon are comparable and give relatively low strength loss. At compressions beyond 4%, which are probably beyond the actual use range, strength loss by flexing begins to occur for Kevlar.

To describe the fatigue response of Kevlar more fully, we see (Fig. 24) that,

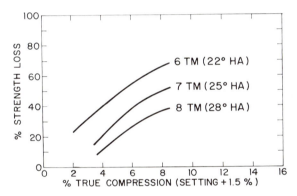

FIG. 24. Disk fatigue strength loss vs. compression for "Kevlar" tire cord at 6% extension setting (0.5 gpd).

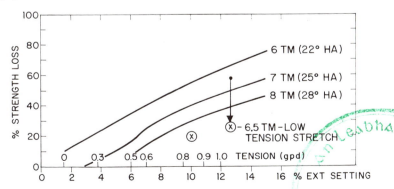

FIG. 25. Disk fatigue strength loss vs. extension for setting "Kevlar" tire cord: 2% compression setting, 3.5% true compression.

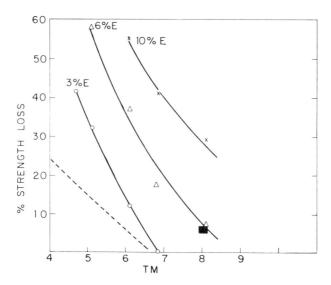

FIG. 26. Disk fatigue strength loss vs. twist multiplier. Sample, Kevlar 1500/1/2 cord at 2% compression setting; (- - -) fleet test radial tire belts with ca. 0.4 lb. of belt fiber (20,00 miles); (■) 4-ply bias tires, 21% defl. wheel test.

TABLE VI
Summary of Kevlar Fatigue Strength Loss in Rubber

• MECHANISM SIMILAR TO NYLON

    – ABRASION AT MODERATE COMPRESSIONS

    – FLEX + ABRASION AT HIGHER COMPRESSIONS

    – POSSIBLE ROLE OF "SNAP-BACK" FATIGUE BEING DEFINED

• UNDER NORMAL COMPRESSIONS (≤4%) AND AT EQUAL CORD HELIX ANGLE AND STRESS, NYLON AND KEVLAR DIFFER LITTLE IN STRENGTH LOSS

    – KEVLAR MORE SENSITIVE AT VERY HIGH COMPRESSIONS

    – FOR EQUAL HELIX ANGLE, KEVLAR REQUIRES 13% HIGHER TM THAN NYLON

• KEVLAR: STRENGTH LOSS CAN BE MINIMIZED BY THE FOLLOWING:

    – USE OF LOW STRETCHING TENSIONS (e.g. 0.2/0.1 vs 1.0/0.3 gpd) CAN REDUCE STRENGTH LOSS BY ABOUT 30%(ABSOLUTE)

    – STRENGTH LOSS IS REDUCED BY ABOUT 14% ABSOLUTE FOR EACH UNIT INCREASE IN TM

    – AVOIDANCE OF PRECOMPRESSION OR WAVY PLIES IN TIRE BUILDING

for a given tension (0.5 gpd in this case), strength loss increases about linearly with compression, but that strength loss at a given compression decreases about 14% per unit of TM. Similarly (Fig. 25), for a given compression (3.5% in this case), strength loss increases about linearly with percent extension. Figure 25 also shows the significant effect of cord-processing tensions on fatigue resistance. A standard set of cord dip-stretching conditions involves two ovens, each at about

250°C, the first of which cures an epoxy subcoat under 1 gpd tension, and the second cures an RFL (resorcinol-formaldehyde/latex) topcoat under 0.3 gpd tension. These stretching conditions result in some further orientation of Kevlar tire yarn, as evidenced by an increase in fiber modulus of about 20%. When this orientation is eliminated by keeping tensions low (e.g., 0.1–0.2 gpd in each oven), fatigue performance is improved substantially, but at some reduction in cord modulus (e.g., to 200 gpd from 350–400 gpd). This kind of cord-stretching process is preferred for fatigue sensitive uses (e.g., bias tires) where the lower modulus is still more than adequate.

The results of disk fatigue testing are compared with several tire test results (Fig. 26). These confirm that in radial tire belts of Kevlar, under realistic operating conditions, very low (but finite) degrees of compression are encountered (estimated at 1% or less). To produce such strength losses in laboratory wheel tests with radial tires, cornering forces must be introduced. Note that at the 6.5 TM multiplier recommended for Kevlar belts, strength loss was only 1% in 32,000 km. In 4-ply bias passenger tires, run in laboratory wheel tests at exaggerated deflections, the strength loss of about 7% with the eight TM cords recommended for bias tires is consistent with a true compression of no more than 2%. In this particular test normal rather than low cord-stretching tensions were used.

The conclusions on fatigue strength loss in rubber for Kevlar are summarized in Table VI.

This paper represents the research of a number of scientists in the Industrial Fibers Division. In particular, the authors wish to acknowledge the contributions of Drs. P. G. Riewald, D. L. G. Sturgeon, W. C. Uy, and C. H. Zweben.

REFERENCES

[1] D. L. G. Sturgeon, paper presented at the 1975 International Conference on Composite Materials, Geneva, Switzerland and Boston, Mass., April 7–11, 14–18, 1975.
[2] C. Zweben, paper presented at the 1975 International Conference on Composite Materials, Geneva, Switzerland and Boston, Mass., April 7–11, 14–18, 1975.
[3] P. G. Riewald and C. Zweben, paper presented at the 30th Annual Conference of the SPI Reinforced Plastics/Composites Institute, Washington, D.C., Feb. 6, 1975.
[4] J. W. S. Hearle, P. Grasberg, and S. Backer, Structural Mechanics of Fibers, Yarns & Fabrics, Wiley-Interscience, New York, 1969.
[5] B. D. Coleman and A. G. Knox, Text. Res. J., 27, 393 (1957).
[6] J. Zimmerman, Text. Mfr., 101, 19 (1974).
[7] A. R. Bunsell and J. W. S. Hearle, J. Mater. Sci., 6, 1303 (1971).
[8] A. R. Bunsell and J. W. S. Hearle, J. Appl. Polym. Sci., 18, 267 (1974).
[9] A. R. Bunsell, J. Mater. Sci., 10, 1300 (1975).
[10] L. Konopasek and J. W. S. Hearle, private communication.
[11] E. I. du Pont de Nemours & Co., Inc., Kevlar 49 Bulletin.
[12] J. Zimmerman, Text. Mfr., 101, 49 (1974).

SYNTHESIS AND PROPERTIES OF NEW REINFORCING AGENTS AND SUPERCONDUCTING FILAMENTS

JAMES ECONOMY

IBM Research Laboratory, San Jose, California 95193

SYNOPSIS

Work on high-strength reinforcing agents is briefly reviewed and new areas of need noted. Three new reinforcing agents including high-modulus boron nitride (BN) fibers, aluminum diboride (AlB_2) flakes, and boron carbide (B_4C) fibers are described which satisfy many of these needs. Preparation of a high-modulus BN fiber which combines good dielectric properties, high thermal stability, and good strength is described. Recent work on synthesis and properties of single-crystal flakes of AlB_2 is considered, with emphasis on their unique planar reinforcing characteristics and resistance to mechanical shock. Progress on improving the strength of B_4C fiber is described along with efforts to develop a high-temperature composite for use at 1200°C.

Problems and goals for superconducting filaments are summarized. A program to develop a continuous multifilament yarn of niobium carbonitride (NbCN) is discussed. The NbCN filaments display a high T_c, acceptable mechanical properties, and inherent stability to flux jumps.

INTRODUCTION

Work on developing new refractory and metallic fibers during the past 15 years has been directed in two distinct areas, namely, for reinforcement of high-performance composites and for use as superconductors. In the early 1960s, considerable impetus to composite development was provided by the aerospace industries because of their need for structural materials which combined high strength and stiffness with light weight and good thermal resistance. As suitable fibers became commercially available in the late 1960s, emphasis shifted away from work on new reinforcing agents and more toward composite development and demonstration of prototype systems. However, at present a number of new needs have emerged for improved reinforcing agents with good dielectric properties, stability at temperatures in excess of 1000°C, high strength to modulus ratios, and ability to reinforce in the off axis direction [1]. In the case of superconducting filaments, research began to intensify in the early 1970s with the hope of developing ductile systems which could operate at temperatures of 15–20°C and would be inherently stable to flux jumps. Considerable progress has been achieved to date; however, no system has been devised which combines all these attributes.

It is the purpose of this paper to describe recent progress on developing new

Journal of Applied Polymer Science: Applied Polymer Symposium 31, 23–35 (1977)

reinforcing agents which could satisfy the needs enumerated above. Also, current problems with superconducting filaments will be briefly reviewed and a new system described which solves several of these difficulties. More specifically, in the section on reinforcing agents, synthesis and properties of BN fibers [2], AlB_2 flakes [3], and B_4C fibers [4] will be described; while in the section on superconducting filaments, work on the NbCN multifilament [5] yarn will be presented. Much of the data on these systems has been reported elsewhere and is therefore only highlighted here. New work and especially interpretations of the data are considered in greater detail in this paper.

DISCUSSION

Background on Reinforcing Agents

During the past decade, three distinct reinforcing fibers have reached the stage of commercial development, namely, boron, graphite, and Aramid fibers. The properties of these fibers are summarized in Table I. Because of their high cost, the boron fibers find use in only selected areas, while the graphite and Aramid fibers find a broader applicability because of their much lower prices and availability in the form of multifilament yarns. All three filaments are used almost exclusively in organic matrices, and practically all of the prototype development has proceeded in the direction of organic composites.

As one looks to the near future, there are several obvious improvements that one would desire, such as lower price and higher strength to modulus ratios. There are also more specific needs which have emerged and cannot be filled by the presently available reinforcing fibers. Thus, there is a requirement for a high-temperature fiber with suitable dielectric properties for reinforcement of electromagnetic windows. There is a need for better alternatives to achieving off-axis or planar reinforcement other than through use of cross-plied laminates or fabrics. Finally, the growing need for more efficient turbine engines which must operate at 1200–1300°C requires structural materials which possess the necessary directional strength at such temperatures.

TABLE I
Properties of Reinforcing Fibers

Fiber Type	Density	Strength	Modulus
	g/cc	x 10^3 p.s.i.	x 10^6 p.s.i.
Boron (5.6 mils)	2.45	515	56
Carbon (Thornel 400)	1.78	425	34
Aramid (Kevlar 49)	1.45	400	20.

BN Fibers

A program to develop a high-modulus reinforcing fiber based on boron nitride had its roots in the earlier work on stress orientation of carbon fibers to yield the high-strength–modulus graphite fibers. Because of the structural similarity between boron nitride and graphite, it seemed reasonable that high-modulus boron nitride fibers could be prepared by a similar hot-stretching process. The preparation of a low-modulus BN fiber [6] had been achieved some years earlier, and it seemed that stress orientation of such a fiber might indeed produce the desired mechanical properties. Unfortunately, the temperature necessary for such a process was shown to be in excess of 2500°C, a temperature at which either carbon or tungsten furnace tubes react with the boron nitride fiber. A more encouraging route was devised by carrying out stress graphitization of a partially nitrided fiber at temperatures of 1400–2000°C. As a result of these preliminary successes, a larger effort was initiated in the early 1970s jointly funded by the Carborundum Co. and AFML at Wright Patterson to develop such a product. The primary impetus to this effort was the need for a high-temperature radome with good mechanical properties, and BN fiber was the only known material which possessed the required dielectric properties and thermal resistance for this application [7]. It was also anticipated that availability of the highly refractory BN fibers would provide for the first time a general purpose fiber for reinforcement of resins and metals.

The process for preparing the high-modulus BN fibers consisted of melt-drawing a multifilament yarn of boric oxide (B_2O_3) in a manner similar to that used in the glass fiber industry. In contrast to fiberglass, B_2O_3 can be melt-drawn at temperatures of only 600–700°C. The drawing process is relatively insensitive to minor temperature fluctuations because of the relatively small change in melt viscosity with temperature.

The reaction of the B_2O_3 yarn with ammonia is carried out by controlled heating over the temperature range of 200–800°C. The fiber diameter is critical in this process, since an internal porosity tends to form during nitriding when the diameter exceeds 10 μ. The internal porosity results from the 30% weight loss in going from B_2O_3 to BN and from the tendency of the nitrided surface to retain the original dimensions of the B_2O_3 filament. Most of the work reported here has been carried out on 6 μ filaments, a range in which internal porosity cannot be detected. Because of the excellent melt-drawing characteristics of B_2O_3 it is possible to also draw a 3–4 μ fiber and this is being emphasized in order to facilitate a continuous nitriding process and to achieve higher strength to modulus ratios.

The optimum fiber for hot-stretching should contain about 48–52% nitrogen (theoretical for BN is 56.7%). The structure of the partially nitrided fiber can be best characterized as a highly turbostratic material with crystallite dimensions of the order of 15–30 Å and an interlayer spacing of about 3.7 Å. These dimensions bear a marked similarity to those of turbostratic carbon formed by heating organic chars to about 800°C. Treatment of the nitrided fiber at a temperature of 1200–1400°C results in a disproportionation reaction in which the poorly formed BN layers coalesce into larger and more ordered crystallites and the B—O and N—H bonds in the grain boundary form $B_2O_3 + N_2 + H_2$. By carrying out this reaction under tension at a temperature of 2000°C, the conversion can be completed in several seconds. The crystallites are oriented parallel to the direction of stress, and the trace amounts of B_2O_3 formed during disproportionation will vaporize at 1550°C. Fibers with tensile strengths in excess of 300,000 psi and modulus values of 30–40 $\times$ 10^6 psi can be prepared in this fashion. Densities of the fibers are in the range of 1.9 g/cc. X-ray diffraction studies indicate that the L_a and L_c dimensions are 140 and 100 Å, respectively, while the interlayer spacing is in the range of 3.38 Å. Most likely the BN crystallites are in the form of extended ribbons similar to what is observed with carbon. Hence the L_a dimension is very likely two orders of magnitude greater. Comparison of the orientation half-widths for oriented carbon and BN fibers indicate that the BN structure has about a 35% lower modulus than carbon for the same degree of orientation.

One feature of the stress graphitized BN fiber is the much smaller crystallite size at the surface of the fibers compared to that of the core. Most likely this difference arises during the nitriding step, since the fiber surface would be expected to display a slightly higher degree of nitriding compared to the core. Hence the surface will not crystallize and orient as easily during thermal stressing at 2000°C. This condition is quite different to that observed in carbon fibers, where the surface appears more highly graphitized and oriented and the core far less ordered. The highly oriented carbon surface results in problems of bonding to the resin matrix, and the surface is routinely etched to enhance bonding. Tokarsky and Diefendorf [8] have noted that the change in anisotropy from surface to core in carbon fibers tends to place the surface in axial and circumferential compression and radial tension. These stresses are sufficient to result in cracks and flaws, which very likely represent the limiting factor in the strength properties of carbon fibers. They estimate that a flaw free carbon fiber could exhibit tensile strength of 1.5 $\times$ 10^6 psi. In the case of BN fiber, it would seem that the surface might display better bonding characteristics to a resin matrix. However, the more crystalline core would tend to place the surface in axial and circumferential tension, leading to a far greater sensitivity to formation of cracks. One approach to minimize this problem is to go to very fine filaments of 3 μ diameter. Presumably, differences in degree of nitriding would be reduced because of the significantly smaller distance necessary for diffusion of ammonia into the fiber. As noted earlier, this possibility is now being tested.

AlB_2 Flakes

A program to develop a planar reinforcing material based on single-crystal flakes had its roots in the mid-1960s in a project to prepare and characterize the various Al–B phases formed during crystallization from an aluminum–boron melt [9]. In these studies, trace quantities of single-crystal flakes of AlB_2 were isolated along with a number of other single crystals of Al–B and Al–B–C. Preliminary experiments with small test bars showed that these flakes did indeed provide planar reinforcement with flexural modulus values significantly higher

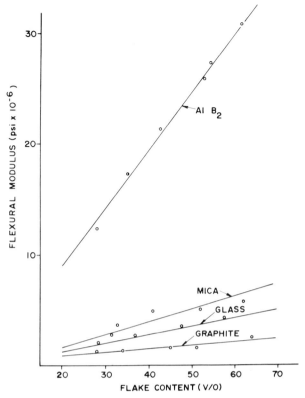

FIG. 1.   Flexural modulus of flake composites (ERLA 4617).

than those achievable with cross-plied boron filaments [10]. As a basis for
comparison, composites were also prepared from flakes of SiC, graphite, mica,
and glass; however, the flexural modulus of such composites was very low, either
because of the low aspect ratios (width/thickness), inherently low modulus, or
presence of lips and curvature in the flakes. The effect of volume loading with
several different flakes on modulus is compared in Figure 1. As can be seen, the
AlB_2 flakes provide an excellent planar reinforcing potential compared to the
other flakes. It appears, in fact, that the AlB_2 flakes may be unique since no other
inherently high-modulus material is known which crystallizes in the form of a
planar crystal with high aspect ratio. However, the flexural strength of the
composites measured only 50,000–56,000 psi, even though bending measurements suggested potential strengths for the flakes of 1×10^6 psi. As shown in
Figure 2, the cause of the low flexural strengths can be traced directly to the
presence in the composite of very blocky single crystals which act as points of
stress concentration. The program to further develop AlB_2 flake composites was
discontinued in the late 1960s because of the inability to achieve high composite
strength and also because of problems in preparing the flakes.

A more concerted effort to develop AlB_2 composites was initiated in the early
1970s with support from the AFML laboratory [11]. A careful analysis of the
various crystals formed along with AlB_2 in the aluminum melt had shown the

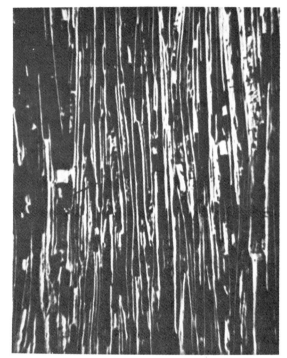

FIG. 2. Cross section showing high concentration of crystalline impurities.

presence of $MgAlB_{14}$, αAlB_{12}, γAlB_{12}, $C_2Al_3B_{48}$, and C_4AlB_{24}. These by-products were significantly reduced by controlling the conditions of crystallization and eliminating the presence of trace amounts of carbon and magnesium. These approaches, coupled with an extensive mechanical separation process involving screening and elutriation, permitted a major reduction of the blocky crystalline impurities. Screening also permitted a separation of flakes into various aspect ratios, permitting a much better insight into the role of aspect ratio on mechanical properties. It was found that the rate of cooling during crystallization also played an important role in determining the size and aspect ratio of the flakes.

Another problem that was encountered involved the actual isolation of the flakes from the aluminum melt. The yield of flakes typically is of the order of 1%, and all the aluminum must be dissolved in order to isolate the 1% yield of flakes. To reduce the excessive losses of aluminum, a filtration procedure was devised to filter the aluminum melt in the temperature range of 750°C in order to concentrate the flakes and permit recovery of most of the aluminum for reuse. By this method, the flakes were concentrated to a level of 20–30%. A cross section of a filtered flake/aluminum solid is shown in Figure 3 prior to dissolution of the aluminum. The fact that the flakes can be concentrated in this fashion demonstrated that flakes were not formed via eutectic solidification, but rather by an actual crystallization from the melt.

A series of composites were prepared using flakes with different aspect ratios

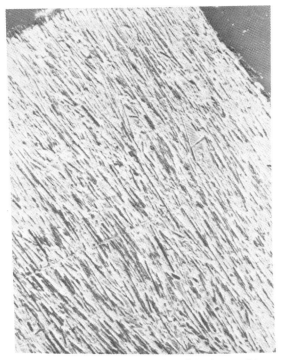

FIG. 3. Cross section of filtered flake/aluminum solid.

(all the values were adjusted to 60 V/o flakes) [3]. It was found that with low aspect ratios of 70/1, the flexural strengths of the composites were in the range of 30,000–40,000 psi. The flexural strengths increased to 100,000–110,000 psi or by a factor of 3 using flakes with aspect ratios of 350/1. The flexural modulus of the composites also increased from 20×10^6 psi (aspect ratio 70/1) to 35×10^6 psi (aspect ratio 350/1). From these results, it can be seen that aspect ratio plays an important role in determining the degree of orientation of flakes and consequently the properties of the composite. As one might expect, low aspect-ratio flakes are not as readily oriented and the degree of misorientation results in the drop in modulus value. Good planarity is apparently achieved at an aspect ratio of 180/1 since the composite modulus values have plateaued at their maximum of 30–35×10^6 psi. The drop in flexural strength with aspect ratio can also be attributed at least partially to the misorientation of flakes. However, at an aspect ratio of 180/1 the flexural strength is only 65,000–80,000 psi, indicating that the flexural strength properties are either more sensitive to the effect of aspect ratios or that the lower aspect-ratio flakes may tend to carry along greater amounts of impurities from the elutriation steps. In fact, the presence of blocky crystal impurities can be detected in practically all cross sections examined, demonstrating that the problem of eliminating impurities has not been solved completely. In spite of this, the composite properties obtained with high aspect-ratio flakes (see Table II) appear to combine the highest planar stiffness with good strength properties compared to any fiber-reinforced system.

TABLE II
Mechanical Properties of AlB_2 Flake Composites[a]

PROPERTY	VALUE
FLEXURAL STRENGTH KSI	109
FLEXURAL MODULUS MSI	35.8
TENSILE STRENGTH KSI	67
TENSILE MODULUS MSI	44
COMPRESSION STRENGTH KSI	66.5
COMPRESSION MODULUS MSI	40
INTERLAMINAR SHEAR PSI	8000

[a] 60 V/o flakes in epoxy; aspect ratio ~325.

The flake-reinforced composites show other characteristics not common to fiber-reinforced systems. For example, AlB_2 flake composites can be cut and machined with diamond tools without appreciably affecting properties. Essentially, no loss in tensile or flexural strengths was observed when $1/8$-in. holes were drilled in a $1/2$-in. test bar using a conventional high-speed drill. Fiber-reinforced composites generally are quite sensitive to any kind of cutting or drilling and considerable delamination is observed. The greater resistance of the flake composite toward delamination appears to be related to the discontinuous nature of the reinforcement. The strength properties of flake composites also appeared to be reasonably insensitive to the presence of voids in the range of 4–12%, although in the 0–4% range the composite appeared to be somewhat more sensitive.

In summary, high aspect-ratio flakes yield composites with unique strength and stiffness properties. Ternary composites consisting of fiber and flake laminates can be fabricated to achieve high directional strength and stiffness combined with significant transverse mechanical properties. There also appears to be a good potential for further enhancement in strength of flake composites by use of higher aspect-ratio flakes and particularly by more effective removal of blocky crystalline impurities. Techniques to reduce or eliminate such impurities will depend on a high-temperature filtration prior to crystallization of the flakes.

B_4C Fiber

Development of B_4C fibers was first reported in the late 1960s as part of an AFML sponsored study [12]; however, work was subsequently discontinued because of the relatively low average strength of the filaments ~70,000–100,000 psi. In the early 1970s, a major effort was initiated to develop ceramic turbine engines that could operate at temperatures of 1200°C since such turbines would permit a far more efficient generation of electrical power. The only candidate materials for this program were SiC and Si_3N_4. Both materials retain their properties at 1200°C; however, methods to design around the inherent brittleness of ceramics have not been forthcoming. Furthermore, ceramic parts that appear identical tend to fail at widely differing loads because of a statistical distribution of imperfections from which cracks can propagate.

A program to improve the mechanical properties of the B_4C fibers was therefore reinitiated with the hope that such fibers could provide the necessary reinforcement for ceramic composites. Presumably, the fiber-reinforced ceramics would have good strength in the direction of principal stress and hopefully would display resistance to crack propagation and to high-temperature creep. It was assumed that the B_4C fibers would show good retention of properties at 1200°C since tests with hot-pressed B_4C specimens had shown that the material retained its strength up to 1600°C. Furthermore, the polycrystalline structure appeared to be very stable at these elevated temperatures and no problems with excessive grain growth were anticipated.

In the study to improve the strength properties of the B_4C fiber, it was found that the strength-limiting feature was related to the formation of crimp during the preparation of the continuous multifilament yarn [4]. The preparation which depends on the reaction of a precursor carbon yarn with $BCl_3 + H_2$ at 1800°C requires that four boron atoms occupy the volume originally occupied by one carbon atom. To accommodate these additional atoms, the fiber tends to expand in diameter and to increase in length; this results in the formation of the crimp. Attempts to eliminate the crimp by heating at elevated temperatures were unsuccessful since the crimped filament yarn had already been extensively damaged in collecting the filaments on a spool. To avoid this problem, a furnace was designed to apply a modest tension at elevated temperatures immediately after the formation of the B_4C fiber. With this type of system, it was possible to establish the appropriate temperature and tension necessary to eliminate the crimp. It was found that at a temperature of 2120°C and a tension of 3000 psi, the crimp was eliminated and filaments with average strengths of 365,000 psi could be obtained (high values of 425,000). In this process, the precursor carbon fibers were converted to about 65% B_4C and hence still retained a carbon core. Development of fibers with even higher tensile strengths, therefore, appears to depend on optimizing the degree of conversion to yield a yarn which can be stressed at slightly higher tension than 3000 psi, and also on minimizing any stresses associated with differences in thermal expansion coefficient of the carbon core and B_4C surface.

The fact that fiber prepared at 2120°C under slight tension displayed good mechanical properties indicates that the crystal structure of B_4C is stable to this temperature; i.e., excessive grain growth did not occur. On the other hand, the carbon core appeared to recrystallize into a graphite phase, and at longer exposures of several hours the carbon core actually dissolves and recrystallizes as discrete small graphite particles dispersed throughout the fiber. One mechanism for achieving this transformation depends on formation of a B/C solid solution with a stoichiometry of less than 4/1. It has, in fact, been postulated that the boron carbide structure can accommodate a much higher percentage of carbon since the icosahedral unit theoretically can contain up to six carbon atoms [13].

Yet it is surprising that such compositions are not observed considering that the fibers are essentially quenched to room temperature after preparation. More likely the dispersed graphite arises from a diffusion and recrystallization of the carbon core with an elimination of the stresses at the B_4C–carbon interface of the fiber as a driving force.

Some experiments were directed toward developing high-temperature composites of B_4C fibers using a carbon matrix. Carbon was selected as the matrix because of its excellent high-temperature stability and compatibility with B_4C. It was found that the fiber–matrix bonding was unusually good since interlaminar shear strengths in excess of 7000 psi were observed. In contrast, composites of carbon fibers in a carbon matrix display interlaminar shear strengths of only 1000–1500 psi. The flexural strength and modulus of the composite at fiber loadings of 50 V/o were 70,000 psi and 23×10^6 psi, respectively. Hence there is considerable room for improvement in strength properties using the higher strength B_4C fibers. The composite had a density of 1.8 g/cc, almost half the values for SiC and Si_3N_4 ceramics. This is of particular significance for turbine blades where strength to weight ratio is critical.

The surface of the composite was protected against oxidation by applying a CVD coating of SiC. The thermal expansion coefficient of the B_4C fiber/carbon composite and of the SiC coating are almost identical, and no microcracks could be detected during thermal cycling up to 1300°C. There was a tendency for the composite to show a small weight pickup in the range of 0.2–0.5% on heating in air to 1000–1300°C. This increase was attributed to the formation of a thin film of SiO_2 which is the mechanism by which SiC is protected against oxidation.

In summary, an approach has been defined for developing structural high-temperature materials for use in a turbine engine operating at 1300°C. Techniques for fabricating complex components of an engine already exist based on advances in prototype development with high performance composites. There is room for considerable improvement in fiber strength and especially in the properties of the composite. Possibilities of oxidative attack during use at elevated temperatures exist especially if pin holes form during the coating operation. Elimination of this problem will depend on the use of thicker SiC coatings or possible infiltration of the carbon matrix with silicon to form SiC as the actual matrix.

NbCN Superconductors

Work on the NbCN superconducting multifilament yarn was undertaken with the intention of preparing a superconducting filament with a high T_c using the technology successfully applied earlier to the preparation of BN and B_4C fibers [14]. Before considering the work on preparing the NbCN filaments, it is instructive to review the problems and present goals for superconducting fibers.

By definition, a superconductor is a material which at low temperatures conducts current with essentially zero resistance. This phenomenon arises when

TABLE III
High T_c Superconducting Materials

MATERIAL	TYPE	T_C	$H_{c2}(4.2°)$
NbTi	A-2	10.2	120
V$_3$Ga	A-15	16.5	220
NbCN	B-1	17.8	130
Nb$_3$Sn	A-15	18.3	220
Nb$_3$Al	A-15	18.9	320
Nb$_3$Ga	A-15	20.3	340
Nb$_3$Ge	A-15	23.2	370

free electrons couple to form electron pairs referred to as Cooper pairs. The significance of superconductivity can be seen from the fact that a coil of superconducting wire generates a magnetic field 10–50 times stronger than conventional magnets or that a superconducting line can transmit 30 times more electricity than standard power transmission cables. However, the costs of refrigeration are too high for systems operating at 4.2°K, and this problem has deterred commercial development of the field. It has been estimated that an increase in operating temperature from 4.2–15°K would reduce refrigeration coats by a factor of four and a further increase to 40°K would cut costs by an additional factor of $2\frac{1}{2}$. Hence emphasis has been placed on developing materials with the highest possible T_c (see Table III).

There are, however, a number of problems [15] with these high T_c superconducting materials. For example, the A-15 type compounds typically are very brittle and difficult to form into strong ductile filaments, and the T_c is very sensitive to slight changes either in stoichiometry or in the long-range crystal order. Another problem with the superconductors indicated in Table III is the tendency for magnetic flux lines to penetrate the superconductor at relatively low fields. These lines become surrounded by whirlpools of super current-creating fluxoids. Movement of these fluxoids can generate heat causing the material to go normal. The tendency for flux jumps would be sharply reduced if the diameter of the superconducting filament were sufficiently small in the range of a few microns.

With the above thoughts in mind, development of a very fine continuous multifilament yarn of NbCN would be very advantageous. The chemistry for preparing NbCN filaments appeared amenable to chemical conversion of a precursor carbon fiber, and the use of fine-diameter filament would act to minimize problems of flux jumps as well as reducing the time for the chemical conversion process. The superconducting properties of NbCN were not sensitive to slight changes in either stoichiometry or long-range order and the material was not considered highly brittle. Hopefully, with a T_c of 17.8°K the NbCN system could be operated at a temperature of 14–15°K, thus permitting a significant reduction in refrigeration costs. On the other hand the H_{c2} value was only one-half to one-third the value of the other systems indicated in Table III, and this shortcoming would obviously limit usage to areas where high fields were not critical.

Fine-diameter carbon filaments were prepared by carbonizing fiber under tension so that the diameter decreased from 10 μ to 5–6 μ. The carbon yarn was converted to NbCN by reaction with a gas stream consisting of $NbCl_5$ + H_2 + N_2 [16]. The reaction was carried out at 1500°C in a tube furnace to which the carbon yarn was fed continuously (contact time ~3 min). Initially, the $NbCl_5$ is reduced to Nb metal on the fiber surface and this reacts to form NbC. The carbon in the core appears fairly mobile and can diffuse through the NbC to react further with the Nb which is depositing on the surface. As the NbC layer increases in thickness, the surface layer of Nb metal can also react with the nitrogen in the gas stream to form NbN. The NbCN then forms as a solid solution between the NbC and NbN. The NbCN composition with the highest T_c is in the range of C ~ 15% and N ~ 85%; hence, the conversion must be driven in the direction of higher nitrogen content. In fact, it is found that the ratio of C/N is strongly dependent on the degree of conversion. For example, the ratio of C/N is 1/1 at a weight increase of 200% and 1/2 at a weight increase of 400%. At a weight increase of 800%, the ratio of C/N is in the desired range of 1/5. This corresponds to approximately a 60% conversion of the carbon fiber to NbCN. The higher degree of conversion is also desirable in order to maximize the current-carrying capacity of the filament. The overall reaction is complicated somewhat by the greater stability of the NbC phase versus the NbN and NbCN. Techniques for enhancing formation of NbCN through use of NH_3 in the gas stream or by quenching to minimize reconversion to NbC have been described earlier [5].

A 720 end multifilament NbCN yarn prepared in the manner described above displayed T_c and H_{c2} properties as indicated in Table III and appeared reasonably flexible. The tensile strengths of the individual filaments were about 70,000 psi; however, no attempts were made to optimize this property by stretching or use of surface finishes to minimize damage arising from contact between fibers. Tests carried out at the AFML laboratory on fine-diameter NbCN filaments showed them to be inherently stable to flux jumps in a magnetic field [17].

The critical current value J_c of the NbCN superconductor was much lower than values observed with the A-15 type systems. It was found that the J_c values could be increased significantly by incorporating trace amounts of silicon (60 ppm). The Si creates points of impurities and dislocations and thus reduces the field dependence of J_c. Presumably, the potential for further improvement in J_c is good as the role of Si and other impurities is clarified.

REFERENCES

[1] J. Economy, Paper presented at the 7th National SAMPE Conference, Albuquerque, New Mexico, Oct. 14, 1975.
[2] R. Y. Lin, J. Economy, R. Ohnsorg, and H. N. Murty, *Appl. Polym. Symp. No. 29*, 175 (1976).
[3] L. C. Wohrer, A. Wosilait, and J. Economy, *SAMPE, 18*, 340 (1973).
[4] J. Economy, W. D. Smith, and R. Y. Lin, *Appl. Polym. Symp. No. 29*, 105 (1976).
[5] W. D. Smith, R. Y. Lin, and J. Economy, *Appl. Polym. Symp. No. 29*, 83 (1976).
[6] J. Economy and R. V. Anderson, *Text. Res. J., 36* (11), 994 (1966); in High Temperature

Resistant Fibers (*J. Polym. C, 19*), A. H. Frazer, Ed., Interscience, New York, 1967, p. 283, J. Economy, R. V. Anderson, and V. I. Matkovich, *Appl. Polym. Symp. No. 9*, 377 (1969).

[7]   R. Y. Lin, J. Economy, and H. D. Batha, *Amer. Ceram. Soc. Bull.*, in press.

[8]   E. W. Tokarsky and R. J. Diefendorf, *Div. Petr. Chem. Preprints, 20* (2), 444 (1975).

[9]   V. I. Matkovich, J. Economy, and R. F. Giese, Jr., *J. Amer. Chem. Soc., 86*, 2337 (1964).

[10] J. Economy, L. C. Wohrer, and V. I. Matkovich, *SAMPE J., 5* (1), 21 (1969).

[11] J. Economy, L. C. Wohrer, and A. Wosilait, AFML Tr-72-74, Part I, June 1972; AFML TR 72-74, Part II, Oct. 1973.

[12] D. J. Beernsten, W. D. Smith, and J. Economy, *Appl. Polym. Symp. No. 9*, 365 (1969).

[13] J. Economy, V. I. Matkovich, and R. F. Giese, Jr., *Z. Kristallogr., 122* (3/4), 248 (1965).

[14] J. Economy, W. D. Smith, and R. Y. Lin, *Appl. Polym. Symp. No. 21*, 131 (1973).

[15] E. Gregory, *Appl. Polym. Symp. No. 29*, 1 (1976).

[16] W. D. Smith, R. Y. Lin, J. A. Coppola, and J. Economy, *IEEE Trans. Magn., 11* (2), 182 (1974).

[17] M. C. Ohmer and W. G. Frederick, *J. Appl. Phys., 45*, 1382 (1974).

PROPERTIES AND USES OF POLYBLEND FIBERS. I. GENERAL REVIEW

S. P. HERSH

School of Textiles, North Carolina State University, Raleigh, North Carolina, 27607

SYNOPSIS

Blends of polymers have long been used to improve the impact resistance and processibility of rigid polymers such as polystyrene and poly(vinyl chloride) and to increase the strength of elastomers. The preparation and study of fibers produced from blends of polymers and their introduction into the commercial marketplace is a much more recent development. A general review of the classes of polyblend polymers and fibers is presented together with a description of techniques used to distinguish between compatible and incompatible blends. A brief description of the composition, phase relations, chief characteristics, and properties of some polyblend fibers is given followed by a discussion of the effects of blend morphology, composition, and compatibility on the mechanical and thermal properties of polyblends. Finally, a number of polyblend fibers (also known as biconstituent, matrix-fibril, or matrix fibers) that have reached the large-scale development stage and/or are available commercially are examined.

INTRODUCTION

Historically, polymer scientists have followed the traditional pattern of beginning their studies using "simple," homogeneous and amorphous systems whenever possible. Indeed, much of the early progress and understanding of polymer behavior was achieved from studies made on such materials. It is becoming more and more evident, however, that most polymers that are considered to be chemically homogeneous become physically heterogeneous when considered on a submicroscopic or even microscopic scale. An example is the presence of a blend of crystalline and amorphous regions in most crystallizable polymers. This heterogeneity can also include different crystalline forms and morphologies such as spherulitic, extended chains, folded chains, etc. In fact, Frank [1] has suggested that it can be useful to consider even chemically homogeneous polymers to be intrinsically composite materials of this type and that such heterogeneity is indeed necessary to produce useful materials.

Although the views of Frank may represent an extreme, there is no doubt that blending of polymers has been used for many years to produce commercial plastics, rubbers, and fibers having useful properties that cannot be obtained as economically with single polymers, and in some cases, cannot be achieved at all.

Journal of Applied Polymer Science: Applied Polymer Symposium 31, 37–53 (1977)
© 1977 by John Wiley & Sons, Inc.

Blended polymers, usually classified as "polyblends," are homogeneous or heterogeneous mixtures of structurally different homo- or copolymers [2]. They must be distinguished from other related composite polymeric systems such as block and graft copolymers, which in themselves are not blends but form solid arrays which physically resemble mixtures, crosslinked polymers and polymers plasticized with monomeric or oligomeric compounds or to which rigid, mostly inorganic, fillers such as silica, salt, clay, aluminum powder, etc., have been added. Recently, Sperling [3] has classified polymers made up of two kinds of mers into 33 different categories. According to Sperling, the distinction between grafts and blends is based on whether the two polymers are joined by chemical bonds (graft or block copolymers) or not (polyblends).

The purpose of blending polymers is either to improve processability, or, more usually, to obtain materials suitable for specific needs by tailoring one or more properties with minimum sacrifice in other properties. The behavior of a poly-blend may be expected to depend on the behaviors of the individual components in the blend, their relative proportions, the degree of heterogeneity, the nature of the interface between the components, and the structure and morphology of the blend. With careful manipulation of these variables, many commercially successful products have been developed.

Perhaps the most important of the heterogeneous blend systems are the high impact plastics such as nitrile rubber blended with polystyrene or styrene–acrylonitrile copolymers and butyl or ethylene–propylene rubber blended with polypropylene. The most important homogeneous systems are combinations of polymeric plasticizers with poly(vinyl chloride) to aid in processing at lower temperatures without reducing the glass transition temperature of the vinyl polymer. Other desirable modifications in polymer properties that can be achieved by blending polymers are to increase stiffness, strength, dimensional stability, toughness, heat distortion temperature, mechanical damping and dyeability; to reduce cost and flammability; and to modify electrical properties. Some specific applications of these principles to fibers will be described later after first considering how polyblends are classified and how blending affects the mechanical and thermal properties of polymers.

CLASSIFICATION OF POLYBLENDS

Three systems are useful for classifying the physical nature of polyblends. These are based on compatibility, on the shape of the heterogeneous phases, and on the relative moduli of the two components. The polyblends can also be classified on the basis of preparation techniques such as melt mixing, solution mixing, and latex or emulsion mixing [2, 4], but this distinction will not be considered here.

Compatibility

In order for two polymers to be compatible with one another, their free energy of interaction must be negative [5]. Since the mixing of two molecules is generally endothermic and the entropy of mixing long polymer chains is small, it is rare that the free energy of mixing is negative. Hence, when polymers are mixed, an incompatible, heterogeneous blend is usually obtained. Compatible polyblends are relatively rare and must result from small or negative heats of mixing. This situation generally occurs as a result of specific interactions between the two polymers, such as hydrogen bonding, stereoisomerization, or strong polar interactions.

In Sperling's classification [3], the incompatible polyblends are divided into amorphous and crystalline subgroups. The only members of the crystalline subgroup are the bicomponent and biconstituent fibers. The difference between these two classes depends on the nature of the two incompatible phases as described in the next section.

Shape of Inclusions

Heterogeneous polyblends, which in certain circumstances are referred to as composite systems, can be further classified in terms of the nature of the two phases and the relative moduli of the components. Three types of phase relationships are generally recognized. First, the blend can consist of a continuous matrix and a discontinuous particulate filler phase; second, the discontinuous phase can be fibrous; and third, the two phases can be continuous. This latter morphology is now generally referred to as an "interpenetrating polymer network" (IPN) blend [6]. As the concentration of the "added" polymer increases in the first two types, phase inversion can occur and the continuous phase can become the discontinuous, filler phase.

As noted above, Sperling [3] considers bicomponent and biconstituent fibers to be different categories of crystalline polyblends. A bicomponent fiber consists of two components divided into two relatively distinct regions within the cross section extending along the length of the fiber. Typical examples are side-by-side, sheath-core, and eccentric. Biconstituent fibers, on the other hand, are composed of a more intimate blend in which one of the components consists of small, discrete fibrils embedded in a more or less continuous matrix. Since the bicomponent fibers do not consist of an intimate mixture of the two polymers, one could reasonably argue that the bicomponent fibers should not be considered to be polyblends at all. (Recently, a new concept for the definition of a biconstituent fiber has been introduced by Saunders et al. [7]. These workers suggested that a biconstituent fiber should be considered to be a mixture of two components, no matter how distributed, which are so chemically dissimilar that they would not normally form a side-by-side bicomponent fiber because of the poor adhesion between the two components. Thus, the distinction between a bicomponent fiber and a biconstituent fiber would depend on the cohesive forces between the components rather than on their distribution. It remains to be seen whether this definition will be accepted.)

Side-by-side and sheath-core bicomponent fibers are frequently referred to as type S/S and type C/C (for centric cover/core), respectively [8, 9]. In a similar fashion, biconstituent fibers are referred to as type M/F (matrix/fibril).

Regulations of the Federal Trade Commission (U.S.A.) permit fibers of the latter type to be identified as a "matrix-fibril fiber" or "matrix fiber" [10].

Moduli of Components

Classifications based on the relative moduli of the two components anticipate to a great extent the properties and use of the blends. For example, the addition of a disperse phase of higher modulus than the continuous phase generally increases the modulus and strength of the polymer and is frequently used to reduce the creep of elastomers. In contrast, the addition of a low-modulus filler is generally employed to increase the impact resistance and elongation to break of rigid plastics or to improve the processability of polymers such as poly(vinyl chloride).

Frequently, rigid inorganic particulate fillers such as carbon black or silica are added to elastomeric polymers for the same reason that high-modulus polymers are added, but the compositions formed should not be considered to be polyblends. In the same fashion, low molecular weight materials are added to serve as plasticizers, lubricants, antistatic agents, light and heat stabilizers, and flame retardants. Because low molecular weight additives can exude to the surface or volatilize out of the composition, much effort has been directed toward developing "permanent" additives which will not be lost. The thrust of much of this work has been to develop polymeric additives, which have low vapor pressures, with improved compatibility [2, 11]. When polymeric materials are added, the product can of course be legitimately referred to as a polyblend.

TECHNIQUES FOR DISTINGUISHING BETWEEN COMPATIBLE AND INCOMPATIBLE POLYBLENDS

Light Scattering and X-Ray Diffraction Effects

The major techniques for determining the compatibility of polymer mixtures are based on optical effects and on changes in the glass transition temperature. A homogeneous two-polymer blend has a refractive index intermediate between those of the separate polymers and will therefore be optically clear. A heterogeneous composite formed by blending polymers having different indices of refraction will of course scatter light depending on the size of the dispersed particles. Hence, light and electron microscopy and small-angle x-ray scattering are useful techniques for determining whether a blend is homogeneous, and if heterogeneous, the size, shape, and extent of incompatible inclusions. As pointed out below, the nature of the inclusions affects the mechanical properties of the blend.

Under certain conditions, however, heterogeneous blends may be optically clear. This situation is possible when (1) the blended polymers have the same index of refraction, (2) the indices of refraction are different but the dispersed particles are so small that they do not scatter visible light, and (3) the indices of refraction are different but the fibers or films are thin enough so that only a few particles are encountered by light passing through.

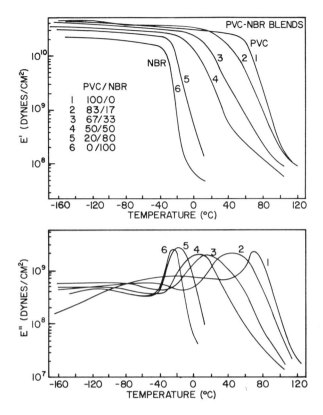

FIG. 1. Elastic (E') and loss (E'') moduli of films made from homogeneous blends of poly(vinyl chloride) (PVC) and butadiene–acrylonitrile copolymer rubber (NBR) at 138 Hz. (Reprinted from ref. [14], p. 67, with deletion of experimental points, courtesy of Kyushu University.)

If any of the components are partially crystalline, wide-angle x-ray diffraction can be used to detect the presence of crystals from the individual components in heterogeneous blends. An example for a nylon 66 fiber containing poly(ethylene glycol) inclusions has been given by Fornes et al. [12].

Thermal Effects

Homogeneous polyblends have a single glass transition temperature T_g intermediate between those of the isolated polymers. The width of the transition zone is broadened somewhat, depending on the degree of miscibility [13]. This behavior is similar to that observed when miscible plasticizers are added to a polymer or for random copolymers, both of which form homogeneous systems. Although the glass–rubber transition as detected in dynamic mechanical tests occurs over a finite temperature range, T_g is usually defined as a single temperature at which the temperature derivative of certain viscoelastic functions (complex modulus E^* and storage modulus E') go through a maximum change and the values of others (loss modulus E'' and loss tangent tan δ) go through a maximum. An example for a blend of poly(vinyl chloride) and butadiene–ac-

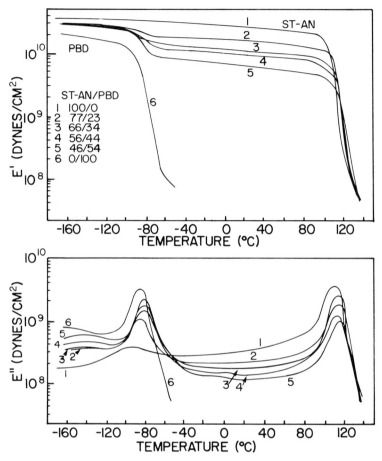

FIG. 2. Elastic (E') and loss (E'') moduli of films made from hetergeneous blends of polybutadiene (PBD) with a styrene–acrylonitrile copolymer (ST-AN) at 138 Hz. (Reprinted from ref. [14], p. 65, with deletion of experimental points, courtesy of Kyushu University.)

rylonitrile copolymer rubber is shown in Figure 1 [14]. In contrast, heterogeneous polyblends characteristically exhibit separate glass transition temperatures for each polymer in the blend. An example for blends of polybutadiene rubber and an acrylonitrile–styrene copolymer is shown in Figure 2 [14]. The modulus–temperature curve shows two steep drops and the tan δ curve shows two maxima, one for each glass transition. The location and breadth of the transitions differ little from those of the pure components. Exceptions have been noted, however, and significant shifts in the T_g have been observed in some two-component polyblends. Such shifts have been suggested to result from filler effects, interaction between the two components, and thermal stresses caused by the difference between thermal expansion coefficients of the two components in the blend [15, 16]. The value of the modulus in the plateau region between the two transitions depends on the moduli of the two components, the ratio of the two components, which material forms the continuous phase, the shape of the dispersed phase and the nature of the interface between the two phases as discussed below.

Since differential scanning calorimetry and other thermal measuring techniques are capable of detecting glass transition temperatures and melting temperatures, these methods can also be used to determine the compatibility of polyblends [17].

MECHANICAL AND THERMAL PROPERTIES OF POLYBLENDS

Moduli

The most successful attempts to calculate the static elastic moduli and dynamic viscoelastic functions of incompatible polyblends from those of the components are based on the Kerner equation [18]. For a two-component system, this equation can be expressed as [19]:

$$\frac{G}{G_m} = \frac{v_m G_m + (\alpha + v_i)G_i}{(1 + \alpha v_i)G_m + \alpha v_m G_i} \tag{1}$$

where subscripts m and i denote matrix (component 1) and inclusion (component 2) respectively, G = elastic shear modulus, v = volume fraction of components, and for spherical inclusions,

$$\alpha = (8 - 10\nu_m)/(7 - 5\nu_m)$$

where ν = Poisson's ratio. This equation, which gives the elastic shear modulus of the blend, assumes strong bonding between the two components.

For certain situations, the Kerner equation can be simplified. For example, for a dispersed phase of spherical particles which are much more rigid than the polymer matrix,

$$\frac{G}{G_m} = 1 + \frac{15(1 - \nu_m)v_i}{(8 - 10\nu_m)v_m} \tag{2}$$

For a low-modulus dispersed phase, again for spherical particles, the Kerner equation becomes

$$\frac{1}{G} = \frac{1}{G_m}\left[1 + \frac{15(1 - \nu_m)}{7 - 5\nu_m}\frac{v_i}{v_m}\right] \tag{3}$$

As the particles change from spherical to elongated, the value of α decreases [20]. For completely oriented rods having high length to diameter ratios, $\alpha \rightarrow 0$ and eq. (1) reduces to

$$G = G_m v_m + G_i v_i \tag{4}$$

the well-known rule of mixtures which gives the modulus expected when two materials are connected in parallel. This combination represents the maximum possible modulus that can be obtained by combining two separate, continuous components. When $\alpha = \infty$, eq. (1) reduces to

$$\frac{1}{G} = \frac{v_m}{G_m} + \frac{v_i}{G_i} \tag{5}$$

and gives the modulus when two continuous materials are connected in series, the lowest possible modulus that can be achieved.

The relationship between eqs. (1), (4), and (5) are given in Figure 3. The modulus of all polyblends theoretically must fall between the upper and lower curves (for a matrix having a Poisson ratio of 0.5). As "rigid" spheres (high modulus) are added to a "rubber" matrix (low modulus), the modulus increases following curve 2 if there is no reinforcement between the two components. If there is interaction between them, the modulus increases from the simple series addition of curve 2 to curve 4. This increase represents a "reinforcement" obtained by dispersing spheres in the matrix. As more rigid spheres are added and the fractional volume approaches 0.5–0.6, a phase inversion can occur during which the rigid phase becomes the matrix and the rubber phase becomes dispersed. At this point, the modulus is given by curve 3, and the system should be treated as a rigid polymer softened by the addition of rubber spheres. The maximum stiffness that can be obtained is then given by curve 1, in which the

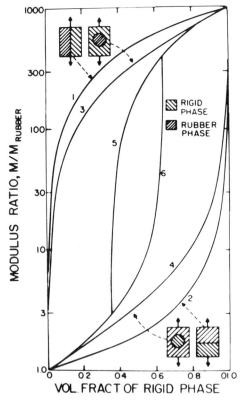

FIG. 3. Relative modulus of blends containing rubber and a rigid polymer having a modulus 1000 times that of the rubber. (1) Polymer and rubber in parallel; (2) polymer and rubber in series; (3) rubber dispersed as spheres in polymer matrix, $v_p = 1$; (4) polymer dispersed as spheres in rubber matrix, $v_p = 1$; (5) rubber dispersed as spheres in polymer matrix, $v_p = 0.64$; (6) polymer dispersed as spheres in rubber matrix, $v_p = 0.64$. (Reprinted from ref. [20], p. 398, courtesy of Marcel Dekker, Inc.)

added particles are no longer spheres but are particles with high length to diameter ratios (ultimately fibers) giving the maximum possible reinforcement between matrix and additive, i.e., with the two phases behaving as continuous elements connected in parallel.

The nature of the discontinuity in the inversion region has been considered by Nielson [21], who modified the relations given by eqs. (1), (2), and (3) by taking into account the packing density of the spherical additives. As the dispersed particles become concentrated enough to contact each other or if they agglomerate, the modulus changes dramatically as phase inversion takes place. Contact occurs when the volume fraction approaches the maximum possible packing density v_p (which is, for example, 0.64 for random close-packed spheres and 0.74 for hexagonal or face-centered cubic spheres). Curves 1–4 in Figure 3 were calculated for $v_p = 1$ (i.e., assuming no free volume in packed spheres!!). Curves 5 and 6 are the moduli ratio calculated for $v_p = 0.64$ and show a more or less realistic transition in modulus as phase inversion occurs when the volume fractions of added polymer v_i approach 0.4–0.5.

An example of the relationship between modulus and composition for a biconstituent fiber (matrix-fibril type) composed of a blend of nylon 6 and poly-(ethylene terephthalate) (PET) in which the modulus ratio of the two components is approximately 1:3 has been given by Papera et al. [22]. Their observations are given in Figure 4.

By representing a polyblend with a "spherical model" and employing an elastic–viscoelastic analogy or correspondence principle ([23], for example), Uemura and Takayanagi [24] derived an equation which gives the complex shear modulus G^* of the polyblend in terms of the complex shear moduli of the components and the volume fraction and Poisson coefficient of the matrix compo-

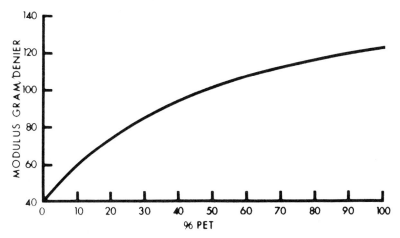

FIG. 4. Modulus of poly(ethylene terephthalate) (PET)–nylon 6 polyblend fiber as a function of percent PET in fiber. (Reprinted from ref. [22], p. 829, courtesy of the Textile Research Institute.)

nent. Christensen [25] obtained the same equation, which may be written as:

$$\frac{G^*}{G^*_m} = \frac{(v_m)G^*_m + (\alpha + v_i)G^*_i}{(1 + \alpha v_i)G^*_m + \alpha(v_m)G^*_i} \tag{6}$$

where the asterisk represents the complex viscoelastic modulus. This equation is identical in form to the Kerner Equation [eq. (1)]. In the derivation of eq. (6), the viscoelastic Poisson coefficient v_m (which is contained in α) is assumed to be a real constant. An expression for the complex Young's modulus E^* was obtained from eq. (6) by Dickie [19]:

$$\frac{E^*}{E^*_m} = \gamma \frac{(v_m)E^*_m + \beta(\alpha + v_i)E^*_i}{(1 + \alpha v_i)E^*_m + \alpha\beta(v_m)E^*_i} \tag{7}$$

where $\beta = (1 + v_m)/(1 + v_i)$, $\gamma = (1 + v)/(1 + v_m)$, and E^* = complex Young's modulus. In deriving this equation, it is assumed that

$$E^* = 2(1 + v^*)G^* \tag{8}$$

and

$$v^* = v' = v \tag{9}$$

which in essence means that the material is isotropic and that the imaginary part of the complex Poisson coefficient v'' is zero. E^* of eq. (7), of course, can be decomposed into its storage (E') and loss (E'') components by algebraic manipulations [19].

The results given by eqs. (1) and (6) may be arrived at by a somewhat different approach first suggested by Takayanagi et al. [26, 27]. In this derivation, the viscoelastic properties of a two-component, incompatible polyblend may be represented by a simple mechanical model of the type shown in Figure 5a. The elements in the model are of two kinds, each of which is assumed to have the viscoelastic property of the blend components. The manner in which the elements are coupled represents the fashion in which the components share the externally applied loads. The model is characterized by two parameters λ and ϕ which describe the extent of series and parallel character, respectively, of the coupling

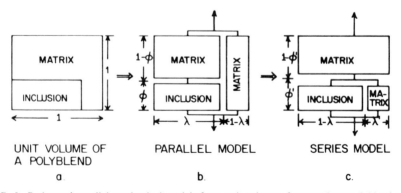

FIG. 5. Series and parallel mechanical models for a unit volume of a two-phase polyblend system.

of the components. The volume fraction of the included component $v_i = \lambda\phi$. If it is assumed that $\nu^* = \nu' = \nu$, the complex shear modulus of the polyblend may be derived from the mechanical model in Figure 5b as

$$G^* = \lambda \left[\frac{\phi}{G^*_i} + \frac{(1 - \phi)}{G^*_m} \right]^{-1} + (1 - \lambda)G^*_m \tag{10}$$

If the two parameters λ and ϕ of eq. (10) are related to the parameter α of eq. (6) by

$$\phi = (1 + \alpha v)/(1 + \alpha) \tag{11}$$

and

$$\lambda = v(1 + \alpha)/(1 + \alpha v) \tag{12}$$

then the two equations are equivalent. Finally, if it is assumed that

$$\nu_m = \nu_i = \nu$$

then eq. (10) can be rewritten, without change of form, to give the complex Young's modulus:

$$E^* = \lambda \left[\frac{\phi}{E^*_i} + \frac{(1 - \phi)}{E^*_m} \right]^{-1} + (1 - \lambda)E^*_m \tag{13}$$

Thus, the models represented by eqs. (1), (6), and (10) are all equivalent.

An alternative parallel-series model as defined by Figure 5c has been derived by Takayanagi et al. [26] which gives the following expression for the complex modulus:

$$E^* = \left[\frac{\phi'}{\lambda'E^*_i + (1 - \lambda')E^*_m} + \frac{1 - \phi'}{E^*_m} \right]^{-1} \tag{14}$$

where $\lambda'\phi' = v_i$.

Although many authors have argued that one model fits a given set of data better than the other [26, 28–30], this contention is incorrect because it has been demonstrated [19, 31] that the two models are exactly equivalent when the parameters are related by eqs. (15) and (16):

$$\lambda' = 1 + v_i - \phi \tag{15}$$

$$\lambda\phi = \lambda'\phi' = v_i \tag{16}$$

Therefore, it is evident that if any of the models expressed by eqs. (6), (10), and (14) fits a given set of data, the other model will fit equally well.

Strength, Breaking Elongation, and Impact Resistance

As stated earlier, one of the major applications of polyblends is to increase toughness and elongation of a brittle polymer by adding a rubbery polymer. A set of stress–strain curves showing the increase in ultimate elongation achieved in biconstituent fibers made from nylon 6/PET is presented in Figure 6 [22]. As soft polymer (nylon 6) is added, the breaking elongation and the energy to

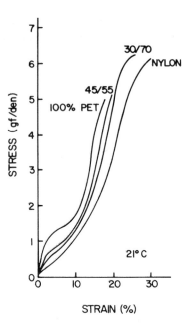

FIG. 6. Stress–strain curves for poly(ethylene terephthalate) (PET), nylon 6, and PET/nylon–6 polyblend fibers. (Reprinted from ref. [22], p. 828, courtesy of the Textile Research Institute.)

break increase, a change which should also increase the impact resistance. Another example showing the effect of composition on the breaking parameters of polyacrylonitrile–cellulose acetate blend fibers is illustrated in Figure 7 [32].

The increase in strength which can be achieved by adding "hard" polymers to "soft" ones is of course implied in Figures 6 and 7. This principle, which is frequently used to stiffen elastomeric polymers, has been utilized in the development of spandex fibers. Although these fibers are block copolymers and therefore are not polyblends, the "hard" urethane segments of the block copolymer serve to stiffen and stabilize the elastomeric segments to form a useful elastic fiber.

Many theories have been offered to explain the mechanism by which the impact resistance of brittle polymers is improved by the addition of a rubbery polymer. Most of these assume that crack propagation is stopped by the dispersed rubber particles. Recent reviews [33–35] may be consulted for details.

Creep and Heat Distortion Temperature

In general, the ratio of the creep compliance $\epsilon(t)$ of a polyblend to the creep of the matrix $\epsilon_m(t)$ varies inversely with the relative elastic moduli E/E_m of the two materials. The relationship can be expressed approximately as [36]:

$$\frac{\epsilon(t)}{\epsilon_m(t)} = \frac{E_m}{E} \tag{17}$$

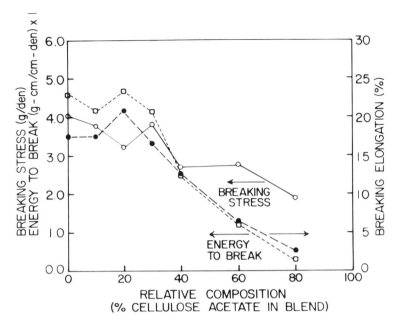

FIG. 7. Breaking stress, breaking elongation, and energy to break of polyacrylonitrile–cellulose acetate polyblend fibers. Data from ref. [32].

When a rigid filler is added to increase the modulus of a polymer, the creep compliance decreases, while the addition of a rubber filler to improve the impact resistance will increase the creep of the polyblend. Since heat distortion tests are based on the temperature at which a defined deflection takes place, the higher the creep, the lower will be the heat distortion temperature. Thus, the above comments describing the influence of blending on creep rates apply also to changes in heat distortion temperatures. It should be noted that an increase in heat distortion temperature, which represents the maximum temperature at which a blend can be usefully employed, results from an increase in modulus and reduced creep rather than from a change in glass transition temperature.

Papero et al. [22] have reported that the creep of a biconstituent fiber formed from blending 30% of PET with nylon 6 is reduced about one-third from that of a nylon 6 fiber.

COMMERCIAL POLYBLEND FIBERS

The first polyblend fibers to be examined were prepared by Cates and White [32, 37] in 1956. These fibers were heterogeneous mixtures of polyacrylonitrile and cellulose acetate, polyacrylonitrile and cellulose (made by hydrolysis of the acetate), and polyacrylonitrile and silk. The first two combinations were prepared by dry-spinning and the third by wet-spinning. Later, in 1963, Morbey and Whitwell [38] studied solution-spun polyacrylonitrile–cellulose acetate blend fibers. Meissner et al. [39] examined the same system in 1968, and in 1973, Duiser and Cooper [40] prepared similar blends by phase separation spinning.

TABLE I

Some Biconstituent Polyblend Fibers that Have Been Introduced Commercially

Trade Name	Composition[a]	Manufacturer	Outstanding property	Reference
Acetokhlorin	85/15 Chlorinated PVC/CA	USSR		8
Antron	Nylon 66/poly(ethylene glycol)	DuPont	Antistatic fiber	45–47
Cadon, Ultron	Nylon 66/poly(alkylene ether)	Monsanto	Antistatic fiber	47, 48
Cordelan	PVC, PVA, and PVC grafted on PVA	Kohjin	Flame retardance	55, 56
EF-121	70/30 Nylon 6/PET	Allied Chemical	Nonflatspotting tire yarn	22
Enkatron	70/30 Nylon 6/PET	AKU	Nonflatspotting tire yarn	58, 60
K-6 (or Chinon)[b]	70/30 PAN/Casein	Toyobo	Silk-like luster & dyeability	8, 50
N-44	80/20 Nylon 66/poly(hexa-methylene isophthalamide)[c]	DuPont	Nonflatspotting tire yarn	42–44
NF-20	70/30 Nylon 6/PET	Firestone	Nonflatspotting tire yarn	42, 58
Mendel (S28)	Nylon 6/PET	Teijin	Silk-like	51, 52
Meraklon DL	Polypropylene/basic polyconden-sate or poly(methyl sorbate)	Montedison	Dyeable polypropylene	4, 50
Source (AC0001)	70/30 Nylon 6/PET	Allied Chemical	Silky luster and hand	22, 41
X-88	Nylon 66/cyclic polyamide	Monsanto	Nonflatspotting tire yarn	8, 42

[a] PVC = poly(vinyl chloride), CA = cellulose acetate, PVA = poly(vinyl alcohol), PET = poly(ethylene terephthalate), PAN = polyacrylonitrile.

[b] Might be 70% graft of PAN on casein with PAN dispersed in casein [54].

[c] Probable composition based on patent literature [43, 44].

The primary interest of the latter authors was to prepare fine fibrils by extracting the matrix with a solvent. They obtained polyacrylonitrile fibrils of 0.1–0.2 μm diameter by extracting the cellulose acetate with acetone from the polyblend. The polyacrylonitrile fibrils obtained could be graphitized. However, no commercial fibers based on any of these mixtures were ever developed.

A list of some of the biconstituent fibers that have reached the large scale development stage and/or have been introduced commercially is given in Table I. The first such fiber was probably Allied Chemical Corporation's EF-121 developed as a nonflatspotting nylon tire yarn and introduced in 1964 [22]. This fiber was composed of a mixture of 70% nylon 6 and 30% poly(ethylene terephthalate). A fiber of the same composition developed for carpet use was introduced in 1968 with the trade name Source [41]. The fibrils in polyblends of this type were reported to range in diameter from 0.06 to about 1 μm and in length from 100 to 200 μm [22]. Another biconstituent polyblend fiber, DuPont's N-44, developed as a nonflatspotting nylon tire yarn, was available in 1965 [42]. This fiber most likely consisted of a 80/20 mixture of nylon 66 and poly(hexamethylene isophthalate) [43, 44]. As shown in Table I, several other polyblend fibers were developed primarily to make nonflatspotting nylon tire yarn.

Another class of biconstituent fibers to be marketed (about 1965) was the permanent antistatic nylon 66 fibers containing up to 10% inclusions of polyethylene glycol or other ethoxylated polymers [45–48]. To be effective, the additive must be uniformly dispersed as a separate phase in the host fiber and elongated during drawing to form fibrils about 0.5 μm in diameter by 20–40 μm long. Fibers of this type, especially their mechanical properties, are discussed further by Bhat and Hersh [49].

Other desirable properties which have been achieved with polyblend matrix-fibril type fibers are improved dyeability (Meraklon DL [4, 50]), a more silk-like hand (Mendel [51, 52], Source [22, 41], and Chinon or K-6 [53, 54]), and flame retardance (Cordelan [55, 56]).

Many other polyblends have been investigated, including one novel system which is a combination of bicomponent and biconstituent fibers [57]. This fiber is a side-by-side bicomponent filament consisting of either a polyamide component and a polyamide-dispersed-in-polyester component or a polyester component and a polyester-dispersed-in-polyamide component. Several comprehensive reviews covering polyblend fiber systems and techniques for preparing them are available [4, 58, 59].

Although some of the polyblend fibers listed in Table I are no longer commercially available, the very fact that so many types have reached the marketplace suggests that many useful new and novel polyblend fibers will be developed in the future.

REFERENCES

[1] F. C. Frank, *Proc. Roy. Soc. London, A319*, 127 (1970).
[2] B. D. Gesner, in *Encyclopedia of Polymer Science and Technology*, Vol. 10, H. F. Mark and N. G. Gaylord, Eds., Interscience, New York, 1969, p. 694.
[3] L. H. Sperling, in *Recent Advances in Polymer Blends, Grafts, and Blocks*, L. H. Sperling, Ed., Plenum Press, New York, 1974, p. 93.

[4] W. Berger and J. Mellentin, *Faserforsch. Textiltech., 21,* 288 (1970).

[5] P. J. Flory, *Principles of Polymer Chemistry,* Cornell Univ. Press, Ithaca, N.Y., 1953, chaps. XII and XIII.

[6] L. H. Sperling and D. W. Friedman, *J. Polym. Sci. A-2, 7,* 425 (1969).

[7] J. H. Saunders, J. A. Burroughs, L. P. Williams, D. H. Martin, J. H. Southern, R. L. Ballman, and K. R. Lea, *J. Appl. Polym. Sci., 19,* 1387)1975).

[8] P. A. Koch, *Textil-Industrie, 72,* 253 (1970).

[9] R. Jeffries, *Ciba Rev., 1,* 12 (1974).

[10] *Code of Federal Regulations, Part 16,* Office of the Federal Register, National Archives and Records Service, General Services Administration, Washington, D.C., Jan. 1, 1975 revision, Section 303.10, p. 484.

[11] M. Combey, in *Plasticizers, Stabilizers, and Fillers,* P. D. Ritchie, Ed., The Plastics Institute, London, 1972, chap. 12.

[12] R. E. Fornes, P. L. Grady, S. P. Hersh, G. R. Bhat, and N. Morosoff, *J. Polym. Sci. Polym. Phys. Ed., 14,* 559 (1976).

[13] S. Manabe, S. Vemura, and M. Takayanagi, *Kogyo Kagaku Zasshi, 70,* 529 (1967).

[14] M. Takayanagi, H. Harima, and Y. Iwata, *Mem. Fac. Eng. Kyushu Univ., 23,* 57 (1963).

[15] S. Manabe, R. Murakami, and M. Takayanagi, *Int. J. Polym. Mater., 1,* 47 (1971).

[16] R. Murakami, S. Manabe, and M. Takayangi, *Rep. Progr. Polym. Phys. Japan, 12,* 271 (1969).

[17] G. R. Bhat, Ph.D. Thesis, North Carolina State University, 1974.

[18] E. H. Kerner, *Proc. Phys. Soc., 69B,* 808 (1956).

[19] R. A. Dickie, *J. Appl. Polym. Sci., 17,* 45 (1973).

[20] L. E. Nielsen, *Mechanical Properties of Polymers and Composites, Vol. 2,* Dekker, New York, 1974, chap. 7.

[21] L. E. Nielsen, *J. Appl. Phys., 41,* 4626 (1970).

[22] P. V. Papero, E. Kubu, and L. Roldan, *Text. Res. J., 37,* 823 (1967).

[23] I. M. Daniel, in *Testing of Polymers,* W. E. Brown, Ed., Interscience, New York, 1969, p. 297.

[24] S. Uemura and M. Takayanagi, *J. Appl. Polym. Sci., 10,* 113 (1966).

[25] R. M. Christensen, *J. Mech. Phys. Solids, 17,* 23 (1969).

[26] M. Takayanagi, S. Uemura, and S. Minami in *Rheo-Optics of Polymers (J. Polym. Sci. C, 5),* R. S. Stein, Ed., Interscience, New York, 1964, p. 113.

[27] M. Takayanagi, in *Proc. 4th Int. Congr. Rheol., Pt. I,* E. H. Lee and A. L. Copley, Eds., Interscience, New York, 1965, p. 161.

[28] T. Horino, Y. Ogawa, T. Soen, and H. Kawai, *J. Appl. Polym. Sci., 9,* 2261 (1965).

[29] M. Siahkolah, M.S. Thesis, North Carolina State University, 1971.

[30] M. Takayanagi, K. Imada, and T. Kajiyama, in *U.S.-Japan Seminar in Polymer Physics (J. Polym. Sci. C, 15),* R. S. Stein and S. Onogi, Eds., Interscience, New York, 1966, p. 263.

[31] D. Kaplin and N. W. Tschoegl, *Polym. Eng. Sci., 14,* 43 (1974).

[32] D. M. Cates and H. J. White, Jr., *J. Polym. Sci., 20,* 155 (1956).

[33] C. G. Bragaw, *Advan. Polym. Ser., 99,* 86 (1971).

[34] S. Newman and S. Strella, *J. Appl. Polym. Sci., 9,* 2297 (1965).

[35] S. Strella, *Appl. Polym. Symp., 7,* 165 (1968).

[36] L. E. Nielsen, *Trans. Soc. Rheol., 13,* 141 (1969).

[37] D. M. Cates and H. J. White, Jr., *J. Polym. Sci., 20,* 181 (1956); *21,* 125 (1956).

[38] G. K. Morbey and J. C. Whitwell, *Text. Res. J., 33,* 673 (1963).

[39] W. Meissner, W. Berger, and H. Hoffman, *Faserforsch. Textiltech., 19,* 407 (1968).

[40] J. A. Duiser and J. A. Copper, *Fibre Sci. Tech., 6,* 119 (1973).

[41] Anonymous, *Chem. Eng. News, 46*(20), 16 (1968).

[42] P. M. Sinclair, *Chem. Eng.,* May 9, 86 (1966).

[43] L. J. Ahles and J. Zimmerman (to E. I. du Pont de Nemours & Co.), U.S. Pat. 3,195,603, July 20, 1965.

[44] J. Zimmerman (to E. I. du Pont de Nemours & Co.), U.S. Pat 3,393,252, July 16, 1968.

[45] E. E. Magat and D. Tanner (to E. I. du Pont de Nemours & Co.), U.S. Pat. 3,329,557, July 4, 1967.

[46] E. E. Magat and W. H. Sharkley (to E. I. du Pont de Nemours & Co.), U.S. Pat. 3,475,898, Nov. 4, 1969.
[47] S. P. Hersh, *Polym. Plast. Technol. Eng., 3*(1), 29 (1974).
[48] G. T. Barnes, *Mod. Text., 51*(7), 45 (1970).
[49] G. R. Bhat and S. P. Hersh, *J. Appl. Polym. Sci. Appl. Polym. Symp., 31,* 55 (1977).
[50] G. Natta and M. Compostella, *Textilind., 69,* 827 (1967).
[51] W. Tsuji, *Senshoku Kogyo, 17*(3), 129 (1969); *Chem. Abstr., 71,* 71765 (1969).
[52] Teijin, Ltd., Brit. Pat 1,126,126, Sept. 9, 1968.
[53] R. Jeffires, *Bicomponent Fibres,* Merrow, Watford, England, 1971, p. 4.
[54] S. Morimoto, *Ind. Eng. Chem., 62*(3), 23 (1970).
[55] T. Koshiro and A. Goldfarb, *Mod. Text., 54*(11), 40 (1973).
[56] T, Koshiro, *Mod. Text., 56*(2), 22 (1975).
[57] S. Etchells, A. Harcolinski, and C. D. Cowell (to ICI Ltd.), Brit. Pat. 1,169,782, Nov. 5, 1969.
[58] K. Dietrich and S. Tolsdorf, *Faserforsch. Textiltech., 21,* 195 (1970).
[59] W. Meissner and W. Berger, *Faserforsch. Textiltech., 16,* 400 (1965).
[60] J. vanGahlen, *Chemiefasern, 17,* 508 (1967).

PROPERTIES AND USES OF POLYBLEND FIBER. II. DYNAMIC MECHANICAL PROPERTIES OF NYLON 66– PEG BLENDS

G. R. BHAT and S. P. HERSH
School of Textiles, North Carolina State University,
Raleigh, North Carolina, 27607

SYNOPSIS

The dynamic mechanical properties of a series of polyblend fibers containing poly(ethylene glycol) fibrils dispersed in a nylon 66 matrix were measured. Fibers of this type impart permanent antistatic properties to nylon 66 fibers and have been commercially available since 1965. Although the electrical characteristics of fibers of this type have been studied rather extensively, little has been published about their mechanical properties. The temperature dependence of the complex modulus and the loss tangent in these isochronal dynamic tensile tests conformed with the predicted behavior of fibrous blends only when the inclusions were assumed to have zero modulus. It was concluded, therefore, that the poly(ethylene glycol) inclusions do not share externally applied stresses and that adhesion between the fibrils and matrix is poor.

INTRODUCTION

In Part I of this paper [1], it was noted that starting about 10 years ago, several synthetic organic fibers made from blended polymers and having durable antistatic properties were marketed. These fibers were blends of polyamides and polyethoxylated compounds such as poly(ethylene oxide glycol). Trade names of some of the permanently antistatic nylons are Cadon and Ultron (Monsanto), Antron (Du Pont), Stataway (Celanese), T-8985 (Akzona), and Body-free and Anso-X (Allied Chemical). These fibers are formed by mixing the polymeric antistatic additive with the molten polymer and extruding the mixture in the conventional manner. The additive remains uniformly admixed within the resultant antistatic fiber as a separate phase. After drawing, the additive is distributed in the fiber as discontinuous, rod-shaped particles about 0.5 μm in diameter and 20–40 μm long with their longest dimension parallel to the fiber axis, thus giving the longitudinal sections a striated appearance [2–5].

The patent and research literature on these materials [2, 3, 6–8] are restricted to the desirable effects of the additives on the static performance of the fibers. They largely ignore the accompanying detrimental effects on other vital physical fiber properties. These detrimental effects are apparently severe in most cases

Journal of Applied Polymer Science: Applied Polymer Symposium 31, 55–61 (1977)

because, in spite of sustained efforts to develop antistatic fibers by blending polymers with polyesters, polyolefins, and acrylics, the few such fibers that have reached commercial success are all blends with polyamides. The mechanical properties of one of the antistatic polyblend fiber compositions will now be examined.

EXPERIMENTAL

Materials

In order to further characterize the nature of the antistatic polyblend fibers, a set of five yarns were prepared by an industrial laboratory by melt-spinning mixtures of nylon 66 and poly(ethylene oxide glycol) (PEG) with a molecular weight of about 20,000 [Union Carbide Corporation's poly(ethylene glycol) compound 20M]. The PEG content of the polymer melts ranged from 0 to 20% by wt in increments of 5%. After melt-spinning, the fibers were drawn five times. Wide-angle x-ray studies of these polyblends, both before and after aqueous extraction [9], indicated that the blend is heterogeneous since the patterns are the superposition of the two polymers in the blend. Before extraction, the c-axis of the PEG is oriented perpendicular to the fiber axis and the b-axis is oriented parallel to the fiber axis. After extraction, the c-axis of the PEG remains perpendicular to the fiber axis, but is now cylindrically distributed about the fiber axis. The b-axis becomes randomly distributed about the c-axis. The c-axis of the nylon 66 is always parallel to the fiber axis. The electrical characteristics of these fibers has been reported by Grady and Hersh [6–8].

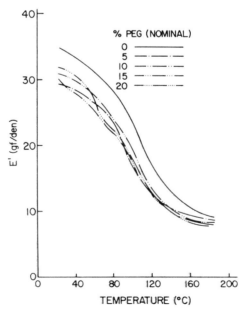

FIG. 1. Elastic modulus E' of PEG–nylon 66 polyblend fibers as a function of temperature at 110 Hz.

The tensile, calorimetric, shrinkage, optical, sonic, and dynamic mechanical properties of these fibers and their composition as determined by infrared, thermal, and chemical analyses have been examined [10]. We wish to now explore the nature of the interface between the two phases in these fibers as implied by the fit of the dynamic mechanical properties of the unextracted fibers to the models described in Part I [1].

Isochronal Viscoelastic Functions

The elastic modulus E' and loss tangent of the fibers measured at a frequency of 110 Hz on a Rheovibron Viscoelastometer Model DDV-II (Toyo Measuring Instruments, Co., Ltd.) are shown in Figures 1 and 2. The moduli E' of the blend fibers are lower than those of the pure fiber. However, the reduction is not proportional to the concentration of inclusions. The blend fibers containing 15 and 20% of PEG show a visible "trough" between 50 and 80°C. Since PEG melts at about 65°C, the fact that there is no sudden drop in modulus at that temperature suggests that PEG particles do not share externally applied stresses. The "trough" may therefore be interpreted as being due to development of internal stresses in the blend fibers caused by increased volume expansion of the PEG particles after melting. A similar effect arising from differences in the coefficient of expansions of matrix and filler has been described quantitatively by Nielsen and Lewis [11].

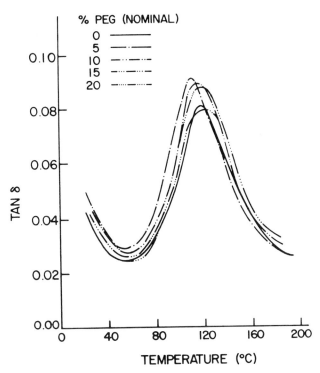

FIG. 2. Loss tangent (tan δ) of PEG–nylon 66 polyblend fibers as a function of temperature at 110 Hz.

The blend fibers have somewhat higher tan δ than the pure fiber. Here again there is no correspondence between PEG concentration and change in tan δ. The maximum in tan δ occurs at about 120°C for the pure nylon 66 fiber and for the blend fibers except for the one containing 5% PEG (112°C). The dynamic mechanical experiments of Dumbleton et al. [12, 13] locate the maximum in tan δ at about 95°C for nylon 66 at a similar frequency which also agrees with the results of Thomas [14].

The change in tan δ in the 20–60°C region is thought to be partly associated with the gradual loss of moisture with heating. The measurements were made by heating the fibers in an ambient atmosphere at 21°C and 65% R.H. The results of Dumbleton et al. [12, 13] show no such change since they used moisture-free samples. The tan δ values exhibit no effect of melting of PEG, which once again suggests that the PEG is discontinuous and not bonded to the matrix in the blend fibers.

Fit of Complex Modulus of Blend Fibers to Mechanical Model

The dependence of the modulus on the volume fraction v_i of PEG inclusions in the blend fiber should be expressed by the rule of mixtures [eq. (4), Part I], since the inclusions have a very high length to diameter ratio. However, the largest actual value of v_i in the fibers available is 0.15, a volume of inclusions too low to verify the fit of eq. (4).

The behavior of nylon 66–PEG blend fibers as a function of temperature, however, might be treated using a model of the type shown in Figure 3b given by eq. (13) in Part I [1]:

$$E^* = \lambda \left[\frac{\phi}{E^*_i} + \frac{1 - \phi}{E^*_m} \right]^{-1} + (1 - \lambda)E^*_m \tag{1}$$

In this model, the complex modulus E^* is expressed in terms of the complex moduli of the matrix E^*_m and inclusion E^*_i and two characteristic parameters λ and ϕ, the product of which equals the volume fraction of the inclusion v_i. (As noted in Part I, this model is equivalent to that of Fig. 3c.) Even though the

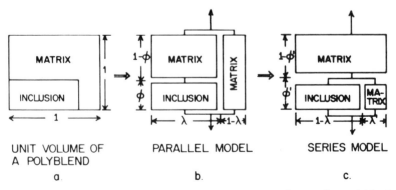

FIG. 3. Series and parallel mechanical models for a unit volume of a two-phase polyblend system.

volume of inclusions is small, the model can be checked since the change in modulus with temperature is determined on a single yarn not subject to sample to sample variability.

The fit to eq. (1) was therefore checked by calculating the parameters λ and ϕ at various temperatures using properly chosen values of E^*_m and E^*_i. To facilitate the computation the equation was re-expressed as

$$\lambda_1 = \frac{v_i(E^*_i - E^*_m)(E^*_m - E^*)}{v_i E^*_m(E^*_i - E^*_m) + E^*_i(E^*_m - E^*)} \tag{2}$$

and

$$\phi_1 = \frac{v_i}{\lambda_1}$$

E^*_m was assumed to be equivalent to the complex modulus of pure nylon 66 fiber. E^*_i was calculated from the results of Takayanagi et al. [15]. The volume fraction v_i of PEG inclusions was calculated for each blend fiber from the mass

TABLE I

Values of λ Calculated for Parallel Model (Fig. 3b) for PEG–Nylon 66 Blend Fiber Using E^*_i from the Data of Takayanagi et al. [15]

Temp. °C	E^*_m gf/den	E^*_i gf/den	E^* gf/den	λ			
				5%[a] 0.052[b]	10%[a] 0.098[b]	15%[a] 0.118[b]	20%[a] 0.152[b]
30	34.0	5.6	29.1	0.314	0.139	0.217	0.092
40	33.1	5.6	28.2	0.348	0.135	0.244	0.095
50	32.0	3.8	27.3	0.235	0.129	0.213	0.102
60	30.6	2.0	26.1	0.183	0.117	0.185	0.146

[a] Nominal PEG content.
[b] v_i determined from chemical analysis [10].

TABLE II

Values of λ Calculated for Parallel Model (Fig. 3b) for PEG–Nylon 66 Blend Fiber with $E^*_i = 0$

Temp. °C	E^*_m gf/den	E^*	λ			
			5%[a] 0.052[b]	10%[a] 0.098[b]	15%[a] 0.118[b]	20%[a] 0.152[b]
30	34.0	29.1	0.144	0.109	0.159	0.082
40	33.1	28.2	0.148	0.106	0.172	0.085
50	32.0	27.3	0.147	0.109	0.172	0.094
60	30.6	26.1	0.147	0.108	0.168	0.137
70	29.2	25.0	0.144	0.120	0.206	0.195
90	25.2	19.0	0.246	0.167	0.226	0.191
110	18.8	13.8	0.266	0.202	0.266	0.245
130	14.0	10.5	0.250	0.214	0.250	0.250
150	10.9	9.0	0.174	0.174	0.174	0.211
170	9.3	8.3	0.108	0.108	0.108	0.161

[a] Nominal PEG content.
[b] v_i determined from chemical analysis [10].

fraction determined from chemical analysis [10] and using 1.14 and 1.20 g/cm^3 as the densities of nylon 66 and PEG, respectively.

λ and ϕ for the blend fibers were calculated in this way for the 30–60°C temperature region since the modulus of PEG above its melting point of about 65°C is zero. The results are shown in Table I. A second set of λ_1 and ϕ_1 values given in Table II were calculated for the 30–170°C region assuming that E^*_i = 0. This assumption, of course, implies that the PEG inclusions occupy volume and add mass but do not share the external load applied to the blend.

DISCUSSION

These results indicate that the behavior of the nylon 66–PEG blend fibers adequately fits the mechanical model when the modulus of the PEG inclusions is set to zero, i.e., if they do not contribute to the modulus. The moduli computed using reasonable values of E^*_i do not fit the mechanical models well in the 30–60°C region. For example, λ decreased from 0.314 at 30°C to 0.183 at 60°C for the fiber containing 5% PEG, whereas for the sample prepared from a melt containing 20% PEG it increased from 0.092 to 0.146 between these two tem-

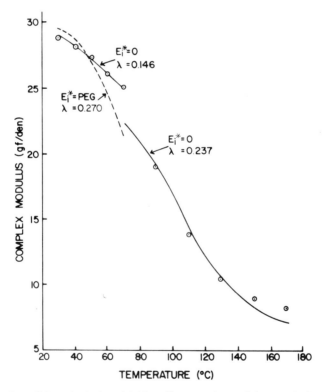

FIG. 4. Fit of parallel mechanical model (Fig. 3b) to complex modulus E^* of 5/95 PEG/nylon 66 polyblend fiber, first with the modulus of the disperse (PEG) phase E^*_i equal to that of PEG (- - -) and then with E^*_i = 0 (——).

peratures. When it is assumed that $E^*_i = 0$ (Table II), the parameters λ and ϕ are reasonably constant from 30 to 60 or 70°C. With one or two discontinuities, λ and ϕ are constant up to 130°C. The fit of the observed moduli for the sample containing 5% PEG to eq. (1) is shown in Figure 4 using both reasonable values for E^*_i and for $E^*_i = 0$. It is believed that the discontinuities in the neighborhood of 60°C result from the melting of PEG. At the lower temperatures, the solid PEG inclusions are soft (rubbery state) and comply with the changes in the contours of the spaces they occupy in the fiber without significant resistance. After melting, the increased volume of these inclusions exerts pressure on the nylon 66 matrix and influences its deformational response. As a result of the constancy of λ and ϕ at temperatures below 70°C, it may be concluded that the PEG inclusions are unattached to the matrix and do not share externally applied stresses because in these calculations E^*_i was assumed to be zero. This conclusion has also been supported by quasi-static tensile tests [10] and the absence of a PEG melting dispersion in the dynamic mechanical spectra of the fibers (Figs. 1 and 2).

One troubling observation, however, is that λ and ϕ do not vary systematically with the volume fraction of added PEG.

REFERENCES

[1] S. P. Hersh, *J. Appl. Polym. Sci. Appl. Polym. Symp., 31*, 37 (1977).
[2] E. E. Magat and D. Tanner (to E. I. du Pont de Nemours & Co.), U.S. Pat. 3,329,557, July 4, 1967.
[3] E. E. Magat and W. H. Sharkey (to E. I. du Pont de Nemours & Co.), U.S. Pat. 3,475,898, Nov. 4, 1969.
[4] Monsanto Co., *22N Anti-Static Nylon,* Monsanto Bulletin, Monsanto Textiles Division, Decatur, Ala., 1969.
[5] Monsanto Co., *Carbon Contract Carpets,* Bulletin N-11, Monsanto Textiles Division, Decatur, Ala., 1971.
[6] S. P. Hersh and P. L. Grady, *DECHEMA (Deut. Ges. Chem. Apparatew.) Monogr., 72*, 251 (1974).
[7] P. L. Grady and S. P. Hersh, *Inst. Phys. Conf. Ser. No. 27*, 141 (1975).
[8] P. L. Grady and S. P. Hersh, *IEEE Trans. Ind. Appl., IA-13*, 379 (1977).
[9] R. E. Fornes, P. L. Grady, S. P. Hersh, G. R. Bhat, and N. Morosoff, *J. Polym. Sci. Polym. Phys. Ed., 14*, 559 (1976).
[10] G. R. Bhat, Ph.D. Thesis, North Carolina State University, 1974.
[11] L. E. Nielsen and T. B. Lewis, *J. Polym. Sci. A-2, 7*, 1705 (1969).
[12] J. H. Dumbleton and T. Murayama, *Kolloid-Z Z. Polym., 238*, 410 (1970).
[13] T. Murayama, J. H. Dumbleton, and M. L. Williams, *J. Macromol. Sci.-1 Phys., B1,* 1 (1967).
[14] A. Thomas, *Nature, 179*, 862 (1957).
[15] M. Takayanagi, K. Imada, and T. Kajiyama, in U.S.–Japan Seminar in Polymer Physics (*J. Polym. Sci. C, 15*), R. S. Stein and S. Onogi, Eds., Interscience, New York, 1966, p. 263.

MATERIAL-PROCESS INTERACTION DURING FALSE TWIST TEXTURING

STANLEY BACKER and DAVID BROOKSTEIN
Department of Mechanical Engineering,
Massachusetts Institute of Technology,
Cambridge, Massachusetts 02139

SYNOPSIS

This paper deals with the experiences of a partially oriented polyester feed yarn as it undergoes false-twist draw texturing.

This paper deals with the succession of actions which a feed yarn of partially oriented polyester experiences as it goes through the false-twist draw texturing process. The sequence of these steps can be stated simply enough. The yarn is twisted, then heated, drawn and twisted some more, cooled, untwisted, and wound up. Or, instead of being wound immediately, it may be relaxed slightly, fed over a second heater, and then wound up in package form. The shorter process provides a twist-lively stretch yarn, the latter a less extensible, but still bulky, yarn.

It turns out that a more complex situation than that described above occurs as the numerous filaments of the typical feed yarn interact with the texturing machine. It is, in fact, necessary to make detailed experimental observations of the nature of these interactions before we can propose a suitable model for analytical treatment of the process.

We shall here report step by step on a study of feed yarn behavior in the false twist texturing process. We first simplify the nature of the process by eliminating many of the features found in modern machines which do not represent the essentials of the texturing system. Figure 1 shows a simplified texturing threadline contained between inlet rollers and outlet rollers. Spaced between these rollers are the heater and the driving spindle. The texturing region between the rollers may be considered to consist of several zones. First, there is the cold entrance zone; second, the hot zone (over the heater); third, the cooling zone after the yarn leaves the heater and before it reaches the spindle; fourth, the post-spindle zone; and, finally, fifth, the take-up zone beyond the rollers.

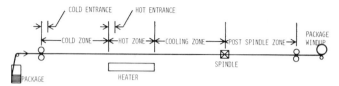

FIG. 1. Single-heater false-twist texturing.

YARN ENTERING THE TEXTURING ZONE

Consider now the behavior of the feed yarn as it enters Zone 1. In the ideal case, this yarn will consist of parallel filaments possessing uniform geometry and uniform properties, i.e., mechanical and thermo-mechanical properties. In addition, this yarn would be entering under negligible tension.

In actual texturing, the filaments of the feed yarn are either purposely or inadvertently entangled to some extent as they enter the front rolls. In addition, the periodic over-end unwinding from the feed yarn package provides a low level of real twist which may be trapped over entering guides until it builds up to a point of surging. This means that periodic real twist levels will be introduced to the entry rolls of the texturing system. In addition, the tension of the yarn entering the front rolls will vary due to changing package size as the feed yarn is unwound over-end. Finally, it may be expected that filament properties will vary depending upon the position of the yarn in the original package. This phenomenon is related to the stress distribution in wound packages and has been studied both theoretically and experimentally.

As each new section of feed yarn protrudes beyond the nip of the feed rollers, its filaments are subjected to threadline tension, torque, and rotation. A simplified picture of yarn conversion from a flat ribbon of seven entering components to a twisted structure is shown in Figure 2. Here, it is seen that yarn components twist around themselves and around each other as they approach the apex of the twist triangle, and then they roll on to the freshly formed end of the yarn. It is seen, also, that with time, the apex of the triangle shifts upward and downward, to the left and right, and forward and backward, in response to the requirement of force equilibrium [1]. We know from many studies on the tensile motivation of filament migration during twisting, that the force level in individual filaments will cycle and this will lead to different force distributions amongst the infeeding components, hence the different geometries of equilibrium apex positions. Also, under the motivation of these force changes, filaments will migrate radially in an effort to minimize the strain energy of yarn twisting. This migration is also evident in Figure 2.

If one considers the rotating yarn at an infinitesimal distance before the triangle apex, it is to be expected that the filaments (or components) are here developing bending moments, twisting moments, and tensile components which sum up to provide the total threadline torque and threadline tension present in the simplified system of Figure 2. It is not unreasonable to expect that similar behavior will occur in the twisting of, say, thirty filaments which have been

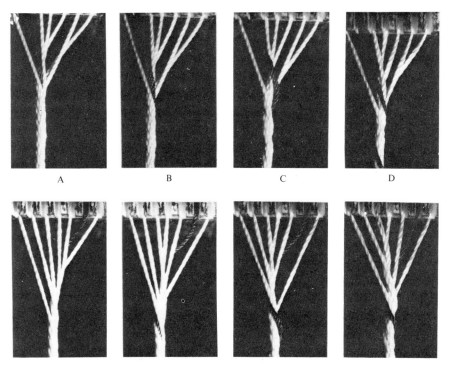

FIG. 2. Migration during twisting of 7-component structure.

flattened into a ribbon-like configuration of zero, or very low twist at the entry roller and then a short distance later are subjected to the threadline torque and tension of the texturing zone. But, of course, interfilament friction and geometric interference will be more intense in the multifilament case.

The actual migration of filaments in the false twist texturing zone is shown in Figure 3, which illustrates a case [2] where a filament migrates from an intermediate radial position to a central position [Fig. 3(b)], a case where a filament maintains its position in the outside helical ring [Fig. 3(a)], and then a case where a yarn goes from the inside position to the outside [Fig. 3(d)].

From such studies of the mechanics of yarn formation we draw certain inferences as a basis for modelling the yarn structure. In particular, we assume that the filaments virtually reach their final bending curvature, twisting torsion, and tensile extension at an infinitesimal distance away from the point of final yarn compaction in the cold twisting zone. This view permits us to make certain simplifications concerning yarn mechanics at this location. In Figure 4 we sketch the componets of filament axial tension which register as components of yarn torque; in addition, we show the components of filament bending moments which contribute to yarn torque, and finally, components of filament twisting torque which contribute to yarn torque. Once we determine the geometry of the filaments, i.e., their curvature, their helix radius, and their helix angle, the calculation of yarn torque becomes a simple case of component determination using suitable sines and cosines, followed by integration of all torque components.

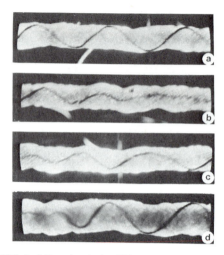

FIG. 3. Migration during FT conventional texturing.

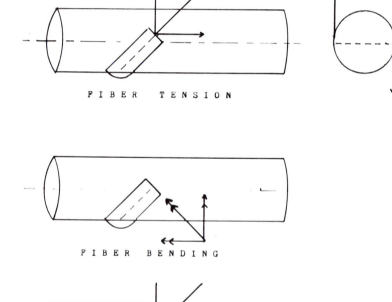

FIBER   TENSION

FIBER   BENDING

FIBER   TORSION

FIG. 4. Sketch of torque components in twisted yarn.

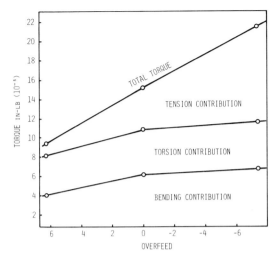

FIG. 5. Torque components in cold-zone threadline.

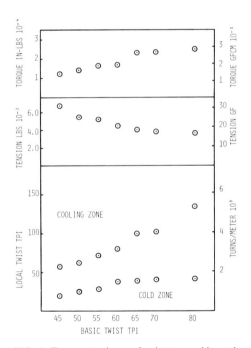

FIG. 6. Torque, tension, and twist vs. machine twist.

It is of interest to note how these components will vary with the tension in the cold zone. Yarn torque components, due to fiber twisting and fiber bending, will vary only slightly with threadline tension for a given machine twist, as shown in Figure 5, but the torque components due to fiber tension will increase significantly as threadline tension is increased [3]. On the other hand, if machine

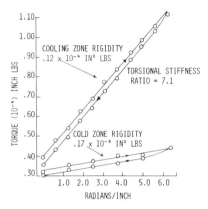

FIG. 7. Incremental torque vs. twist curves.

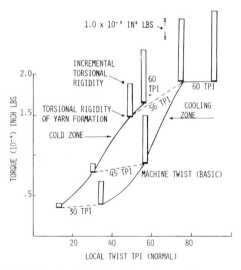

FIG. 8. Torsional rigidities: formation and incremental.

twist is increased, local cold twist will rise with a resultant increase in threadline torque as shown for POY PET draw textured yarn in Figure 6 [4]. Over the commercial range of texturing twists, an increase in machine twist is known to be accompanied by a reduction in threadline tension. Thus, from Figure 6 we can infer that the fiber bending and twisting components contribute more heavily to the cold threadline torque at the high twist levels, while fiber tension contributes more at low twist levels where the twist geometry is less severe and the tension is higher.

At a short distance downstream of the apex of the twist triangle, it is believed that lateral forces induced by fiber twisting and bending, plus filament tension, will develop interfiber frictional forces and increase the resistance of the yarn to further twisting. In other words, the torsional rigidity of the rotating yarn in the cold zone will be considerably higher than that of the rotating collection of filaments at a point just before the yarn is formed at the cold zone entry.

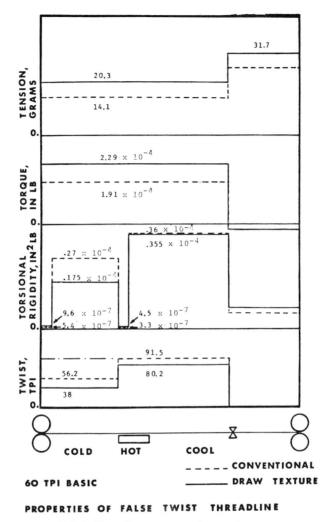

FIG. 9. Threadline parameters in texturing zone.

If the yarn is snatched from the cold zone without loss of twist and with no change in length, it should retain its tension and torque (except for slight stress relaxation). Such a yarn specimen can then be mounted in a torsion-tension tester under a fixed load, and subjected to incremental increasing and decreasing levels of twist. The torque increases which result from these incremental twist increases are plotted in Figure 7 and the slope of these curves represents the torsional rigidity of the yarn as it lies in the rotating cold zone [5]. (Notice, Fig. 7 also includes torsional rigidity measurements of yarns captured from the *cooling* zone.)

Since threadline torque corresponds to the starting points of Figure 7 (noting that more torque relaxation takes place in the cold zone than in the cooling zone), we can plot threadline torques against the actual cold zone twist. The ratio be-

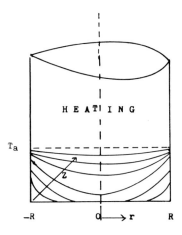

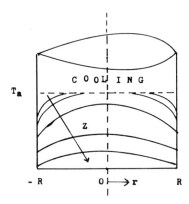

FIG. 10. Temperature gradients during yarn heating.

tween these quantities is, in a sense, the torsional rigidity of formation. Figure
8 illustrates the plotting of threadline torque vs. twist in the cold zone and also
threadlind torque vs. twist in the hot zone. The torsional rigidities of formation,
i.e., the ratios of threadline torque to local twist, are shown in Figure 8 as vertical
(very small) solid bar graphs. These bar graph rigidities are plotted both for
formation of the yarn in the cold zone and for uptwisting of the yarn in the hot
zone.

    The interesting thing about Figure 8 is the comparison of the formation tor-
sional rigidities with the incremental torsional rigidities corresponding to the
slopes of the curves in Figure 7. Note these rigidities are plotted for the cold zone
threadline and for the cooling zone threadline, the incremental torsional rigidities
being plotted as open vertical bar graphs. The level of the incremental rigidities

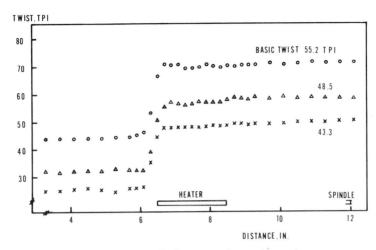

FIG. 11. Twist distributions at varying machine twists.

turns out to be from one to two orders of magnitude above that of the formation torsional rigidities [5]. Thus, our contention that a short distance downstream of the cold formation zone interfiber frictional force causes significant increases in the torsional rigidity of the rotating yarn, is a reasonable one.

Whereas the data of Figure 8 relate to conventional texturing at several different levels of machine twist, the data of Figure 9 are restricted to tests at 60 TPI basic twist for 150 denier texturing output, but they include both the cases of convention texturing and of draw texturing of comparable PET yarns. Here it is seen that the incremental torsional rigidity of both the entry to the cold zone and the entry to the hot zone stand in sharp contrast to the formational torsional rigidity of the threadline in the rest of the cold zone and the rest of the hot and cooling zones. Further, Figure 9 indicates the twist response at a given threadline torque to these differences in torsional rigidity and also records observed differences in tension and torque along the texturing zone and in the post-spindle zone.

APPROACH TO THE HEATING ZONE

At the end of the cold zone and at the approach to the heater, we expect yarn heating, drawing of the partially oriented filaments, and further twisting or uptwisting of the threadline. Seen from a fiber viewpoint, the yarn must first incur heating of its outer layers in accordance with the sketches of Figure 10, which represent the temperature distribution across the radius of the yarn as it proceeds into the heater. The lower curves corresponding to a small value of the dimensionless time parameter Z, show an increase in temperature primarily at the outermost layers. The next curve indicates a further increase of temperature at the outermost and intermediate layers, while the core of the yarn is essentially at room temperature, and so on until finally the yarn is uniformly heated up to the heater temperature.

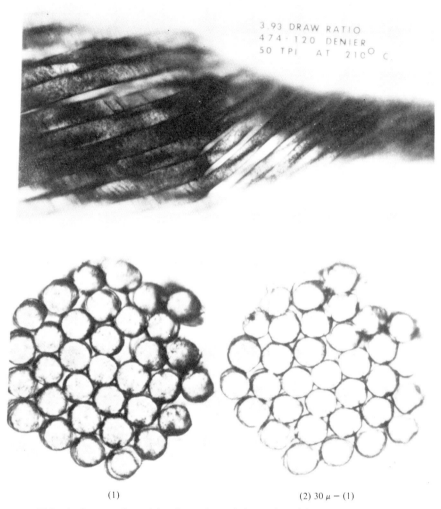

FIG. 12. Cross sections of drawing and uptwisting region of draw textured polyester.

This heat transfer takes place in a fraction of a second, estimated in practical cases to be of the order of $1/10$ or less. But, no matter how short this time is, the curves of Figure 10 indicate that there are intermediate periods in which the outer layers are heated, while the inner layers are at the original room temperature. This means that the resistance of only the outer filaments to deformation will be momentarily reduced as the POY yarns are softened, that is, their yield stress and modulus are reduced—which occurrence will be reflected in a local uptwist of the yarn and in drawing of the filaments in the outer layers. The uptwist which takes place at the entry to the heater zone is well known, not only for draw texturing, but also for conventional texturing, as shown in Figure 11 for three different levels of machine twist [2].

As the yarn is further heated, the intermediate layers become softened and

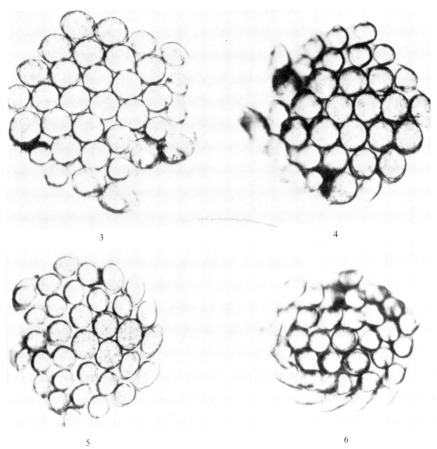

FIG. 13. Cross sections of drawing and uptwisting region. (Each section is 30 μ apart.)

further uptwisting and filament drawing take place, both of outer layers and now of intermediate internal layers. Figure 12 shows the cross section of a virtually unoriented PET yarn during texturing as it approaches the heater. The local drawing and uptwisting is here shown in longitudinal view, along with successive cross sections taken along the cone [6]. Note that initial drawing takes place in the outer layers of the yarn; this is followed by drawing in the intermediate layers of filaments as seen in successive sections approaching the apex of the drawing and twisting cone. By the fourth or fifth section (Fig. 13) the filaments at the center of the cone appear to be fully drawn and from here on, one observes a tendency for these filaments to be compacted laterally and deformed into hexagonal cross sections in the inner portion of the yarn, and part hexagonal and part elliptical in the outer portions of the yarn.

It is of interest to note that even as late as the fifth or sixth cross sections, showing fully drawn filaments virtually over the entire cross section, that there is evidence of space between the filaments in the outer helices, i.e., in the outer ring of the yarn cross section. This would indicate that these filaments have not

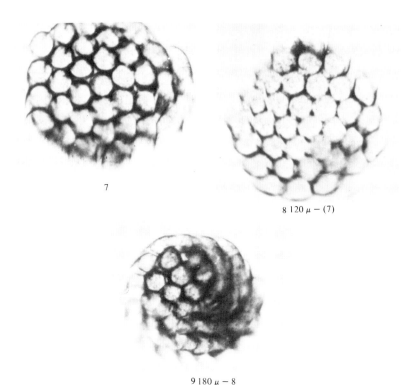

7

8 120 μ − (7)

9 180 μ − 8
FIG. 14. Cross sections of drawing and uptwisting region.

yet been subjected to lateral pressures within that ring. But, as one goes on to the eighth and ninth cross sections (Fig. 14), these interfilament spaces tend to disappear. One is given the impression that lateral forces have now developed in the outer rings, leading to filament cross sectional deformation and to locking of the yarn so as to provide a higher local torsional rigidity than would be expected for the same yarn at the location of the fourth, fifth, or sixth cross sections. Thus, one again sees suggestion of a major difference in torsional behavior within a very short distance along the threadline, this time at the entry to the heater zone.

Still another mechanism is suggested after viewing many cross sections of textured yarns snatched from the cooling zone. It appears that once the structure approaches full packing, there occurs intensive frictional interaction between filaments in the various rings, but also geometric interference to continued twisting and to increases in local helix angles. This geometric interference, and also the additional jamming due to the increased tilting of fibers in any given helical ring, will, on occasion, cause migration as filaments are forced to pop out of their rings seeking a position at a larger radius. Whereas normally a slightly twisted yarn may be expected to have one filament at the center, 6 in the first ring, and 12 in the second ring, it often occurs that the number of filaments in the second ring will be reduced from 12 to 11 as the yarn is twisted up tightly in the texturing zone. This will increase the local yarn radius and it may

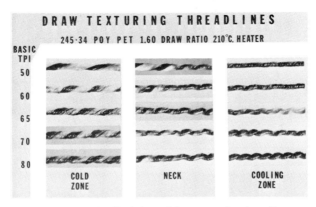

FIG. 15. Longitudinal views of draw-texturing threadline.

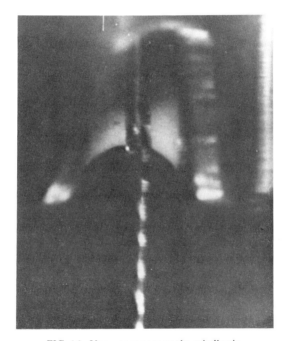

FIG. 16. Yarn movement on the spindle pin.

also cause an increased degree of asymmetry of the yarn cross section. These occurrences should, whenever possible, be taken into account in the structural mechanics of the yarn in the threadline.

Still another occurrence which takes place during uptwisting at the entry to the heater zone is the torsional buckling of the threadline. It has been observed in conventional texturing that if the machine twist is increased to too high a level, the threadline will buckle torsionally, or develop a double helical structure, or corkscrew geometry. Figure 15 shows this kind of behavior for false twist draw texturing of partially oriented polyester. Here it is seen that as the twist level is increased, the configuration of the yarn goes from a fairly straight cylindrical

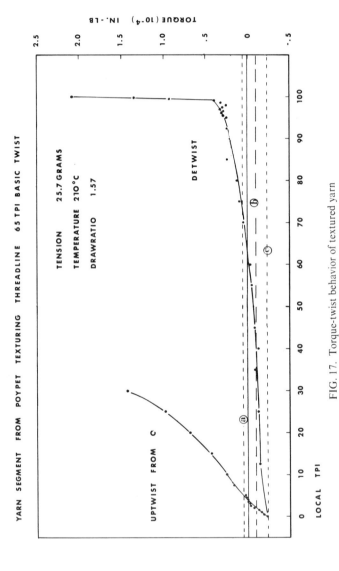

FIG. 17. Torque-twist behavior of textured yarn

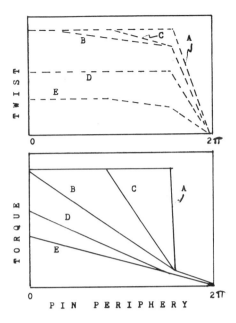

FIG. 18. Yarn behavior at the spindle pin.

configuration for 50 turns/in. basic twist to a heavily buckled and corkscrewed configuration for machine twist of 80 turns/in. Photographs of the threadline taken in the cold zone do not give much evidence of this kind of buckling because the yarns are much heavier and are not so highly twisted. However, at the neck portion representing entry to the hot zone, torsional buckling becomes evident as the basic twist rises to 65 turns/in. There is an occasional buckling apparent in the photograph of the samples at 60 turns/in. basic at the apex of the drawing end and in the cooling zone beyond.

This change in geometric structure of the twisted threadline must be taken into account in any analysis pertaining to threadline behavior, but thus far it has not been considered, and attention has been confined to the case of relatively smooth and unbuckled yarns. Nonetheless, as one measures threadline torque as machine twist is increased, one noted a steady rise at lower twist levels, then the torque vs. twist curve flattens off at twist levels corresponding to the onset of torsional buckling as pictured in Figure 15.

YARN MOVEMENT AROUND THE SPINDLE PIN

As we move down the threadline to the spindle and wrapping of the yarn around the spindle pin, the situation becomes quite complex. We know that the yarn in the post-spindle zone is untwisted and it has, in fact, a slight negative torque compared to the yarn in the texturing zone. The question to be answered is, at what stage does the yarn untwist as it goes around the pin and what is the history of torque change in the circuitous path around the pin? Observations

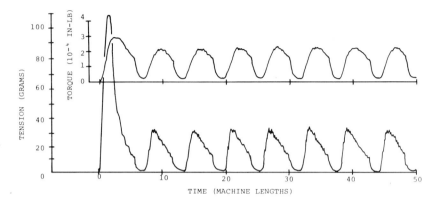

FIG. 19. Cyclic behavior due to yarn rotational slippage at pin.

of yarns moving around a pin during conventional or draw texturing show little evidence of twist change in the first three quadrants of the pin perimeter with virtually all untwisting taking place in the last quadrant. One such example of this kind of observation is shown in Figure 16.

Note well that we stated there is no *evidence* of untwisting of the yarn during its travel through the first three quadrants. However, there is reason to expect some small amount of untwisting to take place in this region even though we cannot record it [7]. This expectation is based upon torque vs. twist tests run on yarns taken from the cooling zone in the texturing threadline. Results of these tests are shown in Figure 17. Here it is seen that the steady torque for a given twist level is at the upper right hand corner of the graph. (The yarn specimen captured from the cooling zone of the threadline has been placed in the torsion tester at the texturing tension, and then the yarn was untwisted.) The thing to note here is the very sharp drop in torque for a relatively little untwisting; in fact, torque drops to about one-fourth of the original threadline torque with removal of but a few turns. With further untwisting, the slope of the curve becomes very flat until all of the turns are removed at a slight negative torque.

These data suggest that there could be a considerable reduction in torque as the yarn moves around the pin and there would be very little evidence of this in terms of a visible reduction in twist. Such behavior is shown schematically in Figure 18. Case A with no twist change through the first three quadrants should show no torque drop during the first three quadrants. On the other hand, case B which shows a considerable torque drop over all three quadrants gives evidence of only a slight twist reduction over those same three quadrants. This is consistent with the data of Figure 17. Case C is another possibility with no torque change and no twist change taking place over the first quadrant and a half, and following this, a slight twist change takes place accompanied by a severe drop in torque. Here, again, most untwisting takes place during the fourth quadrant, as in the earlier cases.

Cases D and E of Figure 18 illustrate an instance where because of low friction, the torque (by which the pin grips the yarn) cannot build up to the level necessary to insert the twist demanded by the machine settings. Under such circumstances, rotational slippage should take place as the yarn translates around the pin, and levels of twist D (or even lower levels of twist E) will be developed. Obviously, materials prepared as in D or E will have lower bulking qualities than for those prepared under conditions A or B or C.

It should be noted that the slippage condition referred to as Cases D or E in Figure 18, may take place on a steady state basis, or in a transient situation. This means that if the spindle is going at 1000 rpm, the slipping yarn will be rotating at a considerably lower velocity in the cooling zone, say, at 900 rpm. This lower level of yarn rotational velocity will dictate the combined rate of twist insertion at the entry to the cold zone and at the entry to the hot zone. And the twist, tension and torque levels of the texturing threadline will ultimately reflect the rotational speed of the *yarn* entering the spindle, and not of the spindle—unless, of course, there is no slippage and their rotational speeds are the same. The situation becomes even more complicated if the rotational slip of the yarn as it passes around the pin is cyclic as reflected in the oscillatory behavior of both threadline torque and threadline tension seen in Figure 19. Such behavior corresponds to a cyclic conversion from condition D or E in Figure 18 to condition A or C, and back.

The mechanics of movement of the yarn around the pin are extremely complicated and we can only cite some of the major facets involved in this action. First, we have the instance of a twisted yarn being bent around a spindle pin, a case which has been given considerable coverage in the textile literature during the last few years. As a highly twisted yarn is bent to a sharp curvature, it tends to develop shear forces in its cross section which form a couple acting in the plane of the bent yarn, and this leads rapidly to a buckling rotation of the loop. This rotation may serve to open the loop or to close it, depending on the direction of wind of the yarn around the spindle pin during string-up. This phenomenon accounts for the very specific instructions given to make sure that a yarn being textured in an S twist is strung up in such a way that the string-up provides for an S helix wind around the pin. If the wrong direction of wind around the pin is encountered, the loop will tend to close on itself and in the ensuing abrasion will cause broken filaments or, indeed, broken yarn.

A second phenomenon present during movement of the yarn around the pin is that, due to the bending of the twisted structure, filaments will want to move relative to each other so as to redistribute local filament sections from high tension regions to low tension zones [8]. This relative motion will result in energy loss which will contribute to the tensile buildup as the yarn moves around the pin from the upstream zone to the post-spindle zone, as has been reported [9]. Friction of the yarn translating around the pin will obviously cause the primary buildup in threadline tension as it moves into the post-spindle zone.

It is recognized that friction between the yarn and the pin in a direction transverse to the yarn axis will provide increments of frictional moment which relate to the letdown of yarn torque from the upstream side of the pin to the downstream. Obviously, these frictional moments must reach the threadline torque requirement for the specified twist, if the yarn is to proceed around the pin without continuous slipping. Rotational slippage leading to untwisting can

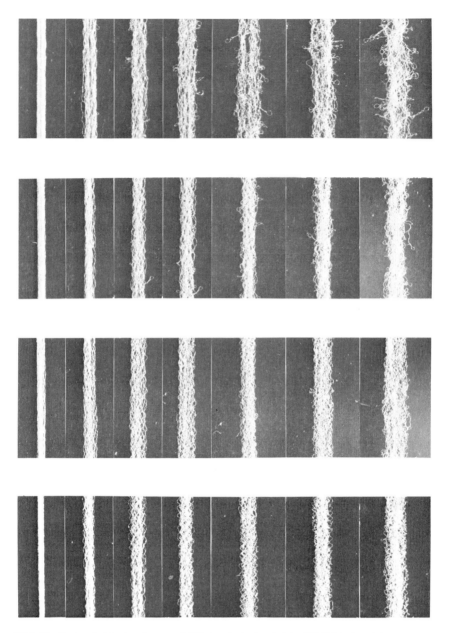

FIG. 20. Contracted textured yarns of differing texturing twists. 150/34 textured dacron yarns, Basic twist (top to bottom); 44.4, 55.2, 66.1, 77.0 turns/in. Contraction (left to right): 0% (at 0.1 G/D), 10%, 20, 30, 40, 50, 60% Texturing conditions: Overfeed, 0%; Temp—228°C.

take place somewhere along the pin, but there should be virtually no such rotation at the early portions of contact with the pin, otherwise the system will be operating at a low efficiency.

It is interesting to note that when conditions D or E apply, namely when the frictional induced moment caused by the pin frictional induced moment caused by the pin friction against the yarn is not sufficient to build up the required threadline torque, then the yarn will rotate and, in some cases, untwist along its entire peripheral contact with the pin. When this takes place, the yarn will be subjected to a cyclic bending and unbending as it moves around the pin and at the same time rotates around its axis. Bending hysteresis of the filaments will result from such cyclic bending [10]. Further, the frictional interaction between fibers in such a bending and unbending test of a highly twisted yarn will constitute an energy loss. Such energy losses must be provided by an additional increment of torque in the revolving threadline beyond the torque managed by yarn pin friction. So, in effect, when slippage starts, there is a "built-in" torque increment provided by the yarn itself, irrespective of friction between the yarn and the pin. This is rather difficult to see at first, but one can anticipate this kind of behavior from analogous situations in which a twisted cable is rotated around a corner without frictional restraints.

POST-SPINDLE BEHAVIOR

In viewing the yarn running through the post-spindle zone, we expect to see untwisted straight filaments, but sometimes one can observe that the crimp which has been heat set into the filaments in the texturing zone is not fully pulled out as the yarn is untwisted around the spindle and moved into the post-spindle zone. This residual crimp in the post-spindle zone should be taken into account in treatment of the overall texturing system.

Finally, in the post-spindle winding zone, or in subsequent second heater zones, it is of interest to observe the behavior and appearance of yarns which have been textured to different twist levels in a single heater system. This is shown in Fig. 20 where four different twist-level textured yarns are contracted by different amounts. Clearly, there is a difference in the character of the contracted yarns for a highly twisted textured yarn vs. a low twisted textured yarn. Of particular interest are the photographs of the yarns contracted 10% or 20%, for this is the range of configurations offered by the system to the second heater zone in a conventional set-texturing process.

SUMMARY

It has been demonstrated through experiments and observations that the rotating threadline of the texturing zone does not behave as a solid cylindrical rod. A significant portion of twist occurs at the very entry of the feed yarn into the cold zone of the texturing machine. Little additional twist takes place until the yarn approaches the heater and the surface filaments are initially heated and softened. Heater uptwisting results as the high temperature line moves inward toward the yarn center. In the case of draw texturing, drastic filament extension also takes place in this region. Beyond the heater entry and through the cooling zone, little additional twisting occurs except as slight twist creep.

At the typical pin spindle virtually all untwisting is observed to occur over the fourth quadrant. If the local frictional grip of the pin is insufficient to provide the required torque for the designated twist level, slippage will take place over the entire periphery of the pin. And in extreme cases the system may oscillate between the no-slip and the slipping regime, with resulting oscillatory effects on threadline tension, torque, and twist.

REFERENCES

[1] A. H. M. El-Shiekh, "on the Mechanics of Twisting," Sc.D. Thesis, Department of Mechanical Engineering, M.I.T., Cambridge, Mass., 1965.

[2] S. Backer and W-L. Yang, *Text. Res. J. 46*, 599–619 (1976).

[3] R. Z. Naar, "Steady State Mechanics of the False Twist Texturing Process," Sc.D. Thesis, Department of Mechanical Engineering, M.I.T., Cambridge, Mass., 1975.

[4] D. Brookstein, "On the Dynamics of Draw Texturing," Sc.D. Thesis, Department of Mechanical Engineering, M.I.T., Cambridge, Mass., 1975.

[5] D. Brookstein and S. Backer *Text. Res. J. 46*, 802–809 (1976).

[6] D. Brookstein, "On the Mechanics of Draw Texturing," M.S. Thesis, Department of Mechanical Engineering, M.I.T., Cambridge, Mass. 1973.

[7] J. J. Thwaites, S. Backer, and D. Brookstein, *J. Text. Inst. 67*, 183–186 (1976).

[8] S. Backer, *Text. Res. J. 22*, 668–681 (1952).

[9] T. Sasaki and K. Kuroda, *J. Text. Mach. Soc. Jpn., 26* (12), T227–232 (1973).

[10] C. Ostertag and A. Trummer, "Man-Made Fiber Technology. Part II: Theoretical Analysis of the Influence of Internal Yarn Friction on the Twist in False Twisting," 70th National Meeting, American Institute of Chemical Engineers, Atlantic City, New Jersey, August 29–September 1, 1971.

THE GEOMETRY AND STRENGTH OF YARNS WITH SPECIAL REFERENCE TO ROTOR-SPUN YARNS

P. GROSBERG and K. H. HO
Department of Textile Industries, University of Leeds,
Leeds, England

SYNOPSIS

This paper describes some work which has been done to determine the number of "wrapper" fibers produced in rotor open-end spinning. It is shown that the number of "wrapper" fibers produced under a wide range of conditions are in accordance with the theoretical predictions. From this geometrical work a simple model of a rotor open-end spun yarn is proposed, and it is shown that the actual strength of rotor open-end spun yarns agrees roughly with the predictions made on the basis of this model. The deviations between the experimentally determined and predicted strengths are discussed.

INTRODUCTION

The basic model on which all previous work on the mechanical properties of yarns has been based is that of a series of helices in which the pitch repeat of the helices of different radius is the same. While various modifications to this model have been made, the mechanical properties of conventional yarns can, on the whole, be explained in terms of this model [1]. It is well known, however, that both the strength and the geometry of yarns produced on open-end spinners are very different.

The geometry of open-end spun yarns differs from conventional yarns in that only some 40–70% of the fibers in an open-end spun yarn are in a helical configuration. Hooked fibers are present in conventional yarns but the proportion found in open-end spun yarns is considerably greater. The presence of random configurations in the open-end spun yarn, however, is unique. These fibers are found mainly on the outside of the yarn and are usually known as "wrapper" fibers. Kasparek [2] showed at an early stage that there was a strong correlation between the number of straight fibers, or, to be more precise, the number of helical fibers and the strength of the yarn. It is the purpose of this paper to investigate the cause of the formation of the wrapper fibers, predict the number of such fibers formed, and hence to estimate the effect of processing parameters on the strength of rotor open-end spun yarns.

Journal of Applied Polymer Science: Applied Polymer Symposium 31, 83–89 (1977)

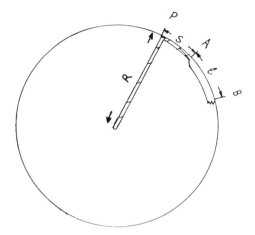

FIG. 1. Diagram of yarn in rotor.

THE CAUSE OF THE FORMATION OF WRAPPER FIBERS

In rotor spinning, fibers are in a groove in the rotor before they are twisted into the yarn which is then withdrawn from the rotor. In the ideal rotor system, the fibers are laid in the groove in such a way that there is an equal probability that a fiber enters any part of the groove. As a result, some fibers must be laid across the point where the fibers in the groove are being withdrawn into the yarn. This point is known as the peeling point (see Fig. 1). Clearly, the probability that a fiber is let into the groove so that the peeling point is within its length is given by $L/2\pi R$, where L is the mean length of the fiber and R is the radius of the rotor.

It has been shown that the yarn twist does not cease at the peeling point P, but extends into the band of fibers in the groove by a distance s which has been called the peripheral twist extent. If a fiber is let into the groove with its right-hand end between the peeling point and the point A shown in Figure 1, then this fiber will not form part of the body of the yarn, since the band of fibers between P and A have already been twisted before the new fiber is deposited into the groove; it will be wrapped loosely around the yarn as at no point is it bound into the yarn structure. It will also subsequently tend to be rubbed into a random form on the surface of the yarn. Clearly, from the argument given above, the probability of a loosely wrapped fiber being produced in this way is given by $s/2\pi R$. Fibers whose right-hand ends are let into the groove between the point A and the point B where the distance between A and B is equal to the mean fiber length of the fibers will, on the other hand, have the right-hand portion of the fiber bound into the structure in the normal way. Only the part of the fiber which initially lies to the left of A will become wrapped round the already formed yarn. By a fairly simple analysis given in the Appendix, one can show that the length of the fibers which will not form part of the structure of the yarn will be equal to $\frac{1}{2}L^2(1 + c^2)N/2\pi R$, where N is the number of fibers in the cross section of the yarn and c the coefficient of variation of the fiber length. The proportion of all the fibers which are not bound into the yarn due to this mechanism is therefore equal to $\frac{1}{2}L(1 + c^2)/2\pi R$. It can therefore be seen that from the above very simple considerations, one would expect that the fraction of wrapper fibers in a yarn should be given by $[s + \frac{1}{2}L(1 + c^2)]/2\pi R$.

The distance s can be varied by altering the true twist in the yarn, the false twist inserted into the yarn inside the rotor by means of various outlet tube de-

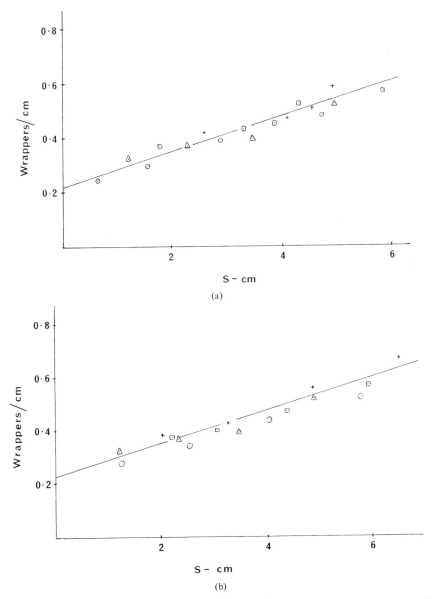

(a)

(b)

FIG. 2. Wrapper fibers/cm vs. peripheral twist extent for (a) varying rotor speeds: (+) 12×10^3 rpm; (□) 10×10^3 rpm; (▲) 8×10^3 rpm; (⊙) 6×10^3 rpm; (b) varying yarn tex: (+) 200 tex; (□) 180 tex; (▲) 150 tex; (⊙) 100 tex; wool yarns.

signs which are commercially available, and by altering the friction in the groove between the yarn and the groove. In a series of experiments, the length s was altered by these means and the number of wrapper fibers per cm of yarn was determined by means of the standard tracer fiber technique. The peripheral twist extent was determined by high-speed photography, by photographing the fiber band inside the groove adjacent to the peeling point.

Figures 2a and 2b show that the number of wrapper fibers per cm of yarn produced increases linearly with the peripheral twist extent; as shown in Figure 2a it is independent of the speed of spinning, and as shown in Figure 2b it is independent of the count of the yarn. Both figures also show that there is an intercept on the number of wrapper fibers per cm axis at zero peripheral extent, as predicted above. From the size of the intercept, it can be shown that $\frac{1}{2}L(1 + c^2)$ is equal to 2.10 cm; the actual value is 1.96 cm. These results show that the predictions made above are fulfilled.

THE RELATIONSHIP BETWEEN THE STRENGTH OF THE YARN AND THE NUMBER OF WRAPPER FIBERS

The work described above leads to the possibility of setting up a simple model of the geometry of a rotor spun yarn. This model consists of a central core with a conventional helical geometry, and wrapped round this core is a layer containing a mixture of fibers which are partially held within the core and partially wrapped loosely round the core, and some fibers which are completely loose and

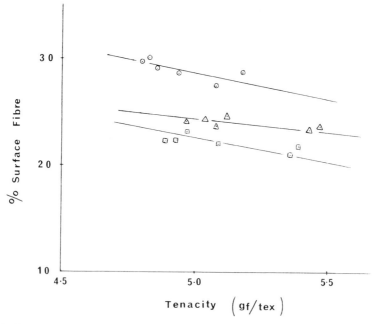

FIG. 3. Strength of wool yarns vs. wrapper fibers/cm: ($\odot$) 14×10^3 rpm; ($\triangle$) 12×10^3 rpm; ($\square$) 10×10^3 rpm.

entangled both with themselves and the first group of fibers. It is obvious that this outer layer of fibers cannot contribute to the strength of the yarn, and one would therefore expect that the strength of the yarn would be directly related to the number of fibers in the core. Such a model has been suggested previously by Shaw [3]. Further work on both long-staple and short-staple fiber yarns have shown that, as expected for such a model, there is a negative correlation between the strength of rotor-spun yarns and the number of wrapper fibers in these yarns. Figure 3 shows a typical example of the effect of wrapper fibers on the strength of rotor spun yarns. Stewart's results and those obtained in our laboratories have shown that for a wide variety of yarns, the strength of rotor-spun yarns can be predicted on the basis of a two-part model of a rotor-spun yarn.

THE RELATIONSHIP BETWEEN THE STRENGTH OF THE YARN AND THE PERIPHERAL TWIST EXTENT

It follows from the above reasoning that the strength of the yarn T should be given in the expression

$$T = T_c[1 - \{s + \tfrac{1}{2}L(1 + c^2)\}/2\pi R]$$

where T_c is the strength of the core when no wrapper fibers are present. A series

TABLE I
Calculated Slopes of Yarn Strengths

Cotton Count	Rotor Speed rev/min	Twist Factor	Actual Slope	S.D. of Slope	Theoretical Slope
12	40	5	-0.1959	0.0971	-0.0748
		7	-0.0961	0.0412	-0.0809
		9	-0.0498	0.0157	-0.0830
		11	-0.0788	0.0133	-0.0826
24	40	5	-0.0466	0.1408	-0.0641
		7	-0.0927	0.0455	-0.0742
		9	-0.1515	0.0136	-0.0736
		11	-0.0783	0.0187	-0.0688
18	40	5	-0.1605	0.0667	-0.0719
		7	-0.1193	0.0118	-0.0787
		9	-0.1029	0.0375	-0.0798
		11	-0.0775	0.0103	-0.0782
18	30	5	-0.2315	0.2455	-0.0622
		7	-0.1470	0.0455	-0.0704
		9	-0.0747	0.0129	-0.0635
		11	-0.1010	0.0081	-0.0753
18	20	5	-0.1559	0.1069	-0.0479
		7	-0.1325	0.0306	-0.0502
		9	-0.0767	0.0177	-0.0631
		11	-0.1085	0.0055	-0.0642

of experiments was carried out on long-staple spinning rotors using mainly wool of 15-cm average length and on short-staple spinning rotors using cotton of 3.1 cm average length. It was found that in all cases, there was a strong negative correlation between the yarn strength and the peripheral twist extent. Table I gives the values of the calculated slopes of the strength against the peripheral twist extent, the experimentally found slopes, and the standard deviation of these slopes.

As can be seen in Table I, these results have been determined for four cotton twist factors, 5, 7, 9, and 11 for three different counts of yarn: 12's, 18's, and 24's cotton count, and three rotor speeds: 20,000, 30,000, and 40,000 rev/min. There does not appear to be any trend in the value of the slope with the speed of spinning and count of the yarn. However, there does seem to be an effect on the slope of the twist factor of the yarn, as shown in Table II which gives the average slope for all the yarns with the same twist factor.

As can be seen from Table II, at lower twist factors the peripheral twist extent has a greater effect on the yarn strength than predicted. This difference decreases as the twist factor increases so that at a twist factor of 11 the actual and predicted values are almost the same, even though the difference is still significant. Possible reasons for this will be discussed in the next session.

DISCUSSION AND CONCLUSIONS

The fact that it has been shown that the peripheral twist extent has a marked effect on the strength of rotor-spun yarns is of considerable practical importance. The size of the peripheral twist extent can be altered by changing the design of the outlet tube or "trumpet" from the rotor. It is clear that to obtain the maximum yarn strength, it is necessary to ensure that the outlet tube does not introduce excessive false twist into the yarn within the rotor.

The size of the correlation coefficient between the yarn strength and the peripheral twist extent, however, is greater than predicted. There are two possible reasons for this effect. First, it has been shown that the twist in the core is not always the same as that calculated from the rotor speed and delivery rate of the yarn. It does vary, also, with the peripheral twist extent. The actual size of the variation is difficult to measure accurately since it is necessary to remove the wrapper fibers before determining the twist in the core, but it would appear that the greater the peripheral twist extent, the higher is the twist in the yarn core.

TABLE II
Average Slopes for Yarns with the Same Twist Factor

Twist Factor	Average Slope	S.D. of Average Slope	Theoretical Slope
5	−0.158	0.0588	−0.064
7	−0.118	0.0154	−0.071
9	−0.091	0.0088	−0.073
11	−0.089	0.0061	−0.074

Since all these yarns have a twist which equals or exceeds the twist required to make a maximum strength yarn, it follows that the higher the peripheral twist extent, the greater will be the obliquity effect on the yarn strength without any improvement due to the slippage effect. It would appear likely, therefore, that with higher peripheral twist extents the yarn strength would be lower than predicted above. It is difficult, however, to see why this effect should be more marked at a twist factor of 5 than at a twist factor of 11.

Another possible cause of this deviation is the fact that the core of a rotor-spun yarn does not consist solely of helically wound fibers. It has been shown by Carnaby and Grosberg [4] that the fibers in rotor-spun yarns in the core region also show marked signs of reversals. These fibers are not only hooked, but they also have double reversals. These reversals are produced by the process of spinning, but the exact cause is not known. It is possible that they are in some way related to the incidence of the peripheral twist extent.

APPENDIX

If $f(l)$ is the fiber length distribution function, then the number of fibers whose length lies between l and $l + dl$ in the cross section of the yarn is $Nf(l)dl$. Of this group of fibers, $l/2\pi R$ have their right-hand ends between A and B. If a fiber end lies a distance x from A, then a length $l - x$ will be wrapped. The total length wrapped is therefore given by

$$\int_0^\infty \int_0^l (l - x) \frac{l}{2\pi R} Nf(l)dl \frac{dx}{l} = \frac{N}{2\pi R} \int_0^\infty \frac{1}{2} l^2 f(l)dl = \frac{1}{2} L^2(1 + c^2)N/2\pi R$$

REFERENCES

[1] J. W. S. Hearle, P. Grosberg, and S. Backer, *Structural Mechanics of Fibers, Yarns and Fabrics,* Wiley, London, 1969.
[2] J. V. Kasparek, Thesis, VTI, Moscow, 1967.
[3] H. V. Shaw, M.Sc. Thesis, Univ. of Manchester, 1972.
[4] G. A. Carnaby and P. Grosberg, *J. Text. Inst.,* submitted.

HIGH-TEMPERATURE COMPOSITES FROM
BISMALEIMIDE RESINS: A BINDER CONCEPT

H. D. STENZENBERGER

Techochemie GmbH, Verfahrenstechnik, 6900 Heidelberg,
West Germany

SYNOPSIS

Bismaleimide resins are frequently used as binders for composites. The processing, thermal, and thermal-oxidative stabilities of the finally cured resins depend on the molecular weight and chemical structure of the prepolymers. Comparative oxidative stabilities for four resins differing in chemical composition are given. The processing for two types of resin, i.e., "hot melt" and "solvent" processable resins, is discussed. Typical laminate properties are given and three prototype applications are presented.

INTRODUCTION

The development of the technology of structural composites has grown rapidly as a consequence of the introduction of new high-modulus, high-strength fibers such as boron and graphite. Graphite fibers are now being used in many aerospace, military, and civilian applications together with epoxy resin binders. In addition, new high-temperature resistant polymers such as polyimides [1], polybenzimidazoles [2], polyphenylquinoxalines [3], and aramides were synthesized and are now used for the fabrication of high-temperature, high-performance composites. These polymers are cured by condensation reactions, liberating low molecular weight material forming voids when used for the fabrication of composites. High-boiling polar solvents such as DMF, DMAc, or NMP are necessary for processing. More recently, the development of addition-type polyimides helped to overcome the aforementioned processing disadvantages of these linear high-temperature polymers.

First, Grundschober [4] reported the homo- and copolymerization of bismaleimides, which can be achieved by a simple heating to temperatures between 150 and 400°C, resulting in highly crosslinked polymers. The great advantage of this approach is that low molecular weight imide prepolymers endcapped with reactive maleimide rings are polymerized into highly crosslinked thermally stable polyimides without the evolution of by-products. Another type of thermosetting polyimides was developed by TRW-System [5]. These compositions contain nadic imide end groups which undergo pyrolitic polymerization when heated to 280–310°C. Resins of this type are commercially available from Ciba-Geigy under the trade name of P13N, but they possess many of the processing problems associated with conventional polyimides because they have to be imidized during processing.

The objectives of this program were (1) the demonstration of the correlation of structure and processibility of various bismaleimides and (2) the development of techniques to fabricate three basic polybismaleimide–carbon fiber structures (filamentary wound structure, honeycomb sandwich panel, autoclave-molded spherical structure).

RESULTS AND DISCUSSION

The usual synthesis for bismaleimides starts from maleic acid anhydride and a diamino compound forming a bismaleamic acid that undergoes cyclodehydration to form the bismaleimide in high yield.

Many variations are possible by changing structure and molecular weight of the diamino compound. The polymerization can be performed easily by heating the monomer (2) to temperatures between 200 and 280°C. The thermal polymerization reaction follows first-order kinetic laws as was recently shown by differential scanning calorimetry [7] and infrared spectroscopic analysis [8].

Resins Investigated

Four bismaleimide resins, differing in structure, were selected to demonstrate the correlation between flow properties and processing characteristics. These structures are given in Table I.

Resin number I was synthesized as outlined in eqs. (1)–(3) using a well-known synthetic route [6]. Resin number II, which was reported to consist of a close

TABLE I
Structures and Melting Behavior of Bismaleimide Resins

RESIN NO.	RESIN STRUCTURE	MELTING BEHAVIOUR
I		Fp. 238 – 245 °C
II	EUTECTIC TERNERY MIXTURE KERIMID 353	LIQUIFYABLE Fp. 70 – 125 °C VISCOSITY 140cP/120°
III	KERIMID 601	MELTABLE Fp. 40 – 110 °C
IV		MELTABLE Fp. 80 – 115 °C

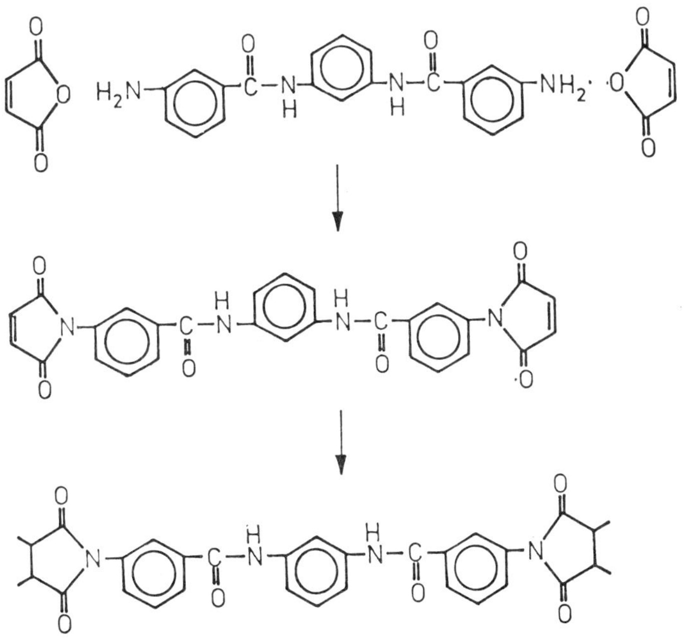

(1)

(2)

(3)

to eutetic mixture of bismaleimides, is marketed by Rhone Poulenc under the trade mark Kerimid 353. Kerimid 601, which was coded resin number III in this work, is also manufactured and marketed by Rhone Poulenc, and is prepared by the reaction of a bismaleimide with an aromatic diamino compound, which forms a maleimide terminated prepolymer that undergoes maleimide type polymerization [eqs. (4) and (5)].

Resin number IV was developed recently, and is prepared by the reaction of a bismaleimide with an amino-terminated aromatic compound [eqs. (6)–(8)]. This polymer was developed with the aim of improving the thermooxidative stability, by increasing the molecular weight of the bridge between the polymerizable maleimide end group without the loss of melting and flow properties.

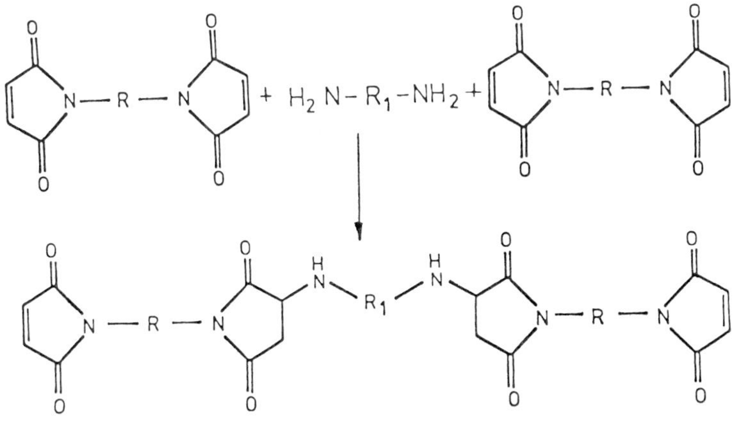

(4)

(5)

$$(6)$$

$$(7)$$

$$(8)$$

Resin Properties

All melting and curing differences of the four bismaleimide resins investigated as a result of prepolymer structural variation (Table I) can be obtained by comparing their DSC traces, as illustrated in Figure 1. Since all the materials investigated cure by addition-type polymerization through the same terminating unsaturated linkages, it was predicted that the cure exotherms would be found in a broadly similar temperature range and reaction rates are undoubtedly a function of the structure and the molecular weight of the bridge between the two terminating maleimide end groups.

Resin number I has a melting point of 238–245°C (Table I) immediately followed by polymerization at a peak temperature of 260°C. No transition is detectable below the melting point. Resin II (Kerimid 353), which is delivered as a resolidified melt, shows a small fusion endotherm at 70–110°C, followed by the crosslinking exotherm in the range of 200–350°C. The peak temperature is 295°C. Resin III (Kerimid 601) also shows a small melting transition of 40–110°C and a broad curing exotherm between 150 and 280°C. The curing proceeds at a lower temperature as compared with resins I and II as a consequence of the asparatimide structure in the prepolymer backbone. The secondary amino groups catalyze crosslinking reaction velocity.

A similar situation is given in resin IV. The resin melts at 80–115°C and polymerizes over a very wide temperature range (150–280°C). This comparative consideration showed that the chemical structure of the prepolymer backbone influences both the melting and polymerization behavior. Polymerization occurs rapidly after the resins are melted, with the exception of resin II, which is melt-stabilized to improve the pot life.

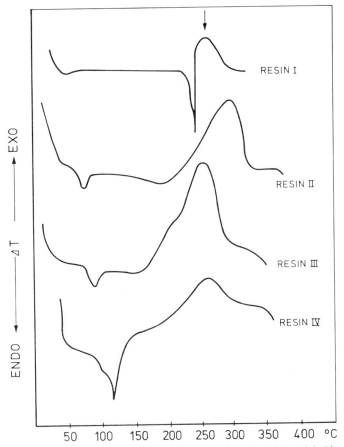

FIG. 1. DSC scans showing melting and polymerization behavior of bismaleimide resins.

TABLE II
Thermal Stability of Bismaleimide Resins

Resin	TGA-Break Temperature °C	Char yield at 750° C	Isothermal aging at 250° C weight loss (%), powderd polymer <0,2 mm			
			100 h	200 h	500 h	1000 h
I	420	60	11,2	17	24,5	--
II	460	44	16,5	29,27	45,70	54,36
III	385	52	14,33	20,64	31,27	40,00
IV	375	54	8,05	11,92	18,99	25,97

Thermal Stability

Programmed thermal gravimetric analysis (TGA) in nitrogen (heating rate 10°C/min) (Fig. 2) and isothermal aging of the powders at 250°C (Fig. 3) were used to evaluate thermal stability (Table II). Kerimid 353 (resin II) had the highest polymer decomposition temperature (PDT) in nitrogen followed by resin I, while Kerimid 601 and resin IV show initial weight loss at 385°C and 375°C, respectively. Resin II had the lowest prepolymer molecular weight, the highest

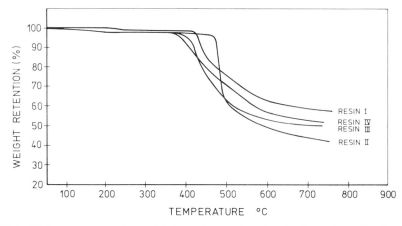

FIG. 2. TGA thermograms of bismaleimide resins used in evaluating thermal stability. Heating rate, 10°C/min.

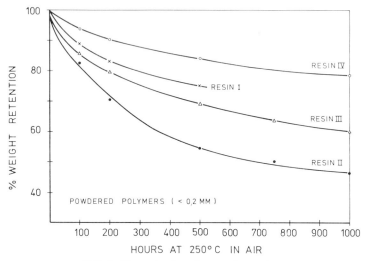

FIG. 3. Thermal aging of neat resin powders.

portion of the aliphatic end groups, and therefore the highest crosslinking density, leading to the highest PDT in nitrogen. As a consequence of the chain-extension reaction during prepolymerization, resins III and IV have higher prepolymer molecular weights, and therefore a lower crosslinking density [eqs. (5) and (7)]. This situation is supported by the char yield data. The main interest lies in thermooxidative stability for applications at high temperatures in air. Weight-loss measurements in air using powdered polymer samples are very useful for comparative investigations. Table II shows that resin II, which had the highest PDT of 460°C, has the lowest thermooxidative stability, while conversely, resin IV, which had the lowest PDT (375°C), possessed the highest thermooxidative stability. This example clearly shows that comparative TGA experiments can be misleading when used alone to predict thermal and/or thermooxidative stability of a polymer sample.

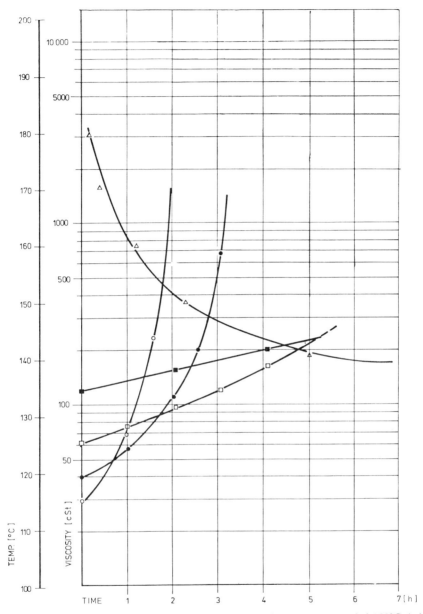

FIG. 4. Viscosity and gelation of Kerimid 353 (resin II) at various temperatures: (O) 155°C; (●) 145°C; (□) 135°C; (■) 125°C.

Processing Considerations

Generally, there are two basic methods for fabricating advanced composite fiber-reinforced structures; i.e., filament winding and hand layup of prepreg materials. The combination of the two methods is sometimes necessary to meet fabrication requirements. Filament winding is the most direct approach for laying the fibers in the desired direction. The direct winding approach in the final structure is limited to cylindrical and conical bodies. Prepregs and broad goods are best fabricated by filament winding techniques.

The choice of the technique for the fabrication of composite materials with bismaleimides depends on the physical properties of the resins. The four bismaleimide resins used in this work were selected to demonstrate the correlation between flow properties and processing characteristics.

The first operation necessary during composite fabrication is the application

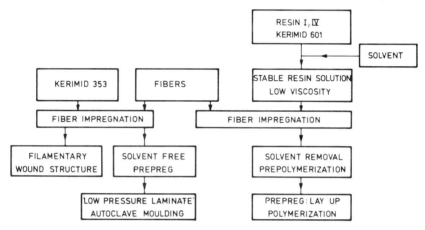

FIG. 5. Processing of bismaleimide resins.

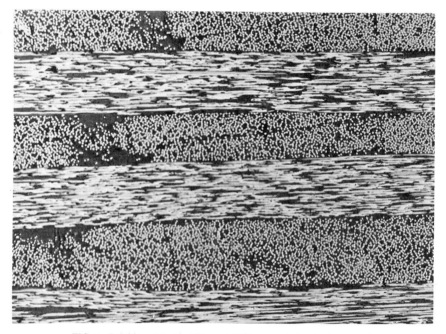

FIG. 6. Polybismaleimide (Thornel 50S) cross-ply laminate (100×).

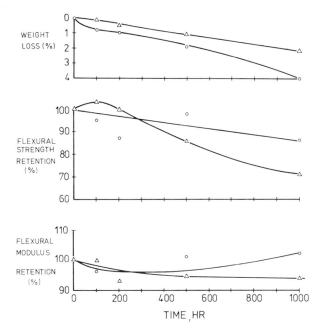

FIG. 7. Changes in flexural properties during aging at 250°C of Kerimid 353 (resin II) in air: (△) unidirectional glass fiber laminates (63 v/o); (O) unidirectional Thornel 50S laminates (55 v/o).

of the selected resin onto the fibers. Liquid or liquefiable binders need no solvent for impregnation (Kerimid 353), while high molecular weight prepolymers that fuse by forming high-viscosity melts need to be dissolved to wet the fibers uniformly (Fig. 4). Resin I and resin III (Kerimid 601) are preferably dissolved in N-methylpyrolidone (NMP), forming low-viscosity stable solutions (Fig. 4). The wetting is easily performed by using impregnation bath techniques, and fiber placement into unidirectional orientation can be done by filament winding of the "on-line" impregnated fibers. The prepregs thus obtained are consolidated to obtain precise fiber orientation and uniform tape thickness. For resin IV, DMAc (dimethylacetamide) was preferred as an impregation solvent. The prepregs contain a large amount of solvent which has to be evaporated prior to composite fabrication. During the drying operation the resin on the prepreg polymerizes simultaneously, forming a "B" stage prepreg. This prepolymerization during solvent removal has to be controlled carefully to obtain the optimal flow properties necessary for laminate consolidation. The advancement of the resin on the prepreg depends on the pressure during composite molding. Kerimid 353, which can be liquefied by heating to 120–130°C, offers the approach of being processed as a wet system without any solvent. The viscosity and gelation characteristics of the resin are given in Figure 5, showing a very extended pot life of 120°C, which was found to be the preferred processing temperature. Because of the low viscosity, prepregs can easily be fabricated within a specified quality by a filament winding operation. The mandrel, which has a diameter of 0.5 m, is heated to 100–125°C to keep the resin flexible during the layup of the filaments. Typical prepreg properties for Thornel 50S fibers are: fiber content, 56 ± 3%; resin content, 44 ± 3%; volatiles, 0; cured ply thickness, 0.2 mm; area weight, 2.9 g/dm². The prepregs are stiff at room temperature but become tacky when heated to 45–60°C, to be plied for molding.

Laminate Fabrication

It was the main objective of a sponsored program performed by Technochemie GmbH and Dornier System GmbH to test the processibility of resin II (Kerimid 353) for the production of three typical carbon fiber structures. As a result of the physical properties of the resin, filament winding has been used to fabricate tubes and tube-like structures (Fig. 8). Filament winding was also used to fabricate NOL rings to evaluate tensile properties (Table III). Basic carbon fiber laminate flexural properties were determined from laminates fabricated by two different methods: the finger-mold technique for unidirectional slabs and the restricted bleed technique for cross-ply laminates. Finger-mold techniques in a compression mold start from prepreg slabs, and excess resin plus entrapped air is allowed to flow longitudinally through the fiber during the heating period and the pressure application. Pressure is applied in increments during the heating period and in such a way that full laminatory pressure is applied just before gelation of the resin. If the pressure is applied too early, resin-poor laminates are obtained. Application of pressure after gelation produces over-sized resin-rich components. Cross-plied laminates were produced by using the restricted bleed technique, which can be scaled up for the production of large laminated sheets, particularly for autoclave use. Entrapped air and excess resin are allowed to flow out laterally to the fibers over the entire area of one face of the laminate into a bleed reservoir. The advantage of this process is that the resin content in the laminate can be easily adjusted by the amount of bleeder fabric.

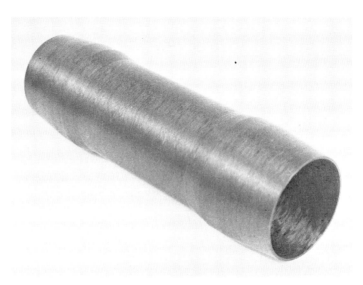

FIG. 8. Filamentary-wound structure: resin, Kerimid 353; fibers, Grafil AS; fabricated by Dornier System GmbH, West Germany.

TABLE III
Mechanical Properties of Poly(bismaleimide) Laminates
Resin: Kerimid 353

Property	E-Glass 25°C	E-Glass 250°C	Modmor II S 25°C	Modmor II S 250°C	Thornel 50 S 25°C	Thornel 50 S 250°C	Thornel 300(+) 25°C	Thornel 300(+) 250°C	Kevlar 49 25°C	Kevlar 49 250°C
60 % by Volume		(60 %)		(60 %)		(60 %)		(55 %)		(60 %)
Density g/cm^3	2,0-2,1		1,54-1,60		1,53		1,543		1,38-1,40	
Tensile Strength (N/mm^2) (NOL)	1050	1020	1100	1050	834	814	--	--	1200	750
Tensile Modulus (KN/mm^2)	42,2	38,0	170	162	176	173	--	--	62,1	50,0
Flexural Strength (N/mm^2)	1205	830	1450	1150	843	746	1284	1201	645	310
Flexural Modulus (KN/mm^2)	42	40	164	169	175	155	112	104	65	49
ILSS (N/mm^2) (LSD:D)	78 (5:1)	49	75 (4:1)	49	43 (4:1)	32	79 (4:1)	--	27 (4:1)	21
Flexural Strength (N/mm^2)	--	--	760	740	510	480	610	--	--	--
Flexural Modulus (KN/mm^2)	--	--	72	69	110	96	60,2	--	--	--

Note: (+) Slightly modified Kerimid 353; LS:D, span to depth ratio; ILSS, short beam shear strength; (NOL), ASTM D 2290-64 T.

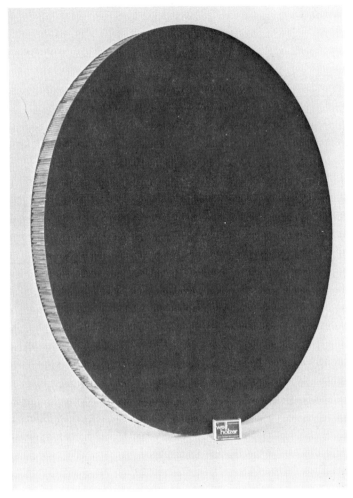

FIG. 9. Sandwich panel composed of facings of Thornel 50S and Kerimid 353. Fabricated by Dornier System GmbH, West Germany.

Laminate Properties

Basic laminate properties from different fibers (E-Glass, graphite, and Kevlar) are given in Table III. Unidirectional samples prepared by using the finger-mold techniques show typical values according to the properties of the selected fibers. Longitudinal tensile strength indicates good retention of properties at 250°C for E-Glass and the three carbon fiber types tested. Kevlar shows a 40% property loss which is due to the fibers property loss of 250°C. Very good flexural property retention is obtained for carbon fibers, up to 90% for Thornel 300 and Thornel 50S. The flexural properties of Kevlar are low because of the low compressive strength of the fiber which leads to a nonlinear flexural stress–strain behavior contrary to carbon fiber laminates.

The interlaminar shear strength depends on the fiber surface properties.

FIG. 10. Graphite/polybismaleimide reflector. Fabricated by Dornier System GmbH, West Germany.

Acceptable values were obtained for E-Glass (78 N/mm^2), Modmer IIS (75 N/mm^2), and Thornel 300 (79 N/mm^2) with a 65% retention up to 250°C. Low shear values were measured for Thornel 50S (43 N/mm^2) and Kevlar (27 N/mm^2), but it must be noted that with Kevlar no real shear failure mode was found.

Bidirectional (cross-ply) laminate properties (Table III) were evaluated from samples prepared by using the aforementioned restricted bleed technique. The advantage of this procedure is that laminates which are nearly void-free can be fabricated (Fig. 6).

Unidirectional glass and Thornel 50S laminates underwent thermooxidative aging at 250°C in air. The flexural property results presented in Figure 7 clearly show excellent retentions for 1000 hr under the conditions tested.

Approximately 90% of the initial flexural strength is retained with Thornel 50S and 75% with E-Glass fiber samples. The flexural modulus for both fibers is almost unaffected. As a point of interest, the weight loss for the Thornel 50S laminates is twice that of the glass fibers; possibly this carbon fiber supports oxidative degradation of the Kerimid 353 resin.

Component Fabrication

The component examples given were fabricated by Dornier System GmbH, West Germany, with the support of the German Space Agency. Filament winding was used to fabricate tube-like structures (Fig. 8). Filament winding was also used to fabricate carbon fiber prepregs which were used for facings of a honeycomb sandwich panel (Fig. 9) and a graphite–poly(bismaleimide) reflector (Fig. 10). The restricted bleed technique was used and the cure was performed by autoclaving in a vacuum bag following a cycle that was adjusted according to the requirements of the tooling and the heat capacity of the autoclave. The maximum pressure necessary to obtain the desired properties was 40 N/cm² and the maximal curing temperature was 210°C.

The author gratefully acknowledges support of this work by the Bundesminister für Forschung und Technologie, Bonn, and H. Stockburger, S. Roth, and F. Frick from Dornier System GmbH, who performed component development.

REFERENCES

[1] E. E. Sroog et al., *J. Polym. Sci. A, 3*, 1373 (1965).
[2] H. Vogel and C. S. Marvel, *J. Polym. Sci., 50*, 511, (1961).
[3] P. M. Hergenrother and H. H. Levine, *J. Polym. Sci. A-1, 5*, 145 (1967).
[4] F. Grundschober and J. Sambeth, U.S. Pat. 3,380,964 (1969).
[5] H. R. Lubowitz (assigned to TRW), U.S. Pat. 3,528,950 (1970).
[6] J. R. Elliott, Ed., *Macromolecular Synthesis, Vol. 2*, 111 (1966).
[7] H. D. Stenzenberger, *Appl. Polym. Symp., 22*, 77 (1973).
[8] D. O. Hummel, K. U. Heinen, H. D. Stenzenberger, and H. Siesler, *J. Appl. Polym. Sci., 18*, 2015 (1974).

ASSESSMENT OF QUALITY IN WOOL/POLYESTER FABRICS

MIRIAM SHILOH

Israel Fiber Institute, Jerusalem, Israel

SYNOPSIS

Studies were carried out on wool/polyester fabrics in their loom state, after heat-setting and after dyeing and final finishing. Properties, such as stress–strain and stress–relaxation performance, were measured on fibers which were withdrawn from the fabrics. Yarn properties were also determined, and then extensive tests were carried out in order to determine as many fabric properties as possible.

Attempts were made to relate fiber properties to fabric wrinkling and bending according to recent theories in which the relative importance of interfiber friction in wrinkling has been in dispute.

The various fabric properties were combined into indices of strength, comfort and appearance. The quality of the fabrics was then calculated by ranking preferences in property groups. By using weighted sums, the changes in fabric quality during finishing could be expressed numerically.

INTRODUCTION

The importance of wool/polyester blend fabrics as a major textile product is maintained through their satisfactory performance in many end-use commodities as well as through the economy of production.

Textile testing methods are applied to gain objective data which are capable of predicting the performance in the future. For instance, data derived from series of physical tests can provide information on a fabric's durability in wear, its contribution to comfort, and its aesthetic appearance after use. A preliminary study has been conducted [1] in which the relative importance of laboratory tested properties was estimated for various clothing items. A high priority was found to be given to the wrinkling and bending performance of fabrics which are designated for outerwear. It is well known that this performance is affected by the characteristic properties of the constituent fibers. Fabric wrinkling and bending are still subjects of many investigations, and they were recently discussed in some extensive studies by Chapman [2–5] and Denby [6–8].

It is known that by the incorporation of the polyester component in a blend with wool, the wrinkling performance is improved under dry and especially under wet conditions. Such improvements can be accomplished even at low polyester percentages [9] when suitable setting procedures are employed. It was also found

Journal of Applied Polymer Science: Applied Polymer Symposium 31, 105–115 (1977)
© 1977 by John Wiley & Sons, Inc.

[10] that for dry wrinkling about 15% of polyester yields good results, whereas for obtaining satisfactory wash-and-wear properties, about 50% polyester is to be recommended. Blends containing 45% wool and 55% polyester are typical products on today's markets, and they are usually considered satisfactory in quality.

In finishing wool/polyester fabrics, however, much attention should be given to utilizing the better qualities of the components so that their contribution to the overall performance would be optimal.

When wool/polyester fabrics are subjected to various finishing treatments, their properties are inclined to change. The measurement and control of properties as related to the processes is of utmost importance in order to avoid damages and to obtain optimal results. Then, when the relative importance of some properties is known, the test results can serve as a guideline for selecting the most suitable finishing procedures. In a recent study [10], attempts were made to use data from groups of measured properties in order to define "quality functions" for various fabrics. With the aid of such assessments, it is possible to generalize the effect of finishing on fabrics.

The present study attempts to assess fabric quality according to the method suggested, with a special emphasis on fabric wrinkling and bending characteristics as affected by the properties of the two fiber types in the blend.

THEORY OF BENDING AND WRINKLING

In his general attempt to predict recovery from bending in fabrics, Chapman [5] measures the residual set and the viscoelastic and frictional limits to recovery. By using a generalized linear viscoelastic relaxation function $B(\tau, t - \tau)$, Chapman relates the bending moment $M(t)$ of the fabric to the curvature $K(t)$ which is formed at a time t by the following Stieltje's integral form:

$$M(t) = \int_0^t B(\tau, t - \tau) \cdot dK(\tau) \pm M_0 \tag{1}$$

where M_0 is a constant frictional moment, $K(\tau)$ is the curvature of the fabric at the bending time τ, $t - \tau$ is the time elapsed after the application of the strain, and $B(\tau, t - \tau)$ can be considered as the generalized time-dependent bending rigidity for a bending moment $M(t)$ per unit fabric thickness.

It is clear that when time effects are negligible, Chapman's equation can be reduced to the simplified cyclic bending moment–curvature relationship, earlier suggested by Grosberg [11]:

$$M = M_0 \pm BK$$

where B is the bending rigidity, which is not time-dependent, and in which the curvature K is linearly related to the bending moment.

From eq. (1), the percentage recovery from bending $R(t)$ was derived and approximated as:

$$R(t) = \frac{B(0,t)}{B(t_R, t - t_R)} \cdot \left[1 - \frac{M_0}{K_0 B(0,t)} \right] = R_{VL}(t) \cdot R_{FL}(t) \tag{2}$$

when a constant creasing curvature K_0 is imposed at $t = 0$ for a period of t_R, and the fabric is allowed to recover freely. In this way, the total recovery can be represented as the product of the viscoelastic recovery $R_{VL}(t)$ and the frictional recovery $R_{FL}(t)$, both being recovery limits.

Test results for various fabrics have been used in order to calculate these recoveries. The calculated values for wool, polyester, and wool/polyester fabrics, after 5 min of creasing and 25 min free recovery, are shown in Table I. It seemed of interest to find out whether fiber stress-relaxation and fabric bending and wrinkling tests results can be compared with Chapman's approximated recoveries as given above.

EXPERIMENTAL

The fabric under study was chosen from a mill (Polgat Textile Mills, Kiryat Gat, Israel) in three stages: (A) in the loom state; (B) after scouring at 45°C, drying at 140°C and heat-setting at 180°C for 30 sec; and (C) after dyeing, decatizing at 115°C and final heat-setting as above. The fabric was a 2/2 Twill, 45/55 wool/polyester, of 64's Merino wool and 3 den Diolen polyester. The fabric consisted of 29 warp yarns (68 tex) and 20 weft yarns (47 tex) per cm, and its finished density was 360 g/m^2.

Fabric samples from the three finishing stages (A, B, and C) were subjected to a wide number of tests, according to procedures specified elsewhere [9]. In addition to fabric tests, fibers and yarns were also tested: original fibers before spinning as well as fiber samples which were withdrawn from the yarns and the fabrics, and also yarns before weaving and yarns which were removed from the three fabric types.

TENSILE TESTS

From the tensile tests (up to breakage), tenacity values (in gf/tex) were obtained for fibers, yarns, and fabrics. These values were then expressed in terms of their percentage change relative to the loom state, as represented in Figure 1.

Another series of tests was then carried out in an attempt to measure the viscoelastic properties of the wool and polyester fibers. Fibers were strained up

TABLE I
Percentage Recovery Limits (%)[a]

Fabric	$R_{VL}(t)$	$R_{FL}(t)$	$R(t)$
100% wool (de-aged)	36.7	90.2	33.1
100% Terylene	93.8	95.5	89.6
45 / 55 Wool/Terylene	71.5	89.4	63.9

[a] From Table I in Chapman [5].

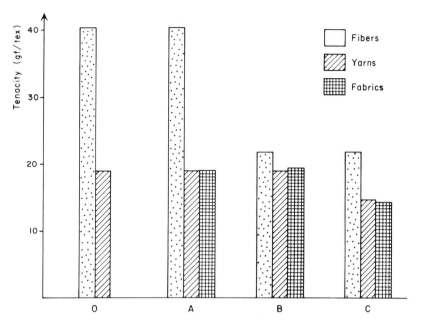

FIG. 1. Tenacity means of fibers, yarns, and fabrics: (O) before weaving; (A) loom state; (B) heat-set; (C) dyed and heat-set.

to 20% on an Instron Tester (at a speed of 0.5 cm/min) and were then left in the strained state for a period of 3 hr. Stress relaxation was measured after 1, 30, 60, and 180 min for each fiber. The identification of the fiber (wool or polyester) was made from the initial extension region. These tests were carried out on the original, unspun fibers, as well as on fibers withdrawn from the three fabrics.

The same experiment was also carried out when the fibers were immersed in water at 20°C for 2 hr before test, and the strain was applied on the fibers while being immersed in a glass tube which was attached to the Instron Tester. No significant differences could be observed between stress relaxation of fibers from the different fabrics and the original fibers. The mean results for stress relaxation in the wool and polyester fibers in air and for those of the wool fibers in water are plotted against relaxation time in Figure 2. The results are also summarized in Table II, together with the relaxation times, which were calculated by assuming a Maxwell unit in which $\mathcal{G} = \mathcal{G}_0 e^{-t/\tau}$.

It appears that the differences between the two fiber types are rather small, much smaller than those that could account for the large differences in recoveries from bending.

Some other mechanical tests, such as flat and flex abrasion and bursting strength tests, were also carried out on the fabrics. The results of the mechanical tests are represented in Figure 3 in percentages of the same values in the loom-state fabric.

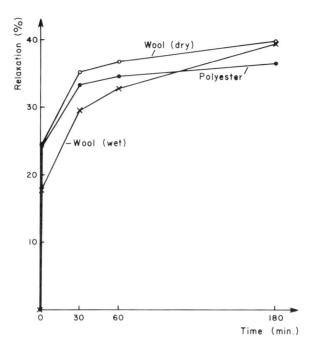

FIG. 2. Fiber stress relaxation (20% extension).

TABLE II
Fiber Stress Relaxation (at 20% Extension)

	Polyester				Wool (dry)				Wool (wet)			
Time (Min)	1	30	60	180	1	30	60	180	1	30	60	180
Mean (%)	24.2	33.4	34.6	36.6	24.4	35.2	36.9	39.9	17.8	29.6	32.9	39.6
S. Dev. (%)	3.0	4.1	4.2	4.4	4.7	4.3	4.0	4.6	1.4	3.0	3.8	5.9
C.V. (%)	12.3	12.3	12.0	11.9	19.1	12.3	10.9	11.6	8.0	10.3	11.6	15.0
τ (hr)	10.5				9.0				7.9			

BENDING AND WRINKLING TESTS

The fabrics were subjected to cantilever bending tests, drape tests, and crease recovery-angle determinations under standard atmospheric conditions. All the results were again expressed as percentages of those of the loom-state fabric, as shown in Figure 4.

In addition to these, cyclic bending tests were carried out on the Shirley Bending Tester [9], from which the bending hysteresis curves were traced for both the warp and the weft directions, as shown in Figures 5a and 5b. The derived values of the bending rigidities (B) and frictional couples (M_0) are summarized in Table III.

In Table III, the results of B and M_0 are given for each fabric direction sep-

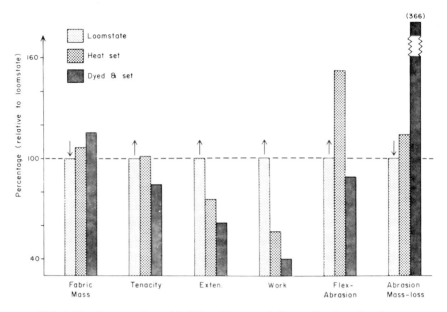

FIG. 3. Tensile properties and finishing. The arrow indicates direction of preference.

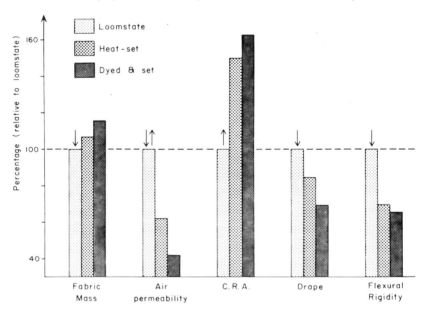

FIG. 4. Performance properties and finishing. The arrow indicates direction of preference.

arately. As could be expected, these parameters differ widely from each other due to the twill structure. Attempts were made to apply these results to theory, and for this purpose, it was necessary to use fabric means. Therefore, in spite of the large differences between the directions, mean values of M_0/B were calculated. These were 2.61, 1.31, and 0.85 cm^{-1} for the fabrics A, B, and C, respectively.

TABLE III
Cyclic Bending Test (W/P Twill)

Fabric:	Warp			Weft		
	A	B	C	A	B	C
Flexural Rigidity (dyn cm^2/cm) - B	600	850	375	100	175	150
Frictional Couple (dyn cm/cm) - Mo	1025	775	200	350	300	175
Residual Couple (cm^{-1}) - Mo/B	1.71	0.91	0.53	3.50	1.71	1.17
C.R.A. (^{o})	119	154	171	89	158	167

Note: (A) loom state; (B) heat-set; (C) dyed and heat-set.

TABLE IV
Recoveries from Bending and Creasing

Fabric	M_o/B (cm^{-1})	$R_{FL}(t)$ (%)	$R(t)$ (%)	Crease Recovery (%)
A	2.61	73.9	52.8	57.8
B	1.31	86.9	62.1	86.7
C	0.85	91.5	65.4	93.9

On the basis of the tensile stress-relaxation data as obtained for the wool and polyester fibers in the present work, it may be assumed that also in their stress relaxation from bending, the viscoelastic properties of these fibers were not affected by the finishing treatments. With this assumption, Chapman's viscoelastic recovery $R_{VL}(t)$ in eq. (2) should remain constant for all three fabrics, while the relaxation due to interfiber friction $R_{FL}(t)$ should vary with the measured B and M_0 values, according to its definition:

$$R_{FL}(t) = 1 - M_0/B(0,t)K_0$$

For the sake of this comparison, let us assume that in crease-recovery tests, the samples were creased at the time $t = 0$ for a period of 5 min to a constant curvature of $K_0 = 10$ cm^{-1} (e.g., the radius of curvature being equal to 1 mm). Then let us assume that the viscoelastic recovery of the wool/polyester fabric in the present study is the same as that of Chapman's fabric, namely, 71.5%. With these assumptions, it is possible to calculate the limiting recoveries of the fabrics, as shown in Table IV.

When crease-recovery angles are expressed as percentage recoveries, these show much larger values than the calculated $R(t)$ recoveries and are more closely related to the recovery due to fiber and yarn friction in the fabrics.

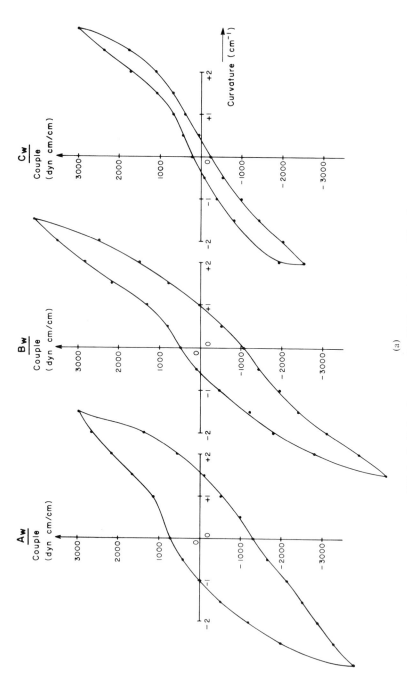

FIG. 5. Cyclic bending tests using W/P twill: (a) warp; (b) weft.

(a)

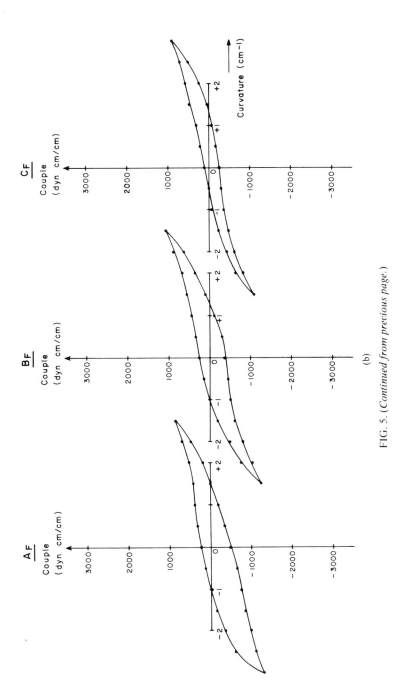

(b)

FIG. 5. (*Continued from previous page.*)

The limiting total recovery of the present fabric in the final finished stage is 65.4%, which is slightly higher than that found by Chapman (63.9%). As no details are given on the structure of the fabrics, however, such a comparison cannot be carried out. At this stage also, any drawing of conclusions cannot be justified. For a better understanding of the role of viscoelastic properties and interfiber friction in bending and wrinkling, well-controlled experiments should be made. Only then would it be possible to test the validity of the mentioned rheological model for the prediction of fabrics wrinkling performance.

ASSESSMENT OF QUALITY IN FINISHING

All the measured data was ranked according to the procedure recommended by Proskuryakov [12] and modified previously [10]. This was based on the results which were presented in Figures 1, 3, and 4. All the properties were summed up in respective groups, the order of preference was listed, and the quality function could be calculated, as illustrated in the following matrix of scores of preferences:

$$Q_m = \sum_{i=1}^{n} \hat{\varsigma}_i \hat{\beta}_i$$

FABRIC	Strength	Comfort	Appearance
A:	1	0	0
B:	2	1	1
C:	0	2	2
$\hat{\beta}_i$:	0.37	0.38	0.25

$$Q_A = 0.04$$
$$Q_B = 0.15$$
$$Q_C = 0.14$$
$$\hat{\varsigma}_i = \varsigma_i / 9$$

From this calculation, it seems that by heat-setting once, quality improved from 0.04 to 0.15. After dyeing and a second heat-setting treatment, it became slightly detrimental, mainly as a result of the tensile losses and in spite of improvements in hand and dimensional stability.

Another illustration of using this method for tracing the effect of finishing is the following: a lighter weight 45/55 wool/polyester was subjected to a stretch modification treatment, attempting to gain better performance through a chemical and mechanical finishing process [13]. The quality function was calculated to be 0.37 before finishing and 0.63 after finishing; thus, the total improvement gain could be quantified as being about 70%. Similar calculations can be made after any processing treatments for the better utilization of fibers in blend fabrics.

REFERENCES

[1] A. Knoll and M. Shiloh, *Text. Inst. Ind.*, *14*(4), 128 (1977).
[2] B. M. Chapman, *J. App. Polym. Sci.*, *18*, 3523 (1974).
[3] B. M. Chapman, *Text. Res. J.*, *45*, 137 (1975).
[4] B. M. Chapman, *Rheol. Acta.*, *14*, 466 (1975).
[5] B. M. Chapman, *Text. Res. J.*, *45*, 825 (1975).
[6] E. F. Denby, *J. Text. Inst.*, *65*, 239 (1974).

[7] E. F. Denby, *J. Text. Inst.*, *65*, 246 (1974).
[8] E. F. Denby, *J. Text. Inst.*, *65*, 250 (1974).
[9] M. Shiloh and R. I. Slinger, *SAWTRI Tech. Rep., 1972*, 181.
[10] M. Shiloh, *Schriftenreihe DWI* (Proceedings of the 5th International Wool Research Conference, Aachen, 1975), *4*, 571 (1977).
[11] P. Grosberg, *Text. Res. J.*, *36*, 205 (1966).
[12] A. Proskuryakov, *Technol. Text. Ind. U.S.S.R.*, *2*, 28 (1975).
[13] I. Kaplan and M. Shiloh, *Amer. Dyest. Rep.*, *64*, 54 (1975).

TENSILE PROPERTIES OF RELAXED AND UNRELAXED PAN FIBERS

J. BANBAJI

Israel Fiber Institute, Jerusalem, Israel

SYNOPSIS

The tensile properties of relaxed and unrelaxed PAN staple fibers prepared by breaking tow filaments, as characterized by their load extension curves, are investigated in water at different temperatures. They are then analyzed and compared with the tensile properties of fibers taken from crimped unbroken tows. In all cases and up to their glass transition temperatures, three defined regions, whose slopes and ultimate parameters are differently dependent on temperature, appeared in the curves. In tow fibers, a "flow" region appeared from about 40°C. This region emerged at a much higher temperature, i.e., 80°C in the relaxed and unrelaxed fibers. Related structures responsible for the various regions in the load extension curve are discussed.

The tensile properties of polyacrylonitrile (PAN) fibers, as characterized by their load extension curves, are generally found in the literature as part of comprehensive studies [1, 2] comparing the tensile properties of various synthetic fibers. Other authors, such as Rosenbaum [3], are more specific and deal with the tensile response of PAN fibers as a function of increasing temperatures. Rosenbaum discusses the various structural components in the fiber. On the other hand, the dynamic mechanical properties are used [4] to explain the relaxation mechanism of polyacrylonitrile as related to its structure.

If a good deal of research work has been done on the as-spun acrylic fiber, little was published on its tensile properties after or during processing treatments. During the bulking process [5], acrylic fibers exist in two structural forms: relaxed and unrelaxed. The unrelaxed or shrinkable fibers, are prepared by extending tow filaments, until breakage, on tow-to-top conversion equipment, under the influence of heat. A portion of these fibers is relaxed in an autoclave and then blended with unrelaxed fibers in the ratio of 60:40 (relaxed:unrelaxed). During the subsequent bulking process, the unrelaxed fibers shrink and compel the relaxed fibers to buckle, producing a high-bulk yarn. The amount of bulkiness acquired by the yarn is obviously a function of the starting structural properties of both types of fibers as well as of the bulking conditions. As a first step in studying the bulking mechanism of an acrylic fiber and its dependence on the above-mentioned parameters, an investigation of the tensile properties of the starting and final structure and its comparison with those of the regular tow fibers would be of interest. This is the aim of the present paper.

Journal of Applied Polymer Science: Applied Polymer Symposium 31, 117–125 (1977)
117

EXPERIMENTAL

The acrylic fiber tow used in this work was a regular production Acrilan 16 Monsanto (3 denier). Tow-to-top conversion and relaxation of the fibers were both carried out in industrial conditions, using a Zeidel converter and conventional autoclaves, respectively. The tensile tests were performed in water at various temperatures ($\pm 0.5°C$) up to 80°C, on a Table Model Instron Tensile Machine, at a rate of extension of 0.2 cm/min using a gauge length of about 4 cm. The diameter of each fiber was measured before the tensile test on a vibroscope at 65% RH and room temperature, and all stress values presented, referred to this measured diameter. Four to six fibers were extended to break on the Instron, and the desired parameters along the load extension curve averaged.

RESULTS AND DISCUSSION

Load Extension Curve of Acrylic Fibers

In water and in air at room temperatures, regular, relaxed, and unrelaxed acrylic fibers exhibit a characteristic load extension curve. This form remains constant for regular fibers tested in air under normal conditions, for several rates of extension, ranging between 0.005 and 10. cm/min. The curve is found to consist of three defined regions (see Fig. 1). The first is a linear region where the load increases rapidly and linearly with the elongation, up to a yield stress value which occurs at about 2% extension. This region defines the initial modulus. The second is a yield region where the load increases very smoothly up to an

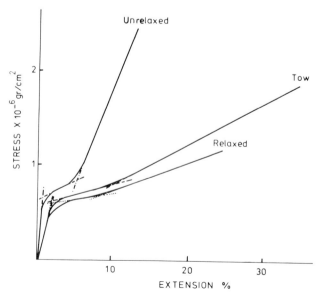

FIG. 1. Load extension curves of tow, relaxed, and unrelaxed polyacrylonitrile fibers.

extension value of about 9%. Finally, there is a postyield region where the load increases more steeply with the extension until breakage of the fibers.

In the regular acrylic fiber, a "flow" region appears at about 40°C (see Fig. 2). In this region, no substantial additional load is needed to extend the fiber in the relaxed and unrelaxed fibers. This region appears at temperatures above the T_g (see below).

It is of interest to note the strong similarity of the load extension curve of wool fibers [6] to that of the acrylic fibers in water at room temperature. In both cases, the linear region which occurs at about the same extension (~2%) is followed by a sharp bending of the curve to a yield region with a slope much lower than that of the linear region. The yield region in the wool fiber extends to a 30% extension, while that of the acrylic extends to 9% only. Moreover, the similarity may go further, to the morphology of both fibers, where it is accepted that the wool fiber consists of a microfibrillar phase embedded in a matrix of different structure [7]. In the acrylic fiber, the fibrillar structure was shown [2, 8] to exist by means of electron microscopy techniques. Actually, the similarity ends here because the wool fiber shows an $\alpha \rightarrow \beta$ crystalline transition [9] when extended to 30%, while acrylic fibers do no show any crystallographic change [10] on extension.

The Linear Region and Its Change with Temperature

Figure 3 shows the behavior of the initial modulus against increasing water temperature, for all three types of fibers. It is seen that the moduli of the regular and relaxed fibers decrease with temperature in a parallel way, up to zero value; this is extrapolated to be at about 70°C. The unrelaxed fiber's modulus starts decreasing from a remarkable higher value to zero at the same extrapolated temperature. This behavior implies that as far as initial modulus is concerned,

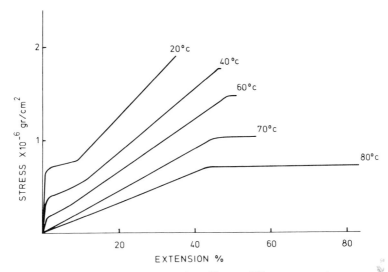

FIG. 2. Load extension curves of tow fibers at different temperatures.

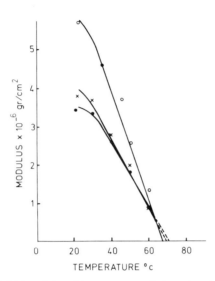

FIG. 3. Change of the initial modulus with temperature: (O) unrelaxed fiber; (×) tow fiber; (●) relaxed fiber.

the unrelaxed fiber, which is in a metastable state, is essentially restored after relaxation to the initial form of the regular as-spun fiber.

The common value of the temperature at which the initial modulus disappears may be attributed to a glass transition. This value is close to that obtained by other workers [3, 11, 12] and may also be determined from the absence of a linear region in the load extension curve.

Figures 4 and 5 show the temperature dependence of the ultimate parameters in the linear region, namely, the stress and extension of the yield point. These parameters are obtained from the intersection of the tangents extended from the linear and the yield regions. It is seen from Figure 4 that for all fibers, the yield stress falls to zero linearly at the same transition temperature as that obtained in the initial modulus case (see Fig. 3). The yield extension behaves differently. This parameter (see lower curves in Fig. 5) remains essentially constant at all temperatures below the T_g, and falls to a zero value just at the transition temperature.

Yield and Postyield Regions and Their Change with Temperature

Figures 5 (upper curves) and 6 show the ultimate extension and stress values of the yield regions of all fibers, respectively. These may be designed as postyield extension and stress points. It is seen from Figure 5 that the yield regions of the tow and relaxed fibers extend up to about 9–10% at room temperature and then reduce to about 7% for both fibers. The tow fibers do not show a postyield point at 70°C, while the relaxed fiber loses it at 60°C. This is also found in Figure 6, where the postyield stress of the relaxed fiber is not distinguishable at 60°C.

The yield region of the unrelaxed fiber behaves differently. The region is shorter (~7%), decreases very slowly to 5.5% and increases abruptly at the T_g, to the initial value of the relaxed fiber. This is peculiar because of the appearance

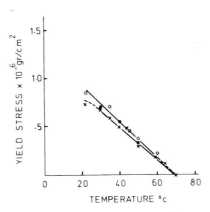

FIG. 4. Change of the yield stress with temperature. Same symbols as in Fig. 3.

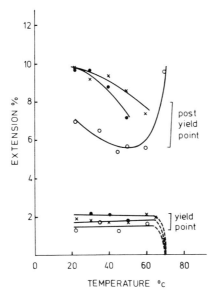

FIG. 5. Change in the extension at the postyield point with temperature. Same symbols as in Fig. 3.

of the postyield region around the T_g in the unrelaxed fiber. It is obvious that a complex system of interactions of time and temperature took place during the testing of the unrelaxed fiber around the T_g. No postyield region was recorded above the T_g, e.g., at 80°C. It is possible that transient time effects are responsible for the appearance of a postyield region around the T_g in the unrelaxed fiber.

Figures 7 and 8 show the slopes of the yield and postyield regions of all fibers against increasing temperature. The slope is expressed by the ratio of the load applied on a corresponding extended length of the fiber. Interesting features may be observed from these figures. It is seen from Figure 7 that in the yield region of the tow and relaxed fibers, there is an increase in the values of the slopes up to about 60°C. Above this temperature, these values decrease and converge

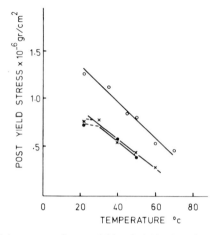

FIG. 6. Dependence of the stress at the postyield and yield points. Same symbols as in Fig. 3.

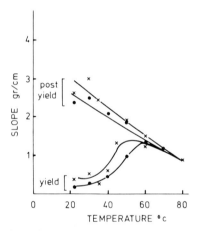

FIG. 7. Change of yield and postyield region slopes with temperature for (×) tow and (●) relaxed fibers.

to the common values of the decreasing postyield slopes. As to the unrelaxed fiber, Figure 8 shows that the slopes of the postyield region are high compared to those of the yield region and fall continually until 80°C.

The increase of the slopes as presented in Figure 7 means, in actuality, an increase of force for different lengths of the fiber. According to the rubber-like elasticity theory [13], an increase in force with temperature at a constant length is equivalent to a decrease of entropy with length at constant temperature. On the other hand, a decrease of entropy is an increase in order at the molecular level. Although the tests were not carried out under reversible conditions, as theoretically required, the results indicate an increase of force with temperature, pointing to an entropy contribution to the deformation. This contribution is likely to exist along the yield portion of the curve in the tow and relaxed fibers.

Along the same line of reasoning, it may be stated that the decrease of force with temperature in the postyield regions of all fibers is due to internal energy

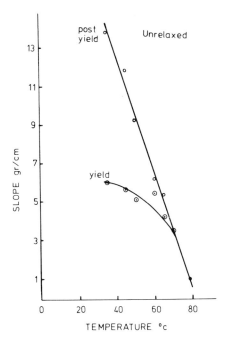

FIG. 8. Change of yield and postyield region slopes with temperature for the unrelaxed fibers.

contribution, which may be attributed to intermolecular stretching of the main chains in the structure. This view is partially supported by Rosenbaum [3], who suggested this behavior for both the yield regions between which there was no differentiation in his results.

The Flow Region and the Breaking Parameters

Figures 9 and 10 show the change of the breaking extensions and stresses versus temperature. The former increase slowly at first, and then at about a temperature considered to be the T_g, rise sharply. The breaking extensions of the relaxed fibers are found to be lower than those of the tow fibers. also, their T_g is lower by some 10°C less than in the tow fiber. This means that during recovery of the unrelaxed fiber, no complete relaxation of the fiber occurred and a higher chain mobility was maintained as evidenced by the lower T_g.

It was already mentioned that a flow region appears in the tow fiber from about 40°C, and only above the T_g in the relaxed fiber. Table I presents the breaking extensions of the tow fibers as according to their components.

It is seen from Table I that the extension in the postyield region remains essentially constant from about 40–50°C, and that the total extension which rises sharply at 70°C is due solely to the flow region. This means that the flow region and not the postyield region is associated with the T_g in tow fibers.

It is worthwhile to mention that the constant extension in the postyield region levels off at 40–50%. On the other hand, Rosenbaum [3] reported high recovery of acrylic fibers after extending them to about 40%. Above this extension, re-

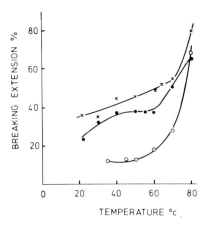

FIG. 9. Change of the breaking extensions with temperature. Same symbols as in Fig. 3.

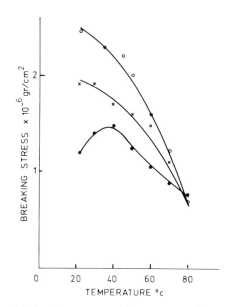

FIG.10.   Change of the breaking stresses with temperature. Same symbols as in Fig. 3.

covery decreases. In the present paper, the flow region which appears in the relaxed fiber only above T_g emerges only at extensions higher than 45%. These features lead to the conclusion that the postyield region is due to stable and recoverable structures, while the flow region is due to nonrecoverable structures which break ultimately.

CONCLUSIONS

The load extension curves of tow, relaxed, and unrelaxed acrylic fibers consist essentially of three regions below the T_g: a linear region, a yield region, and a

TABLE I

Extensions (%) of Tow Fibers in Their Respective Regions at Different Temperatures

Temp °C	Linear	Yield	Postyield	Flow	Tot. Extension
22	1.9	9.8	24.5	–	36.2
30	1.8	9.2	23.6	–	34.6
40	1.7	9.4	34.0	some	45.1
50	1.6	8.6	42.8	3.1	45.9
60	2.1	7.4	43.6	5.3	48.9
70	–	–	44.9	10.2	55.1
80	–	–	41.3	38.7	80.0

postyield region. A flow region, negligible in its extent, starts to appear in the two fiber.

Above the T_g, the linear region in all fibers disappears, the postyield region becomes the continuation of the yield region, and the flow region becomes dominant.

The deformation in the yield regions of the tow and relaxed fibers may be due, up to about 60°C, to entropy contribution. The postyield regions of all fibers are due to internal energy contribution.

Based on the high recoverability occurring up to extensions of 40–45% and to the starting of the flow region thereafter at all temperatures investigated, it is suggested that a well-defined and recoverable phase in the structure is responsible to all extensions up to the postyield region.

The tensile properties of acrylic fibers are not fully restored after industrial relaxation of stretch broken fibers.

The author wishes to express his thanks to Prof. M. Lewin, Director of this Institute, for useful discussions and for his encouragement to publish this paper; to Dr. S. Dershowitz for reviewing the manuscript; and to Mrs. E. Engel for carrying out experimental work.

REFERENCES

[1] I. N. Hall, J. Appl. Polym. Sci., 12, 731 (1968).
[2] R. Meredith, Mechanical Properties of Textile Material, Interscience, New York, 1956.
[3] S. Rosenbaum, J. Appl. Polym. Sci., 9, 2071 (1965).
[4] S. Minami, Appl. Polym. Symp. No. 25, 145 (1975).
[5] F. Malaguzzi and G. C. Sala, in High Bulk Yarns (Fourth Shirley International Seminar), 1971.
[6] M. Feughelman, Text. Res. J., 34, 539 (1964).
[7] M. Feughelman, Text. Res. J., 29, 223 (1959).
[8] J. P. Craig, J. P. Knudsen, and V. F. Holland, Text. Res. J., 32, 435 (1962).
[9] E. C. Bendit, Text. Res. J., 30, 547 (1960).
[10] C. R. Bohn, J. R. Shaefgen, and W. O. Stratton, J. Polym. Sci., 55, 531 (1961).
[11] Z. Gur-Arieh and W. Ingamells, J. Soc. Dyers Col., 90, 12 (1974).
[12] S. Rosenbaum, J. Appl. Polym. Sci., 9, 2084 (1965).
[13] L. R. G. Treloar, The Physics of Rubber Elasticity (2nd ed.), Oxford, 1958.

THERMAL RETRACTION OF POLYESTER POY BY A NOVEL SEMIMICROTECHNIQUE

G. LOPATIN

Israel Fiber Institute, Jerusalem, Israel

SYNOPSIS

A novel and convenient method for measuring thermal retraction of synthetic yarns is described. Short lengths of yarn are confined in capillary tubes and observed while, or after, being heated to the selected temperatures. Shrinkage versus temperature curves for polyester partially oriented yarn (POY) are shown and discussed in terms of crystallinity models and more specifically of POY structure. With respect to their potential as a measure of orientation, the thermal retraction data of three POY samples are compared to the natural draw ratios and sonic moduli of the samples.

INTRODUCTION

Thermal retraction, i.e., shrinkage of filaments, is often used to investigate the structure of oriented systems, especially of synthetic fibers. It has been used to measure the spinning orientation of melt-spun polyethylene filaments [1], elucidate a model of crystallinity [2], and investigate crystallite characteristics of drawn filaments [3]. In general, the techniques are fairly tedious or they require specialized equipment [4], such as a DuPont Versatile Chemical Analyzer. We have used a relatively simple semimicrotechnique which should be convenient for workers in many polymer laboratories.

The technique was applied to the study of partially oriented polyester yarns (POY). POY is a new feed yarn designed for the draw-texturizing process [5]. It is produced without cold-drawing, but enough orientation is imparted by a very high drawdown during spinning to give sufficient stability for further processing. The cold-drawing step, still required for full orientation, is postponed until the yarn is draw-textured. Advantages such as improved dyeability and lower production costs are claimed for POY [6].

From a morphological standpoint, very high drawdown from the melt without cold-drawing puts POY, before it is texturized, into a new metastable structural state, intermediate between as-spun yarns and the usual drawn, crystallized yarns. Therefore, studies of POY structure should be important for their relevance to the fundamental understanding of fiber structure as well as for their technological interest. Our technique of measuring thermal retraction of fibers and some of the more noteworthy results, especially on POY, are described in this paper.

Journal of Applied Polymer Science: Applied Polymer Symposium 31, 127–132 (1977)

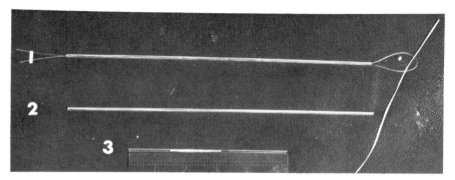

FIG. 1. Capillary technique for thermal shrinkage.

EXPERIMENTAL

Thermal retraction of the yarn was measured by observing its shrinkage within a capillary tube. The capillary tube was of the open-ended thin-walled "melting point" type of about 0.5 mm inner diameter and as long as convenient. The longer the sample, the greater the sensitivity of the measurements. The yarn was put into the capillary by first pushing the two ends of a nylon monofilament (0.2 mm diameter) through the capillary, leaving a loop at one end (Fig. 1, part 1). The yarn was inserted into the loop and it was carried through the capillary by pulling on the monofilament ends until the yarn extended from both ends of the capillary. It was then merely cut off (Fig. 1, part 2). Measurements were made with a millimeter scale through a 10× magnifying glass. Figure 1 (part 3) shows a yarn after shrinking.

To eliminate resistance to shrinkage by yarn–glass friction at the capillary wall, the following means have been used: spraying the yarn with silicone oil, coating the yarn with silicone oil, or even filling the capillary tubes with silicone oil, i.e., a semimicrosilicone bath. The surface tension effects in the small capillary, i.e., capillary action, are sufficient to keep the oil in the tube. For measurements on POY below the melting temperature, spraying the yarn was sufficient for reproducible and reliable measurements. For measurements above T_m, however, the yarns had to be thoroughly coated with oil; even so, data above T_m are difficult to obtain. The molten filament is often broken.

The capillary containing the yarn was heated at about 20°C/min on a Kofler Hot Stage (C, Reichert Optiche Werke AG, Vienna, Austria) and the measurements made at 10°C intervals while heating to obtain the shrinkage curve:

$$\% \text{ Shrinkage} = \frac{L_{\text{final}} - L_{\text{initial}}}{L_{\text{initial}}} \times 100$$

An alternative method which was used to obtain the overall percent shrinkage was to place five tubes containing yarns between the platens of a Carver laboratory press set at 121°C. After 5 min, the tubes were removed and the length of yarn measured. The data in this latter method show a standard error of ±0.3.

The POY yarns tested were obtained from three manufacturers, Monsanto Chemical Co., USA (POY A); SNIA Viscose Ltd., Italy (POY B); and Hoechst GmBh, Germany (POY C); however, these are not necessarily their standard products. The three samples are all 270 denier/34 filaments, and since receipt were maintained at or below 23°C.

The POY flat yarn was a standard product for false twist texturizing of Hoechst—Trevira 150/34. The as-spun polyester yarn was spun at the Israel Fiber Institute (364 den/20 fil) and had been stored longer than 1 year before these shrinkage tests.

The natural draw ratio was obtained at 21°C at 100%/min extension on an Instron Tensile Tester.

The Sonic Moduli of the POY samples were measured on the Morgan Dynamic Modulus Tester, PPM 5 (H. M. Morgan Co., Inc., Cambridge, Mass.).

RESULTS AND DISCUSSION
Morphology

Figure 2 shows shrinkage curves of polyester POY (B), a fully oriented yarn (FOY) (A), and a cold-drawn (1.6× at room temperature) POY sample (C).

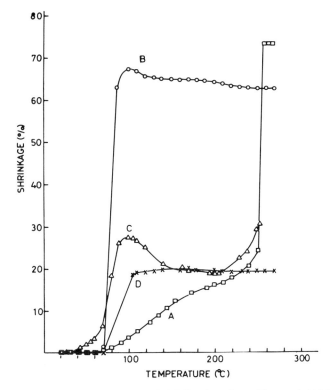

FIG. 2. Thermal retraction of polyester yarns: (A) fully oriented yarn (flat yarn); (O) (B) partially oriented yarn (POY); (△) (C) POY hand-stretched 1.6×; (×) (D) as-spun yarn.

Of the several interesting features of the curves, one is especially noteworthy. These curves are a direct experimental corroboration of the hypothetical curves suggested for polyester yarn by Bosley [2] to help elucidate his concept of functional crystallinity. The curves are remarkably identical to those postulated by Bosley, and his explanation appears very adequate. The POY sample is a relatively highly oriented but substantially amorphous yarn. Thus, at about the T_g (which is somewhat raised here because of the internal stresses arising from the orientation), the segments are freed to move. They randomize, the orientation is eliminated, and the yarn shrinks back to a random morphology (curve B).

The FOY yarn (curve A) will shrink only when sufficient immobilizing crystallites melt [3]. As the temperature rises, more of the oriented structure is freed by the increased melting of crystals. There is increased randomization (loss of orientation), and the yarn shrinks to greater and greater extents.

In the POY yarn drawn cold, the stretching induces partial crystallization so that some oriented amorphous material is removed and immobilized. Thus, although the yarn is overall more highly oriented because of the reduced amount of amorphous (i.e., mobile) structure, there is less retraction in the region above the T_g. The yarn does not shrink further until sufficient of the crystalline structure is melted. In curve C, then, the shrinkage of the amorphous portions is very well separated from the shrinkage of the crystalline portions. The case thus seems a good example of a separation of polymer structure into two relatively distinct regions, i.e., amorphous and crystalline.

With respect to POY, several important points are shown by the data. (1) Above the T_g, the POY structure is substantially completely mobile. (2) There is little, if any, crystallinity. Any crystallinity that may be present is below the limit of interference with segment mobility. From Bosley's view, this would mean zero crystallinity. (3) There is, however, a spontaneous elongation or auto-orientation after the shrinkage above the T_g. The Differential Scanning Calorimeter (DSC) (Differential Thermal Analyzer, E. I. du Pont de Nemours & Co., Inc., Wilmington, Del.) curves in Figure 3 show the crystallization exotherms and melting endotherms of POY and FOY. It is seen that the crystallization temperature range of POY corresponds to the spontaneous lengthening observed in the thermal retraction measurements, showing that the spontaneous lengthening is related to the crystallization of the POY. Thus, since crystallization is taking place, the elongation is probably due to epitaxial crystallization (shish kebab?) on oriented crystal nuclei present in the original POY and persisting above T_g. This spontaneous elongation is seen in the shrinkage curves of all POY samples.

In this connection, the shrinkage curve of an as-spun polyester yarn is also shown in Figure 2. Here again, the T_g is clear and crystallinity is absent, but there is much less shrinkage of the yarn, showing the decreased orientation in ordinary as-spun yarns. Furthermore, in contrast to POY behavior, there is no spontaneous elongation. The relatively high anisotopic morphology of POY therefore probably includes some highly oriented supermolecular structures which are stable above T_g. These serve as nuclei or substrates for further orientation during crystallization even in the absence of a tensile stress.

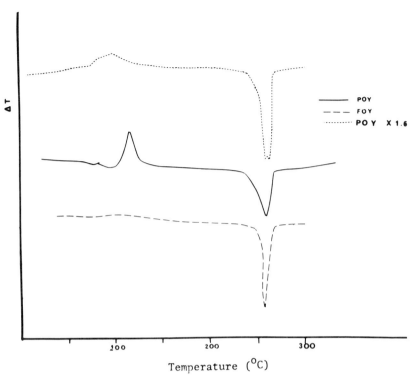

FIG. 3. Differential scanning calorimeter (DSC) traces of polyester yarns.

TABLE I
Relative Orientation Measurements of Partially Oriented Yarn (POY)

Sample	Shrinkage %	Natural Draw Ratio	Sonic Modulus g/den
POY A	67.2 ± 0.3	1.58	35.0
POY B	66.8 ± 0.3	1.60	34.2
POY C	63.3 ± 0.3	1.69	28.9

A Measure of Orientation

Measurements of overall shrinkage relate directly to the degree of orientation of the structural units in the fiber (e.g., crystallites or molecules), provided that there is no viscous flow during the retraction (i.e., so-called nonflow shrinkage) [3]. The shrinkage is an entropy-induced retraction as is the retraction of stretched rubbers [2]. The molecules pass from an oriented state, parallel to the fiber axis, to a random state. Thus, overall shrinkage can give a measure of structural orientation, and the capillary method has been used to compare the orientation of various POY samples.

In Table I, the percent shrinkage of three samples of POY (at 121°C) are shown together with the natural draw ratio of the three samples, an important parameter for draw-texturizing. Sonic modulus data, another measure of orientation for polyester [7], are also shown. An excellent correlation between the shrinkage values, the natural draw ratio, and the sonic moduli is seen, supporting the use of thermal retraction as a measure of orientation.

The author would like to thank Prof. M. Lewin for useful discussions, Mr. Dror Selivansky for some of the data and his helpful suggestions, Portchester Ltd. for encouragement of the project, and Portchester Ltd., Ashkelon, and Nilit Ltd., Migdal Haemek, for supplying the POY samples.

REFERENCES

[1] J. Furukana, T. Kitao, S. Yamashit, and S. Ohya, *J. Polym. Sci. A-1, 9,* 299 (1971).
[2] D. E. Bosley, in *J. Polym. Sci. C, 20,* 77–107 (1967).
[3] D. Prevorsek and A. U. Tobolsky, *Text. Res. J., 33*(10), 795 (1963).
[4] R. Hagege and D. M. Tung, *J. Appl. Polym. Sci., 16,* 1427 (1972).
[5] H. Weinsdoerfer and G. Egbers, *Text. Res. J., 45*(9), 654 (1975).
[6] O. L. Shealy and R. E. Kitson, *Text. Res. J., 45*(2), 112 (1975).
[7] W. W. Moseley, Jr., *J. Appl. Polym. Sci., 3,* 266 (1960).

EXPERIMENTS ON FLY REDUCTION IN CARDING

B. GUTMAN, J. PEREL, and M. SHIMSHONI

Israel Fiber Institute, Jerusalem, Israel

SYNOPSIS

It was demonstrated that in the carding process, the quantity of cotton fly generated at the comb depends, among other parameters, on the electrostatic charge on the web in that area. At low relative humidity, the fly generation was significantly reduced by application of "antistatic" coatings to the comb. Other experiments showed that the efficiency of the coatings depends strongly on the conditions of the combing process.

INTRODUCTION

The cotton industry is presently faced with the need for a drastic reduction of cotton fly and dyes generated in the various stages of cotton processing (ginning, drawing, carding). Although the fly generation in these processes has not been investigated thoroughly, it is known that many factors are involved in it. The problem is complicated still more by the mutual interdependence of their effects. In the present work, we report on some experiments on the effect of electrostatic charges on the fly generation in the zone of the comb in the carding process.

GENERAL

The assumption on which this investigation rests was that fly generation in cotton carding is partly enhanced by the interaction of electrostatic charges generated on the fibers during the process. Therefore, application of "antistatic" coatings to appropriate parts of the card should reduce the quantity of fly produced in this process. This was tested on a pilot carding machine (Shirley miniature spinning plant, Platt Brothers). On this card, the largest portion of the fly is produced at the comb. The vigorous mechanical action of the comb on the web, the bulkiness of the web, and the air currents produced by the comb enhance fly generation in this zone. The nature of the comb–web interaction also favors generation of static electricity. For these reasons and because of the relative ease of handling, the combing zone was chosen for our tests.

The web leaves the doffer positively charged. The metal comb also charges the web with a positive charge. In order to obtain the antistatic effect, the comb

Journal of Applied Polymer Science: Applied Polymer Symposium 31, 133–136 (1977)
133

was coated with a polyurethane-based hard lacquer which charges the web with a negative charge. Since only a small portion of the web comes into contact with the comb, the web is charged positively on leaving the coated comb zone, but the density of the charge is lower than in the case of an uncoated comb. The coating may inhibit fly production by three different mechanisms. (a) The coated comb charges the "potential" fly particle negatively, thus either diminishing the repulsive forces between the potential fly particle and the web or creating attractive forces between them. Obviously, this action diminishes the probability for the particle to leave the web. (b) The coating is positively charged. This produces an "antisticking" effect with a positively charged web, although negative charges induced on the comb proper reduce its value. (c) The electrostatic charge on the web tends to make the web bulky, enhancing fly production. Reduction of the charge on the web produced by the coating inhibits this effect. This is true especially in the comb zone, in which the comb motion produces extensive air currents.

FIG. 1. General view of carding setup: (A, B) laps for coated and uncoated sides, (C) partition, (D, E) glass plates.

FIG. 2. Glass plates with fly: (D) coated side; (E) uncoated side.

EXPERIMENTAL

Two quantities were measured: (a) the electrostatic charge on the web after passing the comb, and (b) the weight of the fly produced at the comb. For this purpose, half of the comb (the left half) was coated with the "antistatic" coating. The cotton was fed into the card, in the form of a narrow lap (about 4 cm wide, compared to the 25-cm width of the comb) on the left and right sides of the feeding belt alternatingly (noted as A and B in Fig. 1). After processing by the card, the cotton was collected into a crude Faraday cage connected to a Keithley 610B electrometer, and the charge was measured. The same amounts of cotton (0.5 g) were processed in each run on the left and right sides alternatingly; the length of the lap was kept constant throughout the investigation. The left-side lap never spread in processing to reach the uncoated part of the comb and vice versa. It was found that the coating applied reduced the charge on the cotton by a factor of 3.5 ± 0.5.

In order to measure the effect of the charge on the fly generation, the following procedure was used. A partition was inserted between the coated and uncoated halves of the comb (labeled C in Fig. 1), so that the fly generated on these two parts of the card could not mix. Then glass plates (labeled D and E in Fig. 1) were mounted horizontally above the left and right sides of the comb, respectively. The undersides of the plates were oiled to enhance sticking of the fly. At each run, a lap of 1 g of cotton was fed on each side of the card simultaneously, and the fly collected on the glass plates was measured. On the coated side of the comb, four times less fly was collected than on the uncoated side (0.25 mg and 1 mg, respectively). A photograph of these plates with the deposited fly is presented in Figure 2. All experiments were done at a temperature of 23 ± 2°C and a relative humidity of 40% ± 2%.

We repeated our experiments at 65% relative humidity values. In this case, our coatings were not found to be effective in fly reduction. This is due to the known fact that at high relative humidity values, the electrostatic charge produced on the cotton web is small.

DISCUSSION

The research demonstrates that in this process of carding, static charges at the comb strongly affect the amount of fly produced and that antistatic coatings are effective inhibitors. In industrial plants, however, this effect may or may not be as significant as in our work because of different values of the relevant parameters (thickness of the web, mechanical parameters of the carding machine, etc.). The effect of "antistatic" coatings on fly generation at other parts of the card (notably the trumpet) was not investigated; yet the fly generated there accounts for a large portion of total fly generated in carding. Another limitation in our experiments was that the glass plates changed the air flow pattern, thus affecting the amount of fly produced and the amount reaching the glass plates.

The quantities of fly obtained in the present experiments are not necessarily representative of the fly quantities that would have been generated in an industrial process and have only a limited relative (coated to uncoated) value. These experiments must be regarded as preliminary only, and further studies are needed concerning the influence of other parameters on the formation of fly in the carding process and its inhibition by antistatic coatings.

ON STRUCTURE AND THERMOMECHANICAL RESPONSES OF FIBERS, AND THE CONCEPT OF A DYNAMIC CRYSTALLINE GEL AS A SEPARATE THERMODYNAMIC STATE

J. W. S. HEARLE

Department of Textile Technology, University of Manchester Institute of Science and Technology, Manchester, England

SYNOPSIS

A general approach to fiber structure, based on six important parameters, is summarized. The explanations of a number of mechanical and thermal properties in terms of a working model of the structure of polyamide and polyester fibers are presented. The major part of the paper is concerned with an explanation of multiple melting phenomena.

Although there are some complications in the experimental data found by Bell and others, the major effects are that rapid quenching leads to form I which shows an almost constant melting point, but annealing leads to a progressive development of form II, initially with a lower melting point but after prolonged annealing with a higher melting point than form I. The various observed effects can be explained by postulating that form I is a separate thermodynamic state, with internal energy and entropy intermediate between a melt and a classical crystalline system, such as form II. The new form is described as a dynamic crystalline gel, because it is assumed to be formed when the density of temporarily paired segments increases to such a level that it gels the structure by preventing relative motion of polymer chains.

INTRODUCTION

Theme of the Paper

A conference presentation concerning the structure and thermomechanical responses of polyamide and polyester fibers should cover a number of topics in order to provoke discussion, and in particular to draw attention to areas of ignorance. Some of these topics have been discussed elsewhere [1–5], and others are early speculations, more suitable for conference discussion than a written record. The present paper concentrates on a particular theme, which has been worked out to a degree suitable for publication. This theme is a possible explanation of the interesting phenomena of multiple melting of polyamide and polyester fibers, and some other polymer systems, observed by Bell and his colleagues [6–9] and others [10–27]. However, it is desirable to set this, briefly, in a more general context by way of introduction and conclusion.

Journal of Applied Polymer Science: Applied Polymer Symposium 31, 137–161 (1977)

A General Approach to Fiber Structure

All the general-purpose textile fibers can be described as partially oriented, partially crystalline, linear polymer systems. The problem in this description, particularly at present for the synthetic fibers because more is known about the natural fibers and rayon, is the large number of possible partially crystalline forms. The theoretical possibilities would range from an assembly of perfect single crystals (which have a well-defined and understood structure) in an amorphous matrix (which is an ill-defined and badly understood form) to a uniform state of partial order, with all sorts of mixtures and intermediate states in between, and all sorts of detailed possibilities for chain paths within the systems (see Fig. 16 in Hearle and Greer [3]).

It is now recognized that there can be no universal, or even widely applicable, model of fiber structure, but that we must endeavor to define models which are useful in application to particular types of fiber subject to particular histories. The recognition of this fact is of special importance in considering the melt-spun synthetic fibers because of their inherent structural instability, compared with other classes of fiber. The two reasons for this are indicated in Table I. First, although all partially crystalline states of homogeneous polymers are metastable, the high-speed industrial formation and modification of man-made fibers are

TABLE I
Structural Instability of Fibers

Fibre type	TYPICAL CONDITIONS		Rate of formation
	for use or processing as fibres	for formation as fluid	
man-made			
melt-spun (e.g.* nylon 66 or PET)	up to 230°C (503°K)	260 to 310°C (533 to 583°K)	milliseconds
solution-spun	no solvent	excess solvent	to
regenerated or synthesised in situ	as required polymeric substance	as precursor	seconds (random)
natural			
biologically laid down	as required polymeric substance	as precursor substances	weeks (controlled)

* For nylon 6 or other melt-spun synthetics, the temperature values will be different, but the differences will be similar.

Note: the real distinction is, of course, not between fibers which happen to be made in these different ways, but in whether they cannot be made by the simpler methods, because their melting point is too high or there is no suitable solvent for the unmodified polymer.

bound to lead to more easily disturbed states than the slow growth of natural fibers under controlled biological conditions in living organisms. Second, the circumstances of use and processing of melt-spun fibers are only slightly lower in absolute temperature than the condition for mobility in the liquid (molten) state. However, in fibers spun from polymer solutions, the solvent would have to be reintroduced to get to the liquid state; and in others, such as cellulose, the liquid state preceding fiber formation may exist only in the form of other chemical substances in solution, such as a cellulose derivative or precursor substances for synthesis *in situ*.

The different forms observed by Bell, which form the main subject of this paper, are illustrative of the diversity of structure which must exist in different samples, depending on whether they have been rapidly quenched from the melt or subject to controlled annealing or drawing treatments in the laboratory. An even greater diversity of structure must occur as a result of the thermomechanical treatments to which typical nylon or polyester fibers are subject in the sequence from the initial supply state, through the melt, extrusion, cooling, annealing, drawing, texturing in one or two stages, dyeing, fabric finishing, garment setting, and washing and drying in their everyday use.

Instead of trying to describe particular geometrical models of molecular packing, an alternative approach is to try and select, from the "Avogadro's number" of parameters required to give a full specification of a partially ordered structure, a few parameters of particular importance. My own list grew from three [2] to five [3, 5] and now, with a previously unpublished addition, stands at six: (I) degree of order; (II) degree of localization of order; (III) length/width ratio of localized units; (IV) degree of orientation; (V) size of localized units; (VI) molecular extent.

The topics discussed in the detailed part of this paper are concerned with changes in I, II, and V as a result of thermomechanical treatments.

The new parameter, molecular extent (introduced orally at the conference and to be published in the future), is difficult because it is not possible to conceive of any way of measuring it, though one can estimate a value in the melt and then speculate on how it might change as a result of changes in the solidifying threadline and in drawing. It describes a generalized folding back of chains, and is important because a fold (a false end) is just as much a source of weakness due to slippage as a real molecular end. This fact is well recognized in the analogous problem in yarns and nonwoven fabrics, where fiber extent, which can be measured by tracer fiber techniques, is recognized to be a more relevant quantity than fiber length in influencing strength.

A Working Model of Polyamide and Polyester Fibers and Its Relation to Some Properties

A number of leading workers on the subject, notably Peterlin [28–30], Prevorsek [31], and Statton et al. [32–34], have now come, through routes depending on various sources of evidence, to rather similar views of structure of melt-spun fibers. A typical representation is shown in Figure 1, which was a

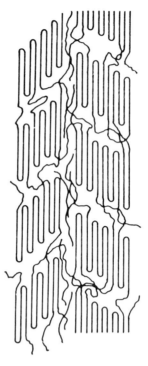

FIG. 1. Model of nylon fiber structure as proposed by Hearle and Greer [35].

particular version proposed by Hearle and Greer [35]. Various terms are used to describe such a structure, but I prefer to call it a modified (because there is a mixture of folding and fringing) fringed-micelle structure, containing pseudofibrils as a result of the stacking of the micelles. Prevorsek lays more stress than other authors on a difference between the disordered material lying between the micelles in a pseudofibril and that lying between the pseudofibrils.

It cannot be too strongly stressed that a representation such as Figure 1 is a grossly idealized simplification because: (a) it is two-dimensional; (b) it is not valid to represent complex polymer molecules, with, for example, 7, 9, or 14 main-chain atoms including different groups along a repeat, by simple lines; (c) it is too neat, tidy, and uniform; and (d) it reflects a personal way of drawing a diagram. There is a need for effort to be put into the creating of more realistic forms of model.

Even within the limitations of the form of Figure 1, more specific information is required on the size and shape of the crystalline regions and of the spaces between them, on the relative extent of fringing and folding, and on the real nature of the disordered regions, all of which will vary with the fiber history. Because of the uncertainties in measurement and interpretation of any experimental technique, such information could only come from the comparison of results from many techniques, carried out on well-characterized samples with the same histories. A difficulty with published experimental data is that they relate to different samples, often of uncertain provenance. New comparative experimental work is needed.

TABLE II
Nylon 66 Crystals[a]

AXIAL, REPEAT			TRANSVERSE, CHAINS		
Number of units	Length Å		Linear number	Area number	Width Å
1	17		1	1	5
			2	4	10
2	34		3	9	15
		2:1	4	16	20
3	51	⟵———————⟶	5	25	25
		⟵——————⟶	6	36	30
		1:1	7	49	35
4	68		8	64	40
5	85		9	81	45
			10	100	50
		3:1			
6	102	⟵———————⟶	11	121	55

[a] Unit cell (1 repeat unit) is 17 Å × 4.9 Å × 5.4 Å.

My present view is that the length-to-width ratio of the micelles is probably greater than suggested by Figure 1, and probably greater in polyester than in nylon fibers. The figures in Table II show that a 1:1 ratio would require, for example, only three repeats of 100 parallel chains: this seems improbable, and six repeats of 50 chains, which would give a 3:1 ratio, seems more likely. But this is no more than ill-informed guessing.

The explanation of some properties in terms of such a working model of structure would be as follows.

(a) The low- and medium-strain room-temperature mechanical properties would be dominated by the behavior of the disordered regions acting as rubbers, comparatively highly crosslinked due to the hydrogen bonding of —CO·NH— groups or the association of benzene rings. The limiting extension values are compatible with the behavior of a rubber with two random links between network points, but a more exact analysis of the properties of the disordered material and the effect of the reinforcement of the crystalline micelles is needed.

(b) The large-strain properties would result from crystalline yielding, with the crack-growth leading to fracture taking place due to cumulative breakage of tie-molecules when the resistance to yielding becomes too great. Scanning electron microscope studies show that the cracks may develop either across the pseudofibers under tensile stress, or between them under shear stress [36, 37].

(c) The change from the glassy to the fully rubbery–liquid state in the dis-ordered regions would occur in steps, with low-temperature transitions giving freedom of rotation of the main-chain bonds in the —CH$_2$— or other aliphatic sequences and higher-temperature transitions giving freedom of rupture to the hydrogen bonds or the benzene ring associations. Both of these transition regions will give rise to viscoelastic effects and to the possibility of temporary set. The hydrogen-bond transitions will be very sensitive to moisture absorption.

(d) An increase in the melting temperature with annealing would be due to the progressive development of larger and more perfect crystalline regions; this leads to a simple theory of permanent heat setting, at some temperature near but below the maximum melting point, as being due to the "melting" of small imperfect crystals within the material and the formation of larger, better crystals.

However, the working model of structure does not offer any explanation of the observation by Bell and others that rapid quenching leads to a form with a rather high melting point in DTA tests, and that annealing then leads to the formation of a lower melting form. This topic will now be examined in more detail. The subject is of practical importance because both forms may be found in commercial fibers, depending on their thermomechanical history.

EXPERIMENTS ON MULTIPLE MELTING

Summary of Data

The work of Bell and his colleagues [6–9] and of others [10–27] has shown that in DTA or DSC studies of nylon 6.6, poly(ethylene terephthalate), and other polymers, endothermic peaks, which are characteristic of melting behavior, occur at different temperatures dependent on the history of the sample. In many instances, two separate endotherms occur on heating a sample.

As a result of his studies, Bell concludes that there are two forms of material involved: form I appears on rapid quenching and shows a constant melting point; form II appears on annealing or on slow cooling, with an initially lower melting point, but as annealing proceeds the melting point rises and the proportion of form II increases. The present paper is concerned with the thermodynamic and structural interpretation of the differences between the two forms.

The original papers need to be consulted for a full account of the experimental results, but a brief summary will be given here. Hearle and Greer [3] have given a longer review of some of the papers.

Figure 2a shows the effect of annealing on the DSC trace of undrawn nylon 6.6 yarn. The rapidly quenched sample, with no annealing, shows an endotherm at 256°C, designated as form I. Annealing at 220°C brings in another endotherm at a lower temperature, designated as form II, and this endotherm becomes larger as annealing proceeds. As shown in Figure 2b, the melting temperature of form I remains constant, but that of form II rises from near 230°C to over 260°C. As annealing proceeds, a point is reached at which there is a single form II endotherm at the same temperature as the original single form I endotherm. However, the difference between the two can be diagnosed since form I will show the appearance of a lower endotherm on annealing while form II will not. Pro-

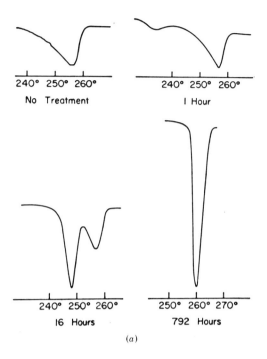

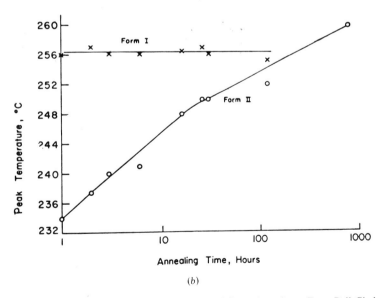

FIG. 2. Effect of annealing undrawn nylon 6.6 at 220°C for various times. From Bell, Slade, and Dumbleton [6]. (a) DSC traces; (b) change of melting temperatures.

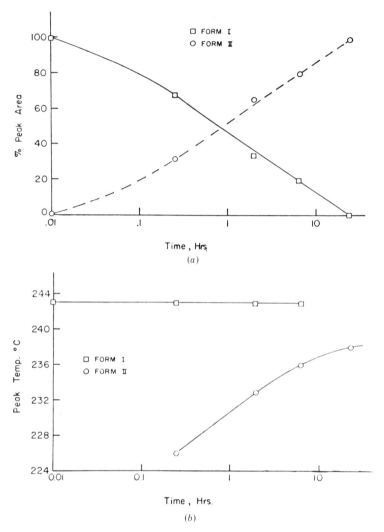

FIG. 3. Effect of annealing PET film at 220°C for various times. From Bell and Murayama [8]. (a) Change in DTA peak area; (b) change in melting temperature.

gressive conversion from form I to form II is also brought about by cold-drawing the yarn, but again the melting temperature of form I remains almost constant while that of form II increases with draw ratio.

Similar behavior has been reported in isotactic polystyrene, but not in polyethylene. In poly(ethylene terephthalate), rapid quenching gives amorphous material. (This can be achieved only with difficulty in nylon by rapid quenching of perfectly dry polymer to a low temperature.) However, the poly(ethylene terephthalate) crystallizes at 110°C and then shows a similar behavior to nylon. Figure 3 shows that the initially crystallized form I, with a constant melting temperature of 243°C, is gradually converted on annealing at 220°C to form II, with a rising melting point.

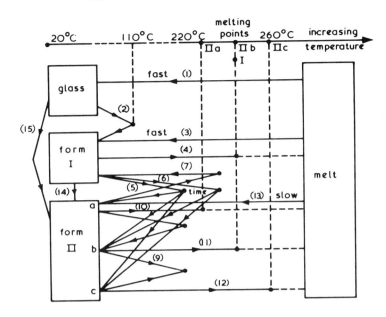

FIG. 4. Summary of observed changes.

Figure 4 summarizes the observed changes. For convenience, three discrete melting points of form II have been shown, though, in reality, there is a continuous gradation. The changes are as follows: (1) rapid quench to glass (typical of PET); (2) crystallization of glass to form I on low-temperature annealing or hot-drawing; (3) rapid quench from melt to form I (typical of nylon); (4) melting of form I on heating; (5, 6) conversion of form I to form II by annealing, going to higher melting form with longer time or higher temperature of annealing; (7) conversion of form II to form I by heat-treating at an intermediate temperature and cooling; (8, 9) continued annealing raising melting point of form II; (10–12) melting of form II; (13) slow cooling from melt to form II; (14) conversion of form I to form II on cold-drawing; (15) conversion of glass to form II on cold drawing.

Other differences between forms I and II have also been reported. Form I will cold-draw (i.e., it will deform plastically at room temperature to very high elongations before breaking) while form II breaks at a low extension, with almost no plastic deformation. The fracture surfaces appear different. There are differences in dyeing behavior. In an earlier work, Statton [38] reported changes in small-angle x-ray diffraction, with the appearance of a distinct four-point diagram on annealing. Bell and Murayama [8] reported changes in the dynamic loss modulus (or tan δ) data. In nylon, form I shows two separate peaks at about $-70°C$ and $-150°C$; but in form II, the lower-temperature peak is absent. In poly(ethylene terephthalate), the form I material shows a peak at about $-40°C$ but this is absent in form II.

It should be stated that the experimental evidence has been criticized on the grounds that the structure changes during the temperature scanning in the calorimeter, and that the results are affected by the rate of heating. But whatever may happen during the test, it must be concluded that rapidly quenched material and annealed material are different because they respond differently. In these systems, we are dealing with metastable states and so there are bound to be rate effects, but it seems likely that these are better regarded as secondary effects superimposed on the primary phenomena. There are also some circumstances in which more than two endotherms are found.

Some Possible Explanations

An increase in melting point on annealing is easily explained as being due to an increase in size and perfection of crystals, or reduction of internal stressing, and this is presumably the explanation of the different melting points observed in polyethylene. However, this is not acceptable as an explanation of the phenomena discussed above since the rapidly quenched material has a higher melting point. It is, however, an explanation of the changes in form II on prolonged annealing.

Another view might be that the annealing leads to some secondary crystallization of the same type as the primary crystallization (i.e., form I), though less perfect and therefore melting at a lower temperature and causing the second endotherm to appear. But this explanation breaks down since form I is different and does not show any change in melting point on annealing; the ultimate highly annealed material derives from form II.

One is thus led to a conclusion that forms I and II are distinct. There have been explanations based on a difference between extended-chain and folded-chain forms but these are difficult to sustain. The present paper gives an alternative explanation.

It has been suggested, for example, by Hobbs and Pratt [26], that form I appears only when recrystallization occurs during the temperature rise in the scanning calorimeter. However, experiments such as those of Sharma [27] indicate that this is unlikely and that form I does develop in the initial crystallization. Even if it is true that recrystallization is a necessary feature, this still leaves the question of why it should lead to a form with a specific melting point which is nevertheless not that of the more stable form found after prolonged annealing. The major features of the thermodynamic and structural arguments given below would remain valid.

THERMODYNAMIC ARGUMENT

A striking fact in the experimental evidence is the constancy of the melting temperature of form I; this suggests that it is a well-defined state. The problem is to reconcile this with its transformation on annealing to an apparently less favorable form with a lower melting point, which on prolonged annealing does reach a higher melting point and then becomes the most favorable form.

A brief indication of a thermodynamic explanation was given by Hearle and Greer [3], but it needs elaborating.

The classical argument on melting is expressed in terms of the free energy F of the system given by:

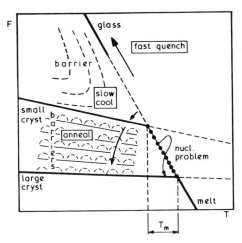

FIG. 5. Classical free-energy diagram showing thermodymamic effects in melting. (*Note.* This and Fig. 6*a* are illustrative and not strictly representative diagrams. The straight lines are plots of free energy against temperature for particular structural forms, with the lowest being the preferred form at any particular temperature. The dotted "contours" are intended merely to give some impression of the barriers to changes between the forms to which the lines refer.)

$$F = U - TS \qquad (1)$$

where U = internal energy, T = temperature, and S = entropy.

If ΔF is the difference in free energy between two states, then at equilibrium between the two states at the melting point T_m we have:

$$\Delta F = \Delta U - T_m \Delta S = 0 \qquad (2)$$

or

$$T_m = \Delta U / \Delta S \qquad (3)$$

Figure 5 shows the application of this to the problem of crystalline melting. The liquid form (the melt) has a high internal energy and a high entropy; the crystalline form has a low internal energy and a low entropy.* As a result, the F–T lines cross (at the melting point) and the liquid is favored at higher temperatures and the crystal at lower temperatures. There are energy barriers to overcome in going from the disordered form of the melt to the ordered form of the crystal, and so it may be possible to quench rapidly to a stable glassy state, with a disordered structure similar to the melt but immobile. Furthermore, small or imperfect crystals will have a higher internal energy (and a higher entropy) and so a lower melting point: annealing, past more energy barriers, will lead to larger, more perfect crystals with a higher melting point. The phenomena of nucleation are associated with the effect of size on melting point. As has been indicated, the behavior of form II on annealing fits this picture: the melting point rises progressively on annealing.

* We can ignore the complication that only about half the material is crystalline and so involved in melting if we assume that the disordered material is still effectively liquid below the fiber melting point, and that the degree of crystallinity is a constant. Some deviation from this simplified view will cause differences in detail, but not in principle.

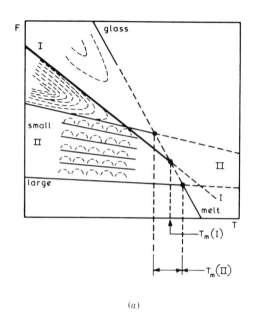

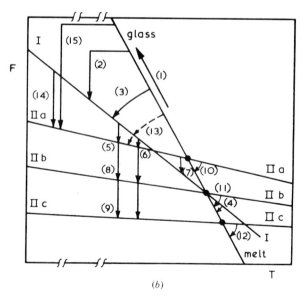

FIG. 6. (a) Free-energy diagram with addition of form I (see note to Fig. 5). (b) Numbered (as Fig. 4) to show observed changes.

It is now postulated: (a) that form I is an intermediate thermodynamic state with values of internal energy and entropy between those of the liquid and the classical (form II) crystals; (b) that the energy barriers between form I and form II are much greater than the barriers between the melt (or glass) and form I; (c) that the free energy of form I is given by a single line, or at least by a narrow band, which crosses the liquid line at the constant melting point of form I.

The free-energy diagram is then as shown in Figure 6a. Given these assumptions, the predicted behavior is as observed, and Figure 6b shows the changes numbered to correspond with Figure 4. Quenching of the melt, unless it is so fast that it leads to a glass (1), will allow the material to pass over the low-energy barrier to form I (3) with its melting point (4), but form II will be blocked off by the higher energy barrier. However, annealing (5, 6) at an appropriate temperature for a long time will allow the barrier to be passed and form IIa to appear. However, although the free energy of form IIa is at a lower level and therefore favored at the annealing temperature, it will, because of the difference in slope of the lines, cross the liquid free-energy line at a lower temperature and so have a lower melting point (10).

Heating form II to an intermediate temperature (7) will convert it back to form I, but continued annealing at a lower temperature (8, 9) will lead to larger, more perfect form II crystals (IIb and IIc) with higher melting points (11, 12). Slow cooling (13) in effect allows annealing to occur and form II appears as well as form I. The relative movement of chain molecules in drawing will also allow the energy barriers to be passed and lead to conversion of form I to form II (14). A glass [of poly(ethylene terephthalate), for example] will crystallize to form I on annealing (2) at a fairly low temperature.

A condition for the occurrence of these effects is that the form I line should cut the liquid line between the crossing points of the imperfect form IIa and the more perfect IIc. This may seem arbitrary, but is plausible in view of the structural situation to be discussed. If the crossing point was beyond that of IIc, which seems unlikely, then as indicated in Figure 7a there would be a potential solid-state transition along PQ below the higher melting point R, though whether this was observed in practice would depend on the effect of the energy barriers. If the crossing point was before that of IIa (the smallest discrete crystallite

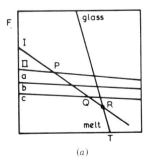
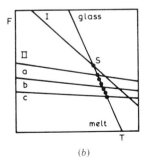

(a)   (b)

FIG. 7. Free-energy diagrams with different levels for form I: (a) with form I at lower internal energy; (b) with form I at higher internal energy.

system), then as illustrated in Figure 7b, rapid quenching would give a material with a lower melting point S and the system would show a commonplace change to higher melting forms on annealing, though with a discontinuity in melting points.

The thermodynamic arguments and the comparison with experiment justify the hypothesis that form I, occurring on rapid quenching, is a separate thermodynamic state, and is not merely the limiting case of a classical crystallite system as the crystallites become vanishingly small. And although the existence of form I depends on kinetics, because it is metastable, its existence is distinct and is not determined solely by kinetics in the sense that it is merely an arbitrary resting place in a continuous sequence of metastable states, such as the various states of form II on continued annealing. The thermodynamic theory leads to the necessity to attempt a structural explanation of the differences between forms I and II.

THE STRUCTURAL ARGUMENT

The Structure of Form II

Form II is relatively easy: it must be a classical structure of well-defined crystallites embedded in, and molecularly continuous with, disordered regions, which are present because of the impossibility of sorting out long-chain molecules into structures with more than a certain degree of order. The precise morphology of the arrangement is not critical to the argument, but a likely form is a modified fringed-micelle structure as shown in figure 1 and discussed as a working model of structure in the introduction. However, the shape of the crystallites may be different: all that is necessary for the present argument is that they are defined as crystalline entities, which become larger and more perfect as annealing proceeds.

The Concept of a Crystalline Gel

The problem is: "What is form I?" It is now proposed that this is what may be called a *crystalline gel*. This structure proposed for form I is a more uniform one, namely, a structure with a low value of *degree of localization of order,* in which no separate crystallites can be identified, but in which a large number of individual pairs of chain segments (or perhaps slightly larger groupings) are in crystallographic register with one another, distributed throughout the material. Linking the crystallographic segments will be other segments which are not in register, and indeed are bent and twisted and not in the right internal conformation to fit in a crystal. The degree of order in nylon and PET would be equivalent to about 50% crystallinity, as measured, for example, by density. Such a structure would have an internal energy lower than an amorphous melt, with little crystallographic register, though not as low as a structure of discrete crystallites with a greater degree of register. The ordering of the structure, namely, the entropy, would also be intermediate between the melt and the system of distinct crystallites. Consequently, it appears to fit the thermodynamic requirements concerning the values of U and S, both of which must lie between the values for a liquid and for a classical crystalline system.

It is not easy to draw models of the proposed structure. A possible view in an unoriented material is shown in Figure 8a, compared with a model of a form II structure in Figure 8b, but these must not be regarded as any more than very inadequate representations as a guide in the effort needed to comprehend the way in which chain molecules pack together.

Another possible representation of a form I structure in a drawn fiber was suggested by Hearle and Greer [3] and is shown in Figure 9. This form could also exist in localized domains of an unoriented system. It differs from Figure 8, and indeed from the description given above, in that there is an element of continuity in the paired crystalline segments, though this is spread through the structure and intermingled with disordered material, and not localized in discrete crystallites. This view of the structure should be compared with the micellar model shown in Figure 1.

Figure 9 is similar to models proposed by Fischer et al. [39] and by Dismore and Statton [32] and shown in Figures 10a and 10b; the corresponding models, after annealing, are shown in Figures 10c and 10d. The point of the present paper is that the form I structure is a separate thermodynamic state, referred to as a crystalline gel.

In polymer science, a gel is an irregular assembly of chains which has changed from the liquid to the solid state by becoming crosslinked into a single network. The present structure is termed a crystalline gel because the crosslinking is due to individual segments in crystallographic register.

The Dynamic View

In the previous section, an over-simplified, static view has been taken. However, it must be remembered that disorder leads to a higher entropy only when there is a continuing interchange between equivalent structural states.

Contrary to the simple view of a liquid, real liquids just above the melting point do contain at any instant many neighboring molecules, or chain segments in polymer melts, which are in crystallographic register. But the situation is dynamic: the pairs are continually breaking and re-forming in new positions, thus giving liquid mobility. It may also be questioned whether in materials like nylon and PET the register extends over the whole lengths of the long repeat units.

As the temperature of a polymer melt falls from some high value, paired segments will fly apart less frequently; at any instant more segments will be paired, and they will stay paired for longer. Nevertheless, provided they break sufficiently often, the material will be liquid and able to flow.

But now, suppose that we get to a state when during the time that pair A in Figure 11a is broken, the neighboring pairs BCDE remain held. Mobility will have been lost, and the segments in pair A will, in due course, come together again. In other words, the structure has become a gel: no liquid-like relative

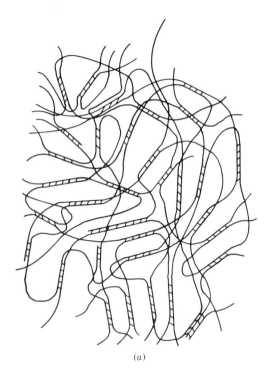

(a)

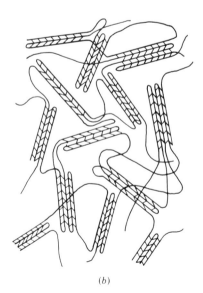

(b)

FIG. 8. Possible view of structure: (a) Form I; (b) Form II.

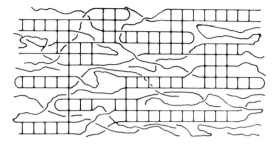

FIG. 9. Another view of a possible form II structure in an oriented system.

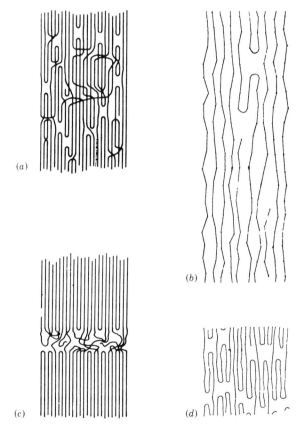

FIG. 10. Models suggested by Fischer et al. [39] and by Dismore and Statton [32] to explain changes on annealing: (a) Fischer, "pre-annealed"; (b) Dismore and Statton, "drawn yarn"; (c) Fischer, "zone-refined"; (d) Dismore and Statton, "annealed."

movement of chains is possible. But it is a gel in dynamic equilibrium in which individual segments are breaking, though never in sufficient numbers at a given time to allow liquid mobility. Thus, the entropy of the dynamic crystalline gel will be high, though not as high as in the liquid; this fits the thermodynamic requirements for form I.

The change between form I and the melt will involve a discontinuous change

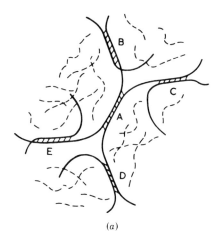

(a)

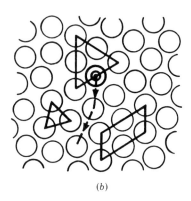

(b)

FIG. 11. Comparison of mobility and local order. (a) Neighboring paired segments in the dynamic crystalline gel (form I). The other neighboring chains are only sketchily indicated by dotted lines in order to avoid distraction in the diagram. (b) A low molecular weight liquid in dynamic equilibrium with clusters in instantaneous crystallographic register. If the molecule 0 was to break loose from the cluster and follow a path like the one indicated (though really pushing through the gaps and allowing other molecules to fill in behind), then it would have lost all relation to the cluster.

in entropy (a latent heat) due to the difference in freedom of the system, which occurs when the relative movement of whole molecules is prevented, and thus it is properly regarded as a first-order phase change (melting). From the mechanistic viewpoint, it would be said that the change is highly cooperative, that the reduction in freedom cumulatively slows down the whole structure, and thus allows a much larger number of segments to get into crystallographic register, and so give the required change in internal energy. Another way of stating this would be to say that in the crystalline gel, as distinct from the liquid, segments which break due to thermal vibrations are not free to move very far and so reform again more quickly, and in the same place.

The formation of a gel is a direct consequence of the long-chain nature of polymers. It could not happen with small molecules where, as indicated in Figure 11*b*, if a crystallographic pair breaks then the separated units are nowhere held and so are free to migrate through the system. There is no intermediate state for small molecules: once they have changed places, any connection with the earlier pairing is lost. But in a polymer, the chains are held at other places. A number of segments must become free together to obtain relative mobility between neighboring molecules. It is difficult to say how many units must become free together, but examination of Figure 12 suggests that it will probably need to approach 10 in order to allow the relative interchange of position of neighboring portions of chains, which would lead in time by successive motions to a change in the relative positions of whole chains in liquid flow. It is also necessary that appreciable lengths of neighboring chains also become free at the same time. Such a high degree of freedom corresponds to a much lower instantaneous occurrence of crystallographic register of segments than would exist in a partially crystalline solid polymer in form I. This is another way of putting the argument that there is a first-order phase change between the melt and form I: although

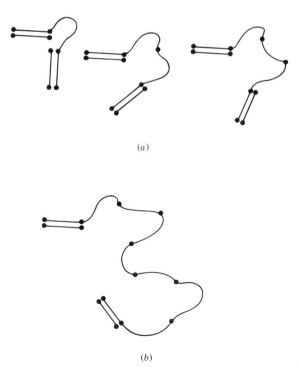

(a)

(b)

FIG. 12. Number of free segments required for liquid mobility. (*a*) With a small number of segments the interchange of position would not be possible. (*b*) With more free segments, interchange would be possible if other neighboring chains also become free.

it needs a large number of segments to be free together to obtain liquid mobility, once this mobility is prevented a much higher proportion of segments will get into register. This is analogous to the conventional argument for the crystallization of a liquid.

Changes between the Forms

The structural ideas being proposed, and their relation to the thermodynamic situation illustrated in Figure 6a may be clarified by summarizing the changes between various forms.

(a) Melt to Form I

As discussed above, this involves a discontinuous change in the number of segments in crystallographic register, freezing out liquid mobility, but with little movement in relative chain position as the segments fall into register. This gives a decrease in internal energy as intermolecular forces are satisfied, a decrease in entropy as the order increases, and low energy barriers to the change.

(b) Form I to Melt

As the temperature rises, the internal vibrations will increase to a level at which the extent of spontaneous separation of chain segments in the dynamic equilibrium is such that the system no longer holds together as a solid. When the separated segments are no longer held by neighboring paired segments, then mobility will be greatly increased; this is the phase change of melting. In some ways this is more similar to the phase change occurring on boiling, when the rate of movement of molecules into the vapor becomes so frequent that the liquid will not hold together, than to a conventional crystalline melting.

(c) Form I to Form II

This is a conventional crystalline growth mechanism, involving the formation of crystallites by packing many chain segments into register over a substantial volume. The internal energy will be lower because of the mutual attraction of many chains, and the entropy will be lower because the system is more ordered. In order to form separate crystallites on suitable nuclei considerable relative chain movement will be needed, and so there will be high energy barriers and a slow change. The differences between form I and form II thus satisfy the thermodynamic requirements.

It should be noted that the paired segments of form I are not stable nuclei: they are continually breaking up and reforming like the clusters in an ordinary liquid, even though they are in sufficient density to gel the structure. Therefore, the existence of form I does not alter the need for nucleation.

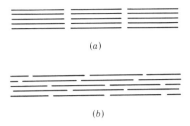

FIG. 13. Changes in packs of cards, analogous to the change from form II to form I.

(d) Form II to Form I

The reversal of the annealing process by the conversion of low-melting (imperfect) form II back to form I is one of the initially surprising features of the situation. Its explanation may be helped by an analogy consisting of the difference between a set of separate packs of cards, as in Figure 13a, and a set in which the cards had been shifted relative to one another as in Figure 13b. In both, attractive forces between the cards would be well satisfied, but the disorder would be greater in the second set. In a similar way, one could imagine small separate crystals of form II breaking up as the temperature rises, with chains shuffling relative to one another to give the form I structure with many segments still in register, but distributed uniformly instead of in organized crystallites. This would be a stable form with the increase in internal energy compensated by an increase in entropy.

(e) Form II to Melt

This is a conventional crystalline melting, and will occur when the form II crystals are so large that the amount of thermal energy needed to disrupt them carries the material into the molten state.

(f) Glass to Form I

The development of form I from glassy PET at about 110°C will be due to individual chain segments falling into lower energy states in crystallographic register with favorably placed near neighbors, with little relative chain movement. The energy barriers to overcome would be low.

Relation to Other Experimental Evidence

The idea that form I is a dynamic crystalline gel has been presented above in relation to the thermal changes, but the structural changes can also be related to other observations.

X-ray diffraction studies of material which would be expected to be in form I show the characteristic spacing, but rather diffusely and not to high orders; the pattern becomes sharper after annealing. This is quite compatible with the structures proposed. Furthermore, the well-defined long spacings, characteristic of separate crystalline regions, appear only on annealing or drawing.

Infrared studies show the appearance of a well-defined fold band on annealing, as would be expected from the structural change.

Dyeing occurs more rapidly in annealed samples; this would correspond to the larger separate noncrystalline regions in form II.

The difference in drawing behavior is easily explained. When the tension is high enough, the separate segment-pairs of form I will come apart and the structure will flow plastically. In effect, this is melting under stress. But the large crystalline units of form II will resist yielding, and chain breakage leading to fracture will occur instead.

The loss modulus data are more difficult to explain in detail. It must be remembered that the peaks reflect the influence of fine structure mobility on large-scale deformation: they can therefore be suppressed if the material giving the peak is prevented from deforming. The peaks would be expected to relate to noncrystalline material, and so they could be blocked by crystallites in parallel. The situation in nylon is difficult to deal with since there are no good explanations of why there are two low-temperature peaks.

The following is a possible explanation of the variations in loss moduli found [40–42] in poly(ethylene terephthalate). The high loss modulus over a wide range of temperature of rapidly quenched amorphous material would be due to the wide variations in local chain arrangements and their internal stresses. The variations are relieved by heat treatment and a concentrated peak appears. Crystallization to form I also allows the mobility to show and the peak is present. But in form II, the crystalline reinforcement prevents the independent mobility of the noncrystalline regions. This implies that the initial modulus is dominated by the crystalline material or by the bulk modulus of the matrix, but this is perfectly possible.

CONCLUSION

Summary

The arguments in this paper are intended to show that the experimental evidence on multiple melting phenomena in polyamide and polyester fibers can be explained if it is postulated that the rapidly quenched form (I) is a separate thermodynamic state of intermediate energy and entropy. This is a state of uniformly distributed partial order, with more than enough crystallographic register of neighboring segments to freeze out liquid mobility. It is formed when segments grab their nearest neighbors and mate together, but are not able to make the larger rearrangement necessary to fit into the serried order within the separate crystals found on annealing to form II. The new form I structure is referred to as a dynamic crystalline gel, because close to the melting point crystallographically paired segments are breaking and reforming in a dynamic equilibrium, although always maintaining enough in register at any instant to leave the material crosslinked as a gel. At a lower temperature, the dynamic equilibrium will be slowed down and the material will be a stiffer solid.

The existence of this separate thermodynamic state is yet another manifestation of the special nature of substances composed of long-chain molecules.

Practical Effects and Scientific Problems

Understanding of the multiple melting phenomena and the structural differences which they indicate is important because of their relation to permanent setting mechanisms and to changes in fiber properties of technological importance. Any second-order transition, such as the transitions above and below room temperature in nylon and polyester fibers, involving change in mobility without structural change, gives a mechanism of temporary set. Any first-order transition, involving changes in structure, gives a mechanism of permanent set, because the movement to a new structure would relieve stresses and stabilize a new form.

The conventional explanation of permanent heat-setting of nylon and polyester fibers has been that it involves the change of small, imperfect crystals into more stable, larger, more perfect ones—the effects discussed as the annealing of form II with a rising melting point. But the recognition of the independent existence of form I means that there are also the mechanisms of the change from form I to form II and from form II to form I.

Undrawn or POY yarns might be in either form I, form II, or glassy states depending on the time–temperature–stress pattern to which the material has been exposed. While cold-drawing will lead to the development of form II from form I, hot-drawing may not; and subsequent hot processes in texturing or finishing may cause more changes between form I and II. We need to understand these changes, first in somewhat simplified laboratory sequences and then in real sequences, more thoroughly and to relate them to changes in structural parameters and to models of fiber structure which are improved, not only in their details but fundamentally in being more realistic representations of complex polymer chain assemblies.

The old rule of thumb that heat-setting gives subsequent stability to lower treatment temperatures but not to higher ones is obviously incorrect and quantitatively inadequate. It can be modified to allow for effects of moisture and stress, and for simultaneous temporary and permanent setting. It must be further modified to allow for form I and form II effects. With much more experimental study, development of understanding, and statement of new rules, a modification of the old ideas might then apply reasonably to nylon fibers, where severe treatments are needed to overcome previous setting.

But polyester fibers seem different. Although there is little relevant evidence from scientific laboratories, commercial practice shows that it is possible to reset polyester yarns and fabrics repeatedly even at temperatures appreciably below previous setting temperatures. Better studies on single fibers are needed to clarify the phenomena, but the industrial evidence does not seem capable of explanation along the conventional lines of structural transitions to more stable forms. There are several possibilities. It could be, though I feel this is unlikely, that the effects are due to higher temperature temporary set in noncrystalline regions. It could be that the setting is chemical and due to the breaking and reforming of main chains. The glycol end groups might attack ester linkages, particularly in strained tie-molecules, leading to new free ends. If further reactions then join free ends in new places, it would stabilize a new structure. Such a change in molecular identity would represent a major change in our thinking about setting, though it would be similar to the rearrangement of cystine crosslinks in wet wool at temperatures above 100°C. Alternatively, and this seems the most likely to me, it could be a thermomechanical effect in the form of a thermally induced plasticity in polyester crystals. This would involve slippage of polyester molecules within crystals, under stress, without a complete melting. The structure could thus repeatedly deform into new states, while still retaining its solid fiber identity. In view of the weak intermolecular bonding in poly(ethylene terephthalate) compared with polyamides, it is plausible that this relative movement could occur. Similar effects probably account for the well-known behavior of polyethylene kitchen articles deformed in hot water. It would not, of course, be required that the same chain segments remained in the crystalline regions; there might be movement both ways between crystalline and noncrystalline regions. But there would always be a stable, though changing, crystalline morphology in the fiber.

Finally, I refer to some interesting thermomechanical effects which we have found recently in my laboratory in studies related to fracture and fatigue. First, in the studies linked to bending fatigue, Jariwala [43] has found that a single bend will cause a visible kink-band in polyester fibers at room temperature, but not above 90°C; while in nylon, visible kink-bands are not generated by bending at room temperature, but are produced by severe bending above 55°C. Second, Konopasek [44] and Mandel [45] have found that tensile fracture at elevated temperatures give complicated stepped crack surfaces, which must reflect some underlying structural features.

The conclusion of this paper cannot be other than a plea for more work on the subject, to give scientific and technological understanding of important and fascinating problems.

REFERENCES

[1]   J. W. S. Hearle, in *Fibre Structure,* J. W. S. Hearle and R. H. Peters, Eds., Textile Institute and Butterworths, London, 1963, chaps. 6 and 19.

[2]   J. W. S. Hearle, in *Supramolecular Structure in Fibers (J. Polym. Sci. C, 20),* P. H. Lindenmoyer, Ed., Interscience, New York, 1967, p. 215.

[3]   J. W. S. Hearle and R. Greer, *Text. Prog. 2*(4), (1970).

[4]   J. W. S. Hearle, in *The Setting of Fibres and Fabrics,* J. W. S. Hearle and L. W. C. Miles, Eds., Merrow, Watford, Hertfordshire, 1971, chaps. 1 and 5.

[5]   W. E. Morton and J. W. S. Hearle, *Physical Properties of Textile Fibres* (2nd ed.), Textile Institute and William Heinemann, London, 1975, chaps. 1, 13–18, 23.

[6]   J. P. Bell, P. E. Slade, and J. H. Dumbleton, *J. Polym. Sci. A-2, 6,* 1773 (1968).

[7]   J. P. Bell and J. H. Dumbleton, *J. Polym. Sci. A-2, 7,* 1033 (1969).

[8]   J. P. Bell and T. Murayama, *J. Polym. Sci. A-2, 7,* 1059 (1969).

[9]   G. E. Sweet and J. P. Bell, *J. Polym. Sci. A-2, 10,* 1273 (1972).

[10] R. C. Wilhoit and M. Dole, *J. Phys. Chem., 51,* 14 (1953).

[11] T. R. White, *Nature, 175,* 895 (1955).

[12] F. J. Hybart and J. D. Platt, *J. Appl. Polym. Sci., 11,* 1449 (1967).

[13] H. Kanetsuna and K. Maeda, *Kogyo Kagaku Zasshi, 69,* 1784 (1966).

[14] Y. Mitsuishi and M. Ikeda, *Kobunshi Kagaku, 23,* 319 (1966).

[15] Y. Mitsuishi and M. Ikeda, *Kobunshi Kagaku, 23,* 310 (1966).

[16] T. Yubayashi, Z. Orito, and N. Yamada, *Kogyo Kagaku Zasshi, 69,* 1798 (1966).

[17] E. L. Lawton and D. M. Cates, paper presented at the annual meeting of the American Chemical Society, San Francisco, March, 1968.
[18] M. Ikeda, *Kobunshi Kagaku, 25,* 87 (1968).
[19] E. L. Lawton and D. M. Cates, *J. Appl. Polym. Sci., 13,* 899 (1969).
[20] R. C. Roberts, *Polymer, 10,* 117 (1969).
[21] R. C. Roberts, *J. Polym. Sci. B, 8,* 381 (1970).
[22] P. J. Holdsworth and A. Turner-Jones, *Polymer, 12,* 195 (1971).
[23] D. L. Nealy, T. G. Davis, and C. J. Kibler, *J. Polym. Sci. A-2, 8,* 2141 (1970).
[24] G. Coppola, P. Fabbri, B. Pallesi, G. C. Alfonso, G. Dondero, and E. Pedemonte, *Makromol. Chem., 176,* 767 (1975).
[25] G. Ceccorulli, F. Manescalchi, and M. Pizzoli, *Makromol. Chem., 176,* 1163 (1975).
[26] S. Y. Hobbs and C. F. Pratt, *Polymer, 16,* 462 (1975).
[27] S. C. Sharma, *Indian J. Chem., 12,* 1297 (1976).
[28] A. Peterlin, *J. Mater. Sci., 6,* 490 (1971) (Fig. 2).
[29] A. Peterlin, *Appl. Polym. Symp. No. 20,* 269 (1973).
[30] K. Sakaoku, N. Morosoff, and A. Peterlin, *J. Polym. Sci. Polym. Phys. Ed., 11,* 31 (1973).
[31] D. C. Prevorsek, P. J. Harget, R. K. Sharma, and A. C. Reimschuessel, *J. Macromol. Sci.-Phys., B-8,* 127 (1973) (Fig. 15).
[32] P. F. Dismore and W. O. Statton, in *Small Angle Scattering from Fibrous and Partially Ordered Systems (J. Polym. Sci. C, 13),* R. H. Marchessault, Ed., Interscience, New York, 1966, p. 133 (Fig. 1).
[33] J. B. Park, W. O. Statton, and K. L. DeVries, unpublished data.
[34] W. O. Statton, in *Molecular Order—Molecular Motion: Their Response to Macroscopic Stresses (J. Polym. Sci. C, 32),* H. H. Kausch, Ed., Interscience, New York, 1971, p. 219.
[35] J. W. S. Hearle and R. Greer, *J. Text. Inst., 61,* 243 (1970).
[36] J. W. S. Hearle, in *Contributions of Science to the Textile Industry,* P. W. Harrison, Ed., Textile Institute, LONDON= [975/
[37] J. W. S. Hearle and B. C. Goswami, *Text. Res. J., 46,* 55 (1976).
[38] W. O. Statton, *J. Polym. Sci., 41,* 193 (1959).
[39] E. W. Fischer, H. Goddar, and G. F. Schmidt, *Makromol. Chem., 119,* 170 (1968).
[40] M. Takayanagi, *Mem. Faculty Eng. Kyushu University, 23*(1), 41 (1963) (Fig. 45).
[41] K. K. Mocherla and J. P. Bell, *J. Polym. Sci. Polym. Phys. Ed., 11,* 1779 (1973).
[42] R. E. Mehta and J. P. Bell, *J. Polym. Sci. Polym. Phys. Ed., 11,* 1793 (1973).
[43] B. C. Jariwala, Ph.D. Thesis, Univ. of Manchester, 1975.
[44] L. Konopasek, M.Sc. Thesis, Univ. of Manchester, 1975.
[45] R. Mandel, unpublished data.

HALOGEN-INDUCED FINE STRUCTURAL CHANGES IN NATURAL AND MAN-MADE CELLULOSE FIBERS

MENACHEM LEWIN, HILDA GUTTMANN, and
DANI SHABTAI

Israel Fiber Institute, Jerusalem, Israel

SYNOPSIS

The effect of Br_2–water on the fine structure of cellulose was studied. The accessibility of cotton was found to decrease and the crystallinity to increase with increase in the time of treatment and in the bromine concentration. The crystallization was evidenced by x-ray, infrared, water absorption, and dye uptake. The crystallization reaction was found to be first-order with respect to the LOR (less ordered regions) of the cotton and to the bromine concentration. At 0.09 mole/l. and 25°C, a rate constant of $3.3 \times 10^{-5} sec^{-1}$ was found. The energy of activation of the crystallization was found to be 8.5 kcal/mole in the range of 10–25°C and 10.9 kcal/mole in the range 25–41°C, indicating a transition point of cotton at ca. 25°C. Measurements of times of sorption equilibrium, rates of oxidation, and energies of activation of oxidation provide further evidence for the transition point.

The crystallization of cellulose appears to depend on the density of packing of the chains in the LOR. This is substantiated by the marked differences in behavior of textile rayon, Vincell and highly oriented high-modulus rayon.

In the case of the rayons first a disordering effect, expressing itself in increased accessibility and moisture regain, is observed. At longer treatment times and higher temperatures crystallization sets in.

Differences between heat- and bromine–water-induced crystallization are discussed. Accessibility–regain plots of cotton and rayon provide indications on their crystallite lateral orders.

INTRODUCTION

In previous communications from this laboratory, the interactions between bromine and cellulose at various pH values were described. The pH was shown to be of primary importance, governing to a large extent the nature of the reactions occurring in the system, their rates and mechanisms [1–3]. This is not surprising, considering the fact that different chemical moieties exist at different pH ranges and the composition of the bromine solutions changes with the pH and with the bromine and bromide concentrations [3, 4]. While in the pH range of 10–14, the bromine appears mainly in the form of the hypobromite ion BrO^-; and in the pH range 6–10 in the form of hypobromous acid, HBrO; at pH values below 4, elementary bromine is the predominant moiety and, depending on the concentration of bromide ions, appears partly in the form of the tribromide, Br_3^-.

Journal of Applied Polymer Science: Applied Polymer Symposium 31, 163–181 (1977)
163

The oxidation of cellulose by hypobromous acid was found [1, 3, 6], based on the low activation energies and on the linear change of the consumption of hypobromite with the square root of time, to proceed through a diffusion mechanism.

Aldehyde, carboxyl, and ketone groups are formed on the cellulose during the oxidation, the relative amounts changing with the change in the pH [1, 6]. It was found that virtually pure keto-cellulose containing negligible amounts of aldehyde and carboxyl groups could be prepared by oxidation with bromine–water at pH 2 at room temperature [3, 6]. The rate of oxidation of cotton cellulose with bromine was found to decrease with decreases in pH [1, 6], and at pH 2 to decrease markedly with the increase in the bromine concentration [5]. This decrease in rate of oxidation coincides with the increase in the reversible sorption of the bromine on the cellulose. The sorbed bromine thus exerts a protective action on the cotton against the oxidation by bromine [5]. The extent of the reversible sorption of the bromine on the cellulose was found to increase with the concentration of the bromine solution and to decrease with increasing temperature. At equilibrium, the bromine–water–cellulose system behaves according to a Freundlich type isotherm with an exponent of 1; i.e., it obeys the distribution law of a solute (bromine) between two solvents. From the isotherm a, distribution constant of 4.23 l./kg and an affinity of ca. 800 cal/mole were calculated for bromine and Deltapine cotton [5]. These values were found to vary slightly for other celluloses [7]. The Langmuir-type isotherms were found to apply to the bromine sorption data of 16 celluloses including several varieties of native, hydrolyzed, and crosslinked cottons, ramie and regenerated cellulose [7]. From these isotherms, accessibility values were calculated by a graphic method. The accessibilities obtained by the bromine method were in the range of 4–70% and were found to give straight-line relationships with crystallinity indices by the infrared method [8, 9] based on the ratio of peaks 1429 cm^{-1}/893 cm^{-1} ($R = 0.984$) and on the ratio of the peaks 1372 cm^{-1}/2900 cm^{-1} ($R = 0.92$) as well as with XRD crystallinity indices ($R = 0.95$) [7]. These correlations indicate that the accessibility measured by the bromine method pertains to the less ordered regions (LOR) and does not include the surfaces of the crystallites. Since the glycosidic oxygens on the surface of the crystallites are buried half a molecule deep under the surface, they are not accessible to hydrogen ions and are therefore not hydrolyzed by acids [10]. The bromine molecule, which is considerably larger than H_3O^+, will not be able to approach the glycosidic linkages on the crystallite surfaces and will therefore not be sorbed on them [7]. Unlike bromine, moisture absorption and deuteration take place at the hydroxyl groups, both of the LOR as well as of the crystallite surfaces, and the accessibility values computed from these measurements are correspondingly higher.

From sorption measurements at several temperatures in the range of 15–40°C, the heat of sorption of bromine to cellulose was found to be 3200 cal/mole, i.e., close to the range of values reported by De Maine [11] for the addition compounds between ethers and bromine, e.g., 3500–6000 cal/mole. Intermediate addition compounds between the halogen and the anomeric oxygen were assumed to be formed during the homogeneous oxidation of aldoses by bromine [12–15] and during the oxidation of cellulose model compounds with chlorine [16–18]. According to Mulliken [19], the long axis of the halogen molecule in the addition compounds is perpendicular to the C—O—C plane. With an ether oxygen, a weak charge-transfer bond is being formed [20] and two possible arrangements may occur: (A) a linear arrangement, donor atom–halogen–halogen, in which one of the halogen atoms is involved in a charge-transfer bond with the donor oxygen atom; (B) a linear halogen–molecule bridge between two oxygen atoms in which both halogen atoms of the same molecule are involved in charge transfer. Such linear bridges extending into regular chains were observed by x-ray studies of the 1:1 crystalline compound formed by 1,4-dioxan with bromine [20]. Bromine bridges or crosslinks between glycosidic oxygens of two neighboring chains formed during the sorption period may be responsible for the stability towards oxidation [7]. The prerequisites for the formation of such bridges are flexibility and mobility of chain segments in the LOR which will enable them to occupy suitable positions. The bridges bring the chains of the LOR closer together and at the same time orient them parallel to each other. Such an orientation may involve a number of chains. When the bromine is reduced or washed out, the position of the chains may be maintained long enough for strong hydrogen bonds to be formed between the chains, yielding nuclei which subsequently bring about crystallization.

Preliminary observations on crystallization of the LOR of cotton cellulose following bromine sorption have been reported upon in a previous communication from this laboratory [7]. It was accompanied by a decrease in moisture absorption and dye uptake. The rate and extent of the crystallization were found to increase with the increase in the bromine concentration. It was found to introduce errors in the accessibility values obtained for cotton by the bromine sorption method. The errors are limited to 1–2% if the bromine concentrations used are below 0.02 mole/l. and the sorption times 1–2 hr at 25°C.

In the following, further data are presented on the changes in the fine structure of cellulose effected by bromine–water solutions.

EXPERIMENTAL

Cellulose Samples

Bleached Acala cotton fabric, supplied by Ata Textiles Co. Ltd., Kfar Ata, Israel; and textile rayon, Vincell 28 and High Modulus Rayon Fibers, supplied by Courtaulds Ltd., England, were used in the present study. The cotton fabric was scoured with nonionic detergent prior to use in the experiments. The rayon samples were extracted with chloroform in order to remove finishes. All samples were dried in air at room temperature after the purification procedures. Measurements of moisture regain at 65% RH was carried out as described earlier [7].

Crystallinity Measurements

The XRD crystallinities were measured on randomly oriented ground fibers (Wiley Mill, Mesh 20) as described earlier [7, 24]. The XRD crystallinity indices and crystallite lateral orders of the high-modulus rayon samples were determined by L. G. Roldan at the J. P. Stevens Technical Center in Garfield, N.J., by the methods described previously [22–24].

Infrared crystallinity indices were determined by the methods of Nelson and O'Connor [8, 9] using a Perkin-Elmer Model 257 spectrophotometer. The pellets used were made from 2 mg of ground fibers (Wiley Mill, Mesh 20) and 300 mg of KBr dried in an oven at 115°C for 2.5 hr. The mixture of KBr with ground fibers was kept in a desiccator over P_2O_5. Two infrared crystallinity indices were calculated: for the rayon samples from the ratio between the intensities of two peaks at 1372 cm^{-1} and 2900 cm^{-1}; and for the cotton samples from the ratio of the peaks at 1.429 cm^{-1} and 893 cm^{-1}.

Measurement of Sorption and Oxidation of Cellulose with Bromine

All the experiments were carried out in duplicate at a consistency of 10 g cellulose per liter. The calculated amounts of aqueous bromine solution were put into a brown flask containing the calculated amount of fibers, and the flask was closed. Bromine-resistant silicone grease was used for the ground glass stoppers which were firmly held in place by stainless steel springs to prevent bromine vapor leakage. The flask was then vigorously hand-shaken and connected to a platform shaker in a thermostatic bath maintaining the desired temperature. After the predetermined time of sorption, the flask was removed from the shaker and its liquid content divided into two equal portions. One half of the solution was transferred by a special pipette to an Erlenmeyer flask containing an accurately measured excess arsenite solution. About 75 ml of 5% $NaHCO_3$ solution was added and the flask was stoppered and placed in the dark for 5 min. The solution was then acidified with 10 ml 4N acetic acid and titrated with iodine solution using starch as indicator. The second half of the solution (still in the reaction flask together with the fibers) was titrated similarly.

The concentrations of bromine in the sorption experiments are expressed in mmole/l. in the solution (C_f) and in mmole/kg in the fibers (C_B). If C_0 is the initial concentration of bromine, C_f is the concentration of bromine in the portion of solution without fibers, and C_2 is the concentration of bromine in the portion of solution containing the fibers, then C_B, the concentration of bromine in fibers, is calculated by the equation:

$$C_B = \frac{1}{2}(C_2 - C_f) \cdot 100 \text{ mmole/kg}$$

The bromine used on oxidation of the fibers, ΔOx, is calculated by the equation:

$$\Delta Ox = \left(C_0 - \frac{C_f + C_2}{2} \right) \cdot 100 \text{ mmole/kg}$$

The values of ΔOx were calculated: ΔOx_1, the rapid oxidation occurring during the sorption; and ΔOx_2, the slow oxidation.

Determination of the Accessibility of Cellulose by the Br_2 Sorption Method

The accessibility of the cellulose samples was determined by applying the Langmuir isotherm to the equilibrium bromine sorption values at several concentrations of bromine in water below 0.02 mole/l. at 25°C. The accessibility in this method is defined by $A = 100/n$, and $n = m/C_{B(s)}$ where $1/n$ is the number of anhydroglucose monomer units (AGU) available for bromine sorption, assuming that 1 mole of bromine corresponds to 1 mole AGU; m is the number of mole AGU's in 1 kg of cellulose and $C_{B(s)}$ is the saturation concentration of bromine in the cellulose [5, 7]. The value of n is calculated from the extrapolation of the straight line obtained by plotting m/C_B against $1/C_f$ to the value of $1/C_f = 0$, where C_B and C_f are equilibrium values [5, 7].

Correlation between the Infrared Crystallinity Index and the Accessibility by the Br_2 Method

The linear correlations described previously [7] between the infrared crystallinity indices and the accessibility by the bromine method were used in the present study. The linear correlation with the $1372 \text{ cm}^{-1}/2900 \text{ cm}^{-1}$ infrared crystallinity indices (I.R.C.I.$_1$) is given by the equation:

$$A = \frac{0.687 - \text{I.R.C.I.}_1}{0.0045}$$

The correlation with $1429 \text{ cm}^{-1}/893 \text{ cm}^{-1}$ infrared crystallinity indices (I.R.C.I.$_2$) is given by the equation:

$$A = \frac{3.136 - \text{I.R.C.I.}_2}{0.0293}$$

The accessibility values were calculated by using the above equations.

Estimation of Orientation

The orientation factor of Hermans [25] as measured by optical birefringence was used to represent the orientation of the samples. Birefringence was measured by the Becke line method as described by Meredith [26]. The immersion media were standard liquids of known refractive indices supplied by the Gargille Corp., U.S.A. These were mixed to provide intermediate values where necessary. The refractive indices of liquids or mixtures that matched a fiber index were further checked with an Abbe-type refractometer, calibrated with 1-bromonaphthalene. All measurements were made with the use of a sodium vapor lamp as a source of monochromatic light.

An optical orientation factor f_0 was calculated for each sample using the relationship $f_0 = N/N_0$, where N is the measured birefringence and N_0 is the theoretical birefringence of a perfectly oriented fiber. N_0 for cellulose II was taken as 0.055, following Hermans [25].

TABLE I

Bromine-Induced Crystallization of Cotton-Cellulose at Several Temperatures[a]

Temp. C^o	Times minutes	C.l. by I.R.	Accessibility %
	untreated control	2.415	24.6
10	30	2.47	22.7
10	60	2.47	22.7
10	90	2.49	22.0
17	30	2.45	23.4
17	60	2.48	22.4
17	90	2.50	21.7
25	30	2.47	22.7
25	60	2.49	22.0
25	90	2.54	20.3
32	30	2.52	21.0
32	60	2.55	20.0
32	90	2.64	16.9
41	30	2.58	19.0
41	60	2.61	17.9
41	90	2.71	14.5

[a] Conc. of Br_2, 0.09 mole/l.; conc. of cotton, 10 g/l.; pH = 2.

TABLE II

Rate Constants of Crystallization of Cotton Cellulose by Br_2–Water at Various Temperatures and Bromine Concentrations

Conc. of Br_2 moles/l	Temp. C^o	k, sec^{-1} x 10^5	Degree of Correlation R
0.09	10	1.83	0.906
0.09	17	2.33	0.99
0.09	25	3.34	0.988
0.09	32	6.5	0.98
0.09	41	9.07	0.974
0.049	25	1.6	0.95
0.098	25	2.7	0.98
0.147	25	5.02	0.97
0.197	25	6.56	0.98

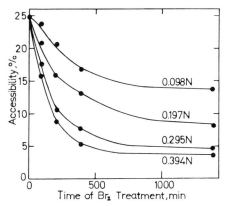

FIG. 1. Change in accessibility of cotton cellulose by aqueous Br_2 treatment; effect of time and Br_2 concentration.

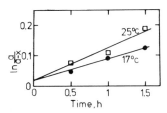

FIG. 2. Kinetics of bromine-induced crystallization of cotton.

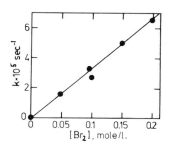

FIG. 3. Effect of Br_2 concentration on kinetics of crystallization of cotton. Slope = 3.38×10^{-4}; $R = 1.00$.

RESULTS AND DISCUSSION

Kinetics of Crystallization of Cotton

In Figure 1, several curves describing the decrease in accessibility of Acala cotton fabric during treatment in a number of concentrations of bromine are shown. The lowest value of 4% accessibility was obtained with a 0.197 molar solution of bromine at 25°C after a 24-hr treatment. In Table I, data of several kinetic experiments carried out with cotton fabric at five temperatures in the range of 10–41°C with a 0.09 mole/l. aqueous bromine solution are presented. Linear plots of ln $(a/a-x)$ against time are obtained for these experiments (see Fig. 2) where a is the initial percentage of accessibility (24.6%) available for crystallization and x is the percentage of the crystalline regions formed after time t. The linearity of the plots indicates that the crystallization process behaves according to first-order kinetics at a given initial concentration of bromine. The first-order reaction rate constants computed from the two series of experiments described in Table I and in Figure 1 (up to 6.5 hr) are listed in Table II.

The rate constant increases linearly with the increase in the bromine concentration (see Fig. 3), showing that the crystallization process is also first-order with regard to the bromine concentration. The overall process appears, therefore, to be a second-order process. The reaction rate increases with temperature (see Fig. 4) and the rate constants for a Br_2 concentration of 0.09 mole/l. vary from 1.8×10^{-5} sec^{-1} at 10°C to 9.1×10^{-5} sec^{-1} at 41°C (see Table II).

The plot of the rate constants against $1/T$ yields two straight lines intersecting at 25°C, indicating a change in the apparent activation energy of the crystallization process. Using Arrhenius plots, two activation energies were calculated: 8.5 and 10.9 kcal/mole for the temperature ranges 10–25°C and 25–41°C, respectively (see Table III).

The rate constants listed in Table II are of the same order of magnitude as the range of values of 10^{-4} to 10^{-5} obtained by Hatakeyama and co-workers from infrared data for the crystallization rates of an amorphous cellulose fiber

TABLE III
Energy of Activation of Crystallization and of Oxidation of Cellulose by Br_2–Water

Temp. °C	$E_{act.}$, Kcal/mole Crystallization	$E_{act.}$, Kcal/mole Oxidation
10–25	8.5	3.3
25–41	10.9	9.3

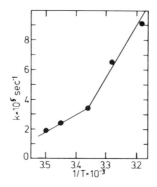

FIG. 4. Temperature dependence of the crystallization rate constant of cotton.

by heat-treating in a dry nitrogen atmosphere at 94–150°C [27]. The activation energies obtained were in the range of 31–46 kcal/mole, i.e., considerably higher than in the present study, indicating basic differences between the mechanisms of the two processes.

In the case of Hatakeyama et al. [27], the crystallization process depends on the alignment of the chains which is governed by the diffusion of the chain segments. This diffusion is hindered by the high stiffness of the chains inherent to cellulose, by entanglements of the chains and by the cohesion between the chains due to Van der Waals forces and to strong hydrogen bonds being formed at random during the heat treatment between the entangled chains. Dehydration may also take place between hydroxyl groups of monomers of two neighboring chains or between C_1 hydroxyls of the chain ends and hydroxyl groups, so as to produce oxygen bridges or crosslinking between chains. This reaction has been postulated earlier for thermal treatments of cellulose [28–31], and its extent and rate were found to increase with increase in LOR [30]. Such crosslinks when formed between entangled chains, may further decrease the probability of their alignment. Since the material used in the experiments of Hatakeyama et al., was designed to be completely amorphous, there are no crystalline regions in which parts of the chains are already aligned. The phenomenon described is thus mainly a primary nucleation, with a high activation energy, producing very small crystallites which could not be detected clearly by x-ray diffraction. The further growth of these crystallites is very slow due to the hindrances listed above, and its rate has in fact not been measured by the authors [27].

The activation energy obtained in the present study is in the range of activation energies of diffusion processes. The substrate used is relatively highly crystalline, and most of the chains are already aligned and ordered in the form of crystalline regions. The crystallization process investigated takes place in the LOR and involves the diffusion of chain segments as well as the diffusion of bromine molecules. The rate-determining step would appear to be the diffusion of the chain segments due to their greater volume as well as to the hindrances to their mobility. Unlike the case of Hatakeyama et al., water is present in the system and serves as a plasticizer; it assists in breaking a part of the hydrogen bonds and in the disentanglement of the chains, thus increasing their mobility. This would account for the lower activation energy measured.

It is of interest to note that an energy of activation of ca. 10 kcal/mole was found for the benzene induced crystallization of poly(ethylene terephthalate) in the temperature range of 25–55°C [32]. This crystallization was thought to be controlled by the rate of diffusion of the solvent through the polymers, and a clear distinction was made between this type of crystallization and the Avrami type thermal crystallizations yielding sigmoidal rate curves. For such a crystallization of amorphous PET, an energy of activation of 37 kcal was reported [33] based on kinetic experiments in the temperature range of 96–132°C, followed by infrared measurements. This value corresponds to the value that Hatakeyama et al. [27] cited above, which was obtained for amorphous cellulose at the same temperature range. The presence of a solvent apparently changes the mechanism of crystallization in both polymers; this is clearly evident in the lower E_{act}.

The bromine molecules being sorbed on the glycosidic oxygen atoms do not compete with water for the hydroxyl groups. It is, however, possible that during the diffusion of the bulky molecules in the LOR to the glycosidic sorption sites, hydrogen bonds between hydroxyl groups or between water molecules and hydroxyl groups will be severed, thus increasing further the mobility of the chain segments.

The alignment of the cellulose chains due to the formation of bromine bridges will also create a tension in the chains and will bring about the breaking of random hydrogen bonds in which various segments of the chains have been involved. The cohesion between the chains will be weakened and a rotation of chain segments will be possible which will facilitate the further alignment of the chains. The alignment will thus spread along the chains in the direction of their axis. Upon removal of the sorbed bromine by washing or reduction, the aligned chains will be in the favored position to form strong hydrogen bonds which may develop into small crystalline regions. If the bromine bridges happened to be formed close to existing crystalline regions, they may bring about an extension of these regions either laterally and increase their thickness or in the main axis direction of the elementary fibrils and increase their length. New small crystalline regions may not be detected by x-ray diffraction, similar to the case of Hatakeyama et al. [27], but may be found and determined by the infrared and bromine accessibility methods.

Since the bromine molecules are not absorbed on the glycosidic oxygens at the walls of the crystalline regions, it cannot be assumed that the crystallites will grow laterally in a regular way. A lateral extension could be obtained, however, if it is assumed that LOR chains exist close and parallel to the walls of the elementary fibrils but not close enough to preclude the penetration of the bromine molecules and consequently the alignment of the chains. Such an arrangement of chains was postulated before by Frey-Wyssling [34] and termed paracrystalline regions. The chains in the paracrystalline regions would align and pack into a new crystalline region which might be close enough to the neighboring crystallite to form an aggregate with it and appear as a widening of the elementary fibril or an increased lateral order. A lateral extension of crystalline regions could also be conceived according to the model of Peterlin and Ingram [35]. The lateral layers of chains, separating the adjacent elementary fibrils from each other and created by the screw dislocation of the consecutive crystalline blocks, could be brought together by bromine bridges and crystallize, thus increasing the order. The cleavage of some chains in these regions might enhance their alignment to a certain extent by further increasing the mobility of the chain ends. Preliminary determinations of the crystallite lateral order from the x-ray diffraction diagrams showed that in the case of high-modulus rayon after several bromine treatment and a pronounced increase in crystallinity, the crystallite width remains essentially constant in the 002 and 101 planes but is increased significantly in the $10\bar{1}$ planes (see Fig. 10 and Table VI). Since in the case of cellulose II the chains in the $10\bar{1}$ crystal plane are strongly hydrogen-bonded, it would indicate that the crystal growth occurred through hydrogen-bonding of additional chains in the $10\bar{1}$ planes.

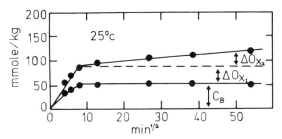

FIG. 5. Interaction of bromine–water with cotton at 25°C. C_B = sorption; Ox_1 = rapid oxidation; Ox_2 = slow oxidation.

An increase in temperature brings about an increase in rate of crystallization chiefly by decreasing hydrogen-bonding and increasing mobility and rate of diffusion of chain segments. As seen in Table III, the activation energy is significantly higher in the temperature range above 25°C.

It is of interest to note that several other changes occur in the system in the vicinity of 25°C. As seen in Table IV, the time needed to reach equilibrium sorption of bromine on cotton decreases sharply from 224 min at 10°C to 64 min at 25°C and 32°C (see Fig. 5). The decrease in sorption equilibrium uptake per degree $\Delta C_B/\Delta T$ is higher (1.7 mmol/kg/deg) in the range 25–41°C than in the range 10–25°C (1.2). Since the equilibrium concentrations of bromine in the solutions (C_f) after completion of the sorption are at all temperatures of Table IV approximately equal, the rates of oxidation can be compared. The rate of the slow oxidation measured as the slope of the straight line of consumed oxidant, ΔOx_2 (see Fig. 5) against $t^{1/2}$ increases sharply above 25°C. The increase in this rate per degree is 1.26 mmole/kg/min$^{1/2}$ in the temperature range 10–25°C as compared to 3.56 in the range 25–41°C, indicating a strong increase in the rate of diffusion of the bromine in the cellulose. This change is evidenced in the change in the energy of activation of the oxidation (see Table III) which is parallel to the change in the E_{act} of the crystallization.

The above changes occurring at 25°C appear to be due to a change in fine structure. They cannot be attributed to changes in crystallinity. At the low concentration of bromine of 11.6 mmole/l. in the experiments of Table IV, the crystallization is slow. However, even if the crystallinity had increased markedly in these experiments, the amount of LOR in the cellulose would have decreased with temperature; therefore, the extent and rate of the sorption and oxidation would have diminished. It appears, therefore, that in the vicinity of 25°C, a second-order transition must take place in the LOR which brings about the effects described here.

The transition near 25°C is clearly evident in Figure 6, in which the accessibility values obtained for cotton by the bromine sorption method using the concentration range 15–65 mmole/l. were plotted against the temperature of sorption. While the accessibility did virtually not change in the temperature range 10–25°C, it dropped sharply above 25°C. This drop coincides with the change in energy of activation of the crystallization (see Table III) and with the sharp increase in the rate of crystallization. A second-order transition at 25°C has been noticed in studies of the change of modulus of elasticity of paper sheets

TABLE IV

Sorption and Oxidation of Cotton by Br_2–Water at Several Temperatures[a]

Temp. °C	Sorption Equilibrium Time, min.	Sorption Equil. Uptake in mmoles/kg.	Cf mmoles/l	ΔOx_1 mmoles/kg	Rate of Oxidation $\Delta Ox_2/\sqrt{t}$ mmoles/kg.min$^{\frac{1}{2}}$
10	240	69.4	10.6	38	0.55
25	64	52.0	10.44	42	0.74
32	64	40.0	10.50	71	1.06

[a] Initial Br_2 conc., 11.61 mM/l.; pH = 2.

TABLE V

Structural Changes in Regenerated Cellulose Fibers by Treatment with Br_2–Water[a]

Fiber	Conc. of Br_2, moles/l	Time, Hours	I.R. Accessibility %	Regain %	fo	XRD C.I.
Textile Rayon	0.0	0.0	48	11.25	0.4	0.53
Textile Rayon	0.170	1.5	59	11.60	0.36	
Textile Rayon	0.170	3.5	64	11.92		
Textile Rayon	0.170	6.5	72	12.27	0.34	
Textile Rayon	0.170	23.0	75	13.00	0.29	0.45
Textile Rayon	0.185	48.0	55			
Textile Rayon	0.147	1.5	55	11.58		
Textile Rayon	0.147	3.5	61	11.81		
Textile Rayon	0.147	6.5	61	12.10		
Textile Rayon (40°C)	0.185	24.0	34.8			
Vincell 28	0.0	0.0	44	11.06	0.76	0.61
Vincell 28	0.170	24.0	59	11.58		0.44
H. M. Rayon	0.0	0.0	32		0.86	
H. M. Rayon	0.170	0.08	41.5			
H. M. Rayon	0.170	0.5	19			
H. M. Rayon	0.170	4.0	4		0.80	

[a] Temp., 25°C; conc. of fibers, 10 g/l.

with temperature [29], in studies of swelling of highly crosslinked cellulose [37], and of sorption of iodine on cellulose [38].

Changes in Fine Structure of Rayon

In Table V, the influence of the bromine treatment on the fine structural parameters of rayon fibers is summarized. It can be seen that for textile rayon and Vincell 28, the accessibility of the fibers increases upon treatment with

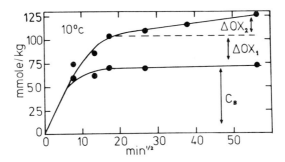

FIG. 6. Interaction of bromine–water with cotton at 10°C. C_B = sorption; Ox_1 = rapid oxidation; Ox_2 = slow oxidation.

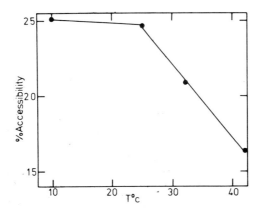

FIG. 7. Temperature dependence of the accessibility by the Br_2 method. Range of $[Br_2]$ = 10–50 mmole/l.

bromine. Longer treatment times and higher concentrations increase the effect. At 25°C, an accessibility of 75% was obtained upon a 23-hr treatment with a 0.17 molar solution. The XRD crystallinity index at the same time decreased and the moisture absorption increased.

A similar phenomenon, although apparently slower, is observed for Vincell 28. The increase in accessibility is accompanied by a marked decrease in orientation of the rayon (Table V) and by a disordered fiber surface as can be seen when comparing the scanning electron micrographs of untreated and bromine-treated rayon fibers (see Figs. 8 and 9).

The disordering effect depends on the time and temperature of the bromine treatment. When prolonging the treatment with 0.18 mole/l. to 48 hr (see Table V), the accessibility increased only to 55% as compared to 75% after 23 hr, while raising the temperature to 40°C reversed the effect and the accessibility decreased to 34.8%, i.e., below the value of the untreated fiber. The reversal is more clearly evident in the case of the more highly oriented and more crystalline high-modulus rayon. As is seen in Table V, the accessibility of high-modulus rayon upon treatment with 0.17 mole/l. of bromine at 35°C first increases and reaches a maximum value, and thereafter decreases and behaves similarly to cotton. The decrease in infrared accessibility is accompanied by an increase in

FIG. 8. Scanning electron micrograph of textile rayon fibers (10,000×).

FIG. 9. Scanning electron micrograph of rayon fibers treated with 0.15 mole/l. Br$_2$–water for 24 hr at 25°C; pH-2 10,000×.

TABLE VI
X-Ray Diffraction Results of Br_2-Treated High-Modulus Rayon[a]

Time of Treatment hours	XRD C.I., %	XRD Crystallite Lateral Order, Å Planes		
		002	101	10$\bar{1}$
0.0	75.0	43 ± 2	50 ± 2	46 ± 2
1.0	81.0	47 ± 2	50 ± 2	54 ± 2
2.0	82.5	47 ± 2	50 ± 2	57 ± 2
3.0	85.0	47 ± 2	50 ± 2	64 ± 2
24.0	87.0	52 ± 2	50 ± 2	79 ± 2

[a] Conc. of Br_2, 0.095 mole/l.; pH = 2; temp., 25°C.

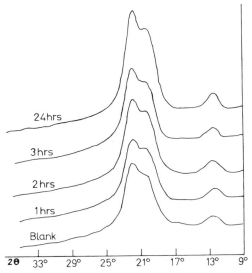

FIG. 10. X-ray diffractograms of high-modulus rayon treated with 0.095 mole/l. Br_2 at 25°C.

the XRD crystallinity index (see Fig. 12 and Table VI) and a decrease in moisture absorption.

Based on the XRD data of Table VI, a first-order kinetic rate constant was calculated for the crystallization reaction of high-modulus rayon in the first 3 hr of the bromine treatment by putting a, the initial percentage of accessible polymer, as 100 minus the XRD crystallinity index, and x, the increase in crystallinity after time t. The constant obtained, 4.48×10^{-5} sec^{-1}, although higher, is of the same order of magnitude as in cotton under similar conditions. The initial decrystallizing effect was disregarded in this preliminary calculation.

The most pronounced difference between the three rayons of Table V appears to be in their orientation factor. It appears that while the decrystallizing as well as the crystallizing effects are typical for all the rayons, their relative magnitudes and rates depend to a large extent on the orientation and consequently on the distance between the chains, since higher orientation is usually accompanied by a higher density of packing of the molecules in the polymer. Larger distances will decrease the probability of the chains to align themselves to positions favorable for the bromine molecules to produce the bridges necessary for the subsequent crystallization. Longer times will thus be needed for the segments to approach each other and the crystallization rate will decrease. A similar effect of orientation was observed earlier in this laboratory in the case of pyrolysis of rayon [30, 31]. The rates of the thermal crosslinking as well as the bulk decomposition reactions were both found to increase in a rectilinear way with the orientation factor [30]. The present results do not yet permit to draw quantitative conclusions, especially when taking into consideration the complexity of the phenomena occurring in the system and the additional morphological factors which may influence them. The degree of crystallinity and the size of the crystalline regions and their distribution in the polymer may have a profound influence on the segment mobility in the LOR, especially in the proximity of the crystalline regions. The "chain immobilization factor" due to the crystalline regions is known to hinder diffusion into the polymer [39, 40] but at the same time introduces an element of order into the LOR, contributes to the decrease in the distance between chains, and therefore may facilitate the alignment of the chains by the bromine bridges. This effect would superimpose upon the effect of the orientation and augment it.

In view of the above results, it appears that the bromine solution exerts two opposing effects, both being influenced by the state of the cellulose treated. The disordering effect appears to be due to the hydrogen bond-breaking activity of the bromine molecules, while the crystallizing effect is due to the bromine–glycoside bridges. In celluloses with relatively large distances between the chains, the former effect will predominate since it depends only on the diffusion of the bromine molecules in the water-swollen fiber. The rate of this diffusion increases with decrease in orientation and with decrease in crystallinity. The regions in which the disordering effect takes place cannot yet be defined, but they may be associated with disaggregation of microfibrils. The rate of the crystallization, on the other hand, appears to increase with the increase in order and in orientation, and it depends on the rate of diffusion of the monomeric units.

Depending on the conditions, both processes, ultimately yielding opposing results, may occur simultaneously. They may be partly overlapping processes or even appear as consecutive stages. In the case of textile rayon and Vincell 28 the processes appear to be separated, the state of the polymer being such as to favor the disordering effect while the crystallization becomes noticeable only after prolonged periods of time and at higher temperatures. In the case of high-modulus rayon, the disordering process, although apparently setting in earlier, may partly overlap with the crystallization which sets in at a much earlier stage than in textile rayon. An indication that the disordering effect occurs to a small extent also in the case of the highly ordered cotton fibers can be seen in the initially slower crystallization when working with a relatively low concentration of bromine (see uppermost curve in Fig. 1). It is proposed to investigate these phenomena further with other celluloses in the future.

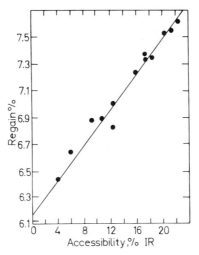

FIG. 11. Moisture regain vs. infrared accessibility of Br$_2$-treated cotton fibers: $R = 0.98$; intercept $= 6.17$; slope $= 0.065$.

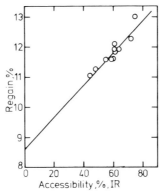

FIG. 12. Moisture regain vs. infrared accessibility of untreated and Br$_2$-treated rayon fibers: $R = 0.94$; intercept $= 8.55$; slope $= 0.054$.

Regain–Accessibility Plots of Bromine-Treated Celluloses

In Figure 11, the moisture regain at 65% of a series of cotton samples both untreated and bromine-treated is plotted against the infrared accessibility values. The straight line obtained when extrapolated to zero accessibility shows an intercept of 6.1% regain. These data corroborate the preliminary results reported earlier [7] which indicated the existence of two types of sorption sites in the accessible part of the fiber. The bromine molecules appear to be sorbed on a smaller number of sites, i.e., the glycosidic oxygens of the LOR.

The relatively large size of the bromine molecules prevent their approach and sorption on the glycosidic oxygens located on the surfaces of the crystallites, since they are buried half a molecule deep within the crystallite [10]. The bromine accessibility will therefore refer to the LOR only and will yield values of a category similar to those obtained from acid hydrolysis, which does not occur on the crystallite surfaces. Zero bromine accessibility would, therefore, indicate a cellulose which does not have accessible glycosidic linkages in the LOR. This would mean essentially that all the LOR have crystallized or aggregated or become ordered in such a way as not to exhibit any bromine or infrared accessibility.

Moisture absorption as well as deuteration take place on the hydroxyl groups of the LOR as well as of the crystallite faces. The crystallization prevents the absorption of moisture in the LOR. Consequently, the moisture absorption will take place only on the crystallite faces and its value will give an indication of the size of the crystallites. The larger the size, the smaller the moisture regain value obtained from extrapolation of the regain–accessibility plots. Extrapolation of the corresponding plot for rayons (see Fig. 12) yields a value of 8.55% moisture regain in keeping with the well-known smaller crystallite size of rayons as compared to cotton. The fact that straight-line relationships with high degrees of correlation are obtained between regain and accessibility values would seem to indicate that the crystallite widths remain essentially constant throughout the bromine treatments. The construction of such plots might therefore be considered as a method for the determination of crystallite sizes in various celluloses.

The authors wish to thank the J. P. Stevens Co., Technical Center in Garfield, N.J., and Dr. L. G. Roldan for the XRD measurements of the high-modulus rayon fibers (Table VI and Fig. 10); Ata Textiles, Co., Ltd., Kfar Ata, Israel for the cotton; and Courtaulds, Ltd., Coventry, England for the rayon samples used in the present study.

REFERENCES

[1] M. Lewin, *Tappi, 48,* 333 (1965).
[2] M. Lewin, in *Bromine and its Compounds,* Z. E. Jolles, Ed., Ernest Benn Ltd., London, 1966, pp. 703–736.
[3] M. Lewin, *Mild Oxidation of Cotton.* Project FG-Is-109, Final Report submitted to Agricultural Research Service, U.S.D.A., Jerusalem, Israel, 1968; Project FG-Is-58, Final Report submitted to Agricultural Research Service, U.S.D.A., Jerusalem, 1963.
[4] M. Lewin, *Bull. Res. Counc. Israel, 2,* 101 (1952).
[5] M. Lewin and A. Ben-Bassat, Sirtec, paper presented at the 1st International Symposium on Cotton Textile Research, Paris, April 1969, Institute Textile de France; Preprints, pp. 535–556.
[6] M. Albeck, A. Ben-Bassat and M. Lewin, *Text. Res. J., 35,* 935 (1965).
[7] M. Lewin, H. Guttmann, and N. Saar, *Appl. Polym. Symp. No. 28,* 791 (1976).
[8] M. L. Nelson and R. T. O'Connor, *J. Appl. Polym. Sci., 8,* 1325 (1964); *ibid., 8,* 1311 (1969).
[9] R. T. O'Connor, E. F. Dupre, and D. Mitcham, *Text. Res. J., 28,* 382 (1958).
[10] A. M. Scallan, *Text. Res. J., 41,* 647 (1975).
[11] P. A. D. DeMaine, *J. Chem. Phys., 26,* 1192, 1199 (1957).
[12] B. Perlmutter-Hayman and A. Persky, *J. Amer. Chem. Soc., 82,* 276 (1960).
[13] C. Brian, *Chem. Rev. 4,* 493 (1964).
[14] J. R. L. Barker, W. G. Overend, and C. W. Rees, *Chem. Ind.* (London), *1960,* 1297; *J. Chem. Soc., 1964,* 3254.
[15] H. S. Isbell, *J. Res. N.B.S., 66A,* 233 (1962).
[16] A. Dyverman, B. Lindberg, and D. Wood, *Acta Chem. Scand., 5,* 253 (1951).
[17] O. Theander, *Svensk Papersbidn., 61,* 581 (1958).
[18] L. S. Levitt, *J. Org. Chem., 20,* 1297 (1955).

[19] J. Mulliken, *J. Amer. Chem. Soc., 76*, 3869 (1958).

[20] O. Hassel and Chr. Rømming, *Quart. Rev., 16*, 1 (1962).

[21] L. Segal, *Amer. Dyest. Rep., 46*, 637 (1957).

[22] L. G. Roldan, F. Rahl, and A. R. Paterson, in *Analysis and Functionation of Polymers (J. Polym. Sci. C, 8)*, J. Mitchell, Jr. and F. W. Billmeyer, Jr., Eds., Interscience, New York, 1965, p. 145.

[23] M. Lewin and L. G. Roldan, *Text. Res. J., 45*, 308 (1975); in *Proceedings of the Seventh Cellulose Conference (J. Polym. Sci. C., 36)*, E. C. Jahn, Ed., Interscience, New York, 1971, pp. 213–229.

[24] K. Y. Heritage, Y. Marin, and L. G. Roldan, *J. Polym. Sci. A, 1*, 671 (1963).

[25] P. H. Hermans, *Physics and Chemistry of Cellulose Fibers*, Elsevier, New York, 1949.

[26] R. Meredith, *J. Text. Inst., 37*, T205 (1946).

[27] H. Hatekeyama, T. Katakeyama, and J. Nakano, *Cellulose Chem. Technol., 8*, 495 (1974); *Appl. Polym. Symp. No. 28*, 743 (1975).

[28] E. L. Back and L. O. Klinga, *Svensk Papperstidn., 66*, 745 (1963).

[29] E. L. Back and E. I. Didriksen, *Svensk Papperstidn., 72*, 687 (1969).

[30] A. Basch and M. Lewin, *J. Polym. Sci. Polym. Chem. Ed., 12*, 2053 (1974).

[31] M. Lewin, A. Basch and C. Roderig, in *Proceedings of the International Symposium on Macromolecules, Rio de Janeiro*, Elsevier, Amsterdam, 1974, p. 225.

[32] R. P. Sheldon, *Polymer, 3*, 27 (1962).

[33] K. G. Layhan, W. J. James, and W. Bosch, *J. Appl. Polym. Sci., 9*, 3605 (1965).

[34] A. Frey-Wyssling, *Science, 119*, 80 (1954).

[35] H. Peterlin and P. Ingram, *Text. Res. J., 40*, 345 (1970).

[36] M. Lewin and L. G. Roldan, *Text. Res. J., 45*, 308 (1975).

[37] K. Chitumbo, W. Brown, and A. Deruvo, in *Transformations of Functional Groups on Polymers (J. Polym. Sci. Polym. Symp. Ed., 47)*, C. G. Overberger and B. Sedláček, Eds., Interscience, New York, 1974, p. 261.

[38] K. Aziz and M. A. Abn-State, *Cellulose Chem. Technol., 8*, 443 (1974).

[39] A. S. Michaels and M. J. Bixler, *J. Polym. Sci., 50*, 413 (1961).

[40] S. W. Lasosky and W. H. Cobbs Jr., *J. Polym. Sci., 36*, 21 (1959).

STRESS–STRAIN BEHAVIOR OF ORIENTED POLY(ETHYLENE TEREPHTHALATE) BY DYNAMIC INFRARED STUDIES. I. LOAD-BEARING OF THE AVERAGE MOLECULAR BACKBONE BONDS

K. K. R. MOCHERLA and W. O. STATTON

Department of Materials Science & Engineering,
University of Utah, Salt Lake City, Utah 84112

SYNOPSIS

A new dynamic infrared technique is used to study the influence of morphology on the response of backbone bonds to external stress in poly(ethylene terephthalate) films. Stress on skeletal atomic bonds causes a peak shift which is linearly related and is shown to be morphology-dependent. It is suggested that the shift measures the tautness distribution of load-bearing chain segments in the polymer. The drastic effects of heat relaxation treatments are now shown to be molecular changes which affect the load-bearing ability of the system. No shift is observed at low loads, indicating that in the elastic, Young's modulus region of deformation the *majority* of the backbone bonds are not measurably distorted.

INTRODUCTION

Stress–strain behavior of crystalline polymers has been shown to be greatly influenced by the morphological state of their samples. However, there has been a lack of a technique that can directly relate the state of molecular bonds in the polymer chain to the macroscopic stress behavior. This paper is a preliminary part of new research being carried out to study such changes that occur in individual molecular segments during application of stress to a polymer sample.

A relatively new infrared technique, first developed by Zhurkov and coworkers [1], is being employed for this purpose. The method consists of examining a sample while it is held under stress in an IR-spectrophotometer. Since the infrared vibrational frequencies of a polymer molecule are dependent on the interatomic bond strengths and valence angles, if the bonds are distorted by stress, the frequency of an absorption bond involving them will shift to longer wavelengths, and the band itself will undergo shape distortion. This peak frequency shift can be explained on the basis of the simplified model which treats a molecular chain as a series of coupled oscillators, each oscillator representing a repeat unit of molecule [2]. If the oscillator linkage is deformed by broadening

Journal of Applied Polymer Science: Applied Polymer Symposium 31, 183–191 (1977)
© 1977 by John Wiley & Sons, Inc.

of valence angles and stretching of interoscillator links such that the overall length of the assembly is increased, then for a given amount of input energy the linkage will vibrate at a lower frequency as compared to the unstrained state. If the distortions are removed, the assembly would now be expected to vibrate at its original frequency. Zhurkov et al. [1] demonstrated that for skeletal or backbone bands, such a frequency shift does occur upon application of external stress. This behavior was demonstrated for polypropylene, poly(ethylene terephthalate), and nylon 6. The frequency shift was shown to have a linear relationship with stress, σ:

$$\Delta\nu = \alpha\sigma$$

$\Delta\nu$ is the change in peak frequency of the particular band, and α is a proportionality constant. This paper deals with the new work done to study the effect of morphology on this α value, as well as establishing that the above equation is actually $\Delta\nu = \alpha(\sigma - \sigma_0)$, where σ_0 is a threshold stress.

An interatomic bond in the molecular chain that is infrared-active is expected to sharply absorb infrared radiation at a particular frequency. However, a smearing and broadening of absorption bands in polymers is always caused by the random configuration and distribution of chain lengths in the noncrystalline regions, and in the crystalline regions due to the presence of crystal defects. In spite of this broadening, the peak of an infrared absorption band represents the majority of those particular interatomic bonds which are the absorbing elements. Thus, the peak frequency shift indicates that the *majority* of the absorbing elements are so affected that they are now absorbing at longer wavelengths.

The band-shape distortion mentioned earlier appears as a broadening on the lower-frequency side of the absorption peak. This "tail-end" distortion, as it is called, is attributed to the presence of highly overstressed bonds [1]. Thus, in addition to the majority of atomic bonds that experience a certain "average" stress on application of external stress, there are some bonds that have much higher stresses imposed on them. These bonds with "overstress" on them now absorb at still lower frequencies than the average bonds, resulting in the "tail-end" absorption on the lower-frequency side of the peak. By a deconvolution technique the deformed band can be used to provide a stress distribution on a molecular level [1, 3, 4]. The results of such deconvolutions on our materials will be the subject of another paper. It will suffice to say here that we are interpreting the main peak to represent the "majority or average bond stresses" which in turn represents the *trans* segments in the crystalline domains, whereas the "overstressed or tail-end" region represents the *trans* segments in the disordered, amorphous domains. The full logic and bases for these interpretations are in preparation [4].

Poly(ethylene terephthalate) (PET) was selected for the present research as it can be easily obtained in a variety of morphological states and also because of its commercial importance. Of all the skeletal bands in the infrared spectrum of PET, it was found that the 973 cm^{-1} band was the most convenient to study. Hence, the data reported here pertains to the behavior of this band on stress application. Several other areas of the spectrum show stress effects and will be reported later as the work progresses.

EXPERIMENTAL

All infrared spectra were recorded on a Perkin Elmer Model 521 Spectro-photometer. The film samples were uniaxially loaded in the infrared beam cavity by an air-operated loading device built especially for this purpose [5]. A strip chart recorder provided a continuous record of both load and extension. The tensile load applied to the sample was accurately and directly measured by a previously calibrated transducer read-out. Samples were loaded at a uniform rate of 10 lb./min.

Highly uniaxially oriented PET films of different molecular weights and thicknesses were obtained from two sources. To distinguish between them, they will be henceforth referred to as films A, B, and C.

Film A, a 3-mil thick film with $\overline{M}_n$ = 22,000 and $\overline{M}_w$ = 56,000, was supplied by Dr. C. Heffelfinger, Film Department, E. I. du Pont de Nemours & Co., Circleville, Ohio.

Film B, a 5-mil thick film with $\overline{M}_n$ = 29,000 and $\overline{M}_w$ = 84,000, was supplied by Dr. G. Adams, E. I. du Pont de Nemours & Co., Wilmington, Del.

Film C, a 2-mil thick film with $\overline{M}_n$ = 22,000 and $\overline{M}_w$ = 60,000, was supplied by Dr. C. Heffelfinger, Film Department, E. I. du Pont de Nemours & Co., Circleville, Ohio.

Specimens of a particular film were heat-treated to provide different morphological states. The heat treatment consisted of annealing the films either slack or constrained in an inert N_2 atmosphere for different lengths of time at various temperature levels.

Samples 0.5 in. wide and 4.0 in. long were cut from the roll such that the draw direction was parallel to the length of the film specimen.

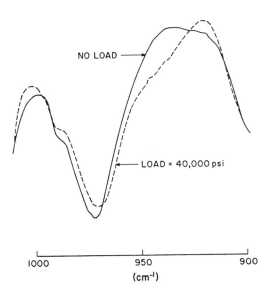

FIG. 1. Highly uniaxially oriented PET films: (——) normal sample; (- - -) sample examined under stress, exhibiting longer wavelength and tail-end distortion.

RESULTS AND DISCUSSION

The 973 cm^{-1} band of PET is a skeletal absorption assigned to motions in the *trans* isomer of the ethylene glycol (OCH$_2$CH$_2$O) linkage of the molecule [6–8]. This isomer exists in the crystalline as well as in the noncrystalline segments when oriented [7]. Heat treatment or orientation increase the intensity of this absorption peak since both processes increase the *trans* content [7].

Figure 1 shows the changes induced in the 973 cm^{-1} absorption band when the sample is examined under stress. The two distinct effects of the stressing phenomena are that the peak has shifted to lower frequency (longer wavelength), and the lower-frequency side of the band has been deformed (tail-end distortion). The amounts of peak frequency shift and tail-end distortion are dependent on the applied external stress. This shift in peak frequency is directly and linearly related to stress and is characterized by the quantity α, the slope of the shift versus stress curve. The band deformation in the low-frequency region has no simple direct relation to the applied stress, but can be used to determine a stress distribution on the atomic bonds, as mentioned earlier and to be reported separately.

It is important to note that the peak frequency shifts are completely reversible at all stress levels; i.e., they totally disappear when the load is removed. This indicates then that the shift is entirely due to the application of external stress and is not a permanent deformation effect.

α is measured from the shift in peak frequency and is thereby a measure of the response to external stress of the majority of those atomic bonds in the polymer that give rise to that particular absorption band. Different values of

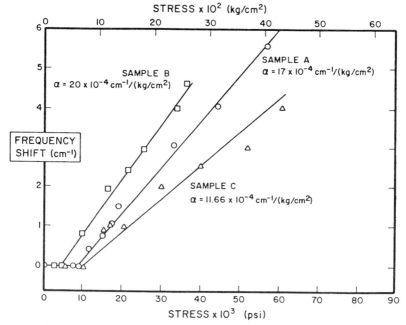

FIG 2. Comparison of α values for three different samples of PET films.

α can be expected for polymer samples with different morphologies, since the external load will be distributed differently so that the atomic bonds will respond differently in each case. For example, if two polymer morphologies are compared, one with more uniform lengths of load-bearing segments than the other, the sample with the uniform matrix should show the larger α value since more of the interatomic bonds will bear the stress at any moment. In the less uniform case, fewer chains will bear the greater part of the load, allowing the majority to be stressed to a lower level.

Such logic seems to be demonstrated by the results shown in Figure 2, which compares the response of the three different samples of PET films. Sample B has the highest α value, followed by A and C, in that order. The three samples show different stress–strain curves (Fig. 3), indicating different morphologies. On comparing the α values, it can be concluded from the above logic that sample B has a more uniform matrix than A, which itself has a more uniform matrix than C. Furthermore, by comparing Figures 2 and 3, it is seen that the slopes of the stress–strain curves for the three samples follow the same order as the α

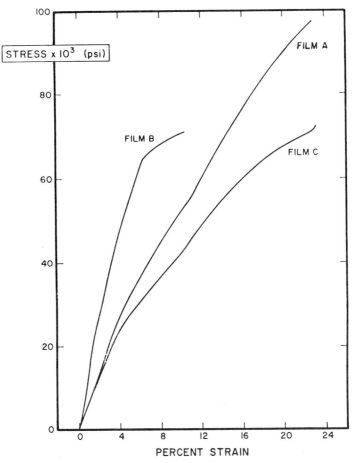

FIG. 3. Stress–strain curves for three samples of PET films.

values. This suggests that the uniformity of the matrix is important in the stress–strain curve.

Another new feature of these results is the existence of a region for all samples in which there is no peak frequency shift, despite an externally applied load. The absence of any frequency shift indicates that the majority of the interatomic bonds are not participating in the stress-bearing process in that load range. A close examination of Figures 2 and 3 reveals that the upper limits of this "no-shift" region are in the neighborhood of the elastic limits (or end of the region of linear stress to strain relationship) of the samples. The conclusion is, therefore, that within the elastic behavior region, the majority of the chains are not involved in the load-bearing process.

From Figure 2, it can be seen that sample B has its no-shift region ending earlier than the other two samples. From Figure 3, one can see that sample B has also the highest Young's modulus of the three samples. Thus, there appears to be a definite relationship between the initial modulus and the beginning of the shift. The other two samples follow the same pattern.

Slack annealing of a crystalline-oriented polymer, such as PET, has been shown to result in recrystallization to folded-chain crystals and relaxation of some of the oriented amorphous segments [9–15]. Such a morphology should change the load-bearing situation; this is indeed found, as shown in Figure 4. It is envisaged that heat treatment of the sample without constraints causes the oriented noncrystalline chain segments to relax partially and some of them to participate in the formation of larger chain-folded crystals. This results in the matrix becoming less uniform than before, since the frozen-in drawing stresses which homogenized the structure are relieved. The load-bearing in such a sit-

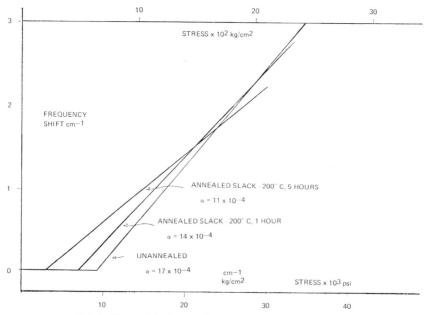

FIG. 4. Effect of slack annealing on load-bearing in PET films.

uation is now mainly carried out by tie molecules that have not relaxed or folded. Thus, the stress on the "average" chains in annealed samples is lower for a given unit of applied external stress, as compared to an unannealed sample.

Figure 5 compares samples of film A annealed slack at different temperatures: 200°C and 230°C. The films annealed at 230°C will have greater recrystallization and chain relaxation. The lower value of α obtained by the experiments indicates a less uniform matrix for the sample annealed at the higher temperature, as would be expected from the above discussion. Similarly the threshold stress value σ_0 is decreased.

The above changes produced on annealing are possible if there is sufficient chain segment mobility. If the chains are constrained during their heating, their decreased mobility will allow the rearrangements to occur only to a limited extent. Thus, if shrinkage is restricted, the amount of nonuniformity introduced into the matrix by heat treatment is also controlled, as shown in Figure 6. The samples were fixed in a specially constructed frame so that changes in two directions were restricted during annealing. The amount of allowed shrinkage was determined by the amount of restraint put on the sample.

SUMMARY AND CONCLUSIONS

Application of an external load to an oriented polymer sample results in the measurable stressing of the backbone atomic bonds in the chains. Since the chains are well oriented, the majority of the bonds experience a uniform stress for a fixed external load.

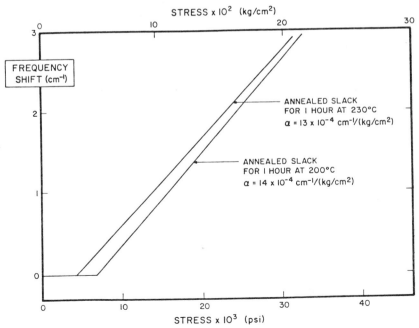

FIG. 5. Comparison of samples of film A annealed slack at two different temperatures.

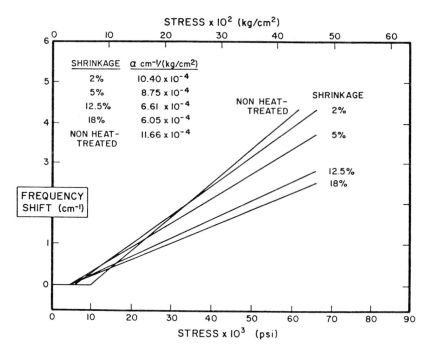

FIG. 6. Control of nonuniformity in PET films annealed at 155°C for 17 hr by restricting shrinkage.

α, a quantity determined as the slope of the curve of the peak frequency shift of a backbone band versus stress, characterizes the response of the majority of the atomic bonds to the external stress. α is shown to be morphology-dependent, having a higher value for a uniformly oriented matrix.

Annealing highly oriented films with no constraints causes chain segment relaxation and chain folding. The matrix is transformed to a less uniform state by formation of folded-chain crystals and disorientation of noncrystalline tie chains. This results in a few chains bearing the larger part of the applied external stress, while the majority of chains experience lower stresses.

The extent to which the value of α decreases is determined by the amount of relaxation introduced in the matrix by heat treatments. Thus, controlling shrinkage controls the amount of folded-chain crystal formation and chain relaxation and thereby controls the response of the sample to external stress. α was shown to decrease with increasing amount of shrinkage allowed.

Corresponding to the elastic region in the stress–strain curves of the polymer samples, there is a region of no frequency shift for low loads in the stressed infrared spectrum. This suggests that in the range of elastic behavior of the polymer, the *majority* of chain segments are not involved in the load-bearing. The threshold stress required for load-bearing by the majority of backbone segments was also found to be morphology dependent.

The authors are grateful to the National Science Foundation for the support of this research.

REFERENCES

[1] S. N. Zhurkov et al., *Fracture, 1969,* Chapman and Hall, London, 1969, p. 545.

[2] R. Zbinden, *Infra-red Spectroscopy of High Polymers,* Academic Press, New York and London, 1964.

[3] V. A. Kosobukin, *Sov. Phys. Solid State, 14,* 2246 (1973); V. I. Vettegren and I. J. Novak, *J. Polym. Sci. Polym. Phys. Ed., 11,* 2135 (1973).

[4] K. K. R. Mocherla, Ph.D. thesis in preparation.

[5] R. P. Wool and W. O. Statton, *J. Polym. Sci. Polym. Phys. Ed., 12,* 1575 (1974).

[6] A. Miyake, *J. Polym. Sci., 38,* 479 (1959).

[7] P. G. Schmidt, *J. Polym. Sci. A-2, 1,* 1271 (1963).

[8] S. K. Bahl, D. D. Cornell, and F. J. Boerio, *J. Polym. Sci. Polym. Letters Ed., 12,* 13 (1974).

[9] J. L. Koenig and M. J. Hannon, *J. Macromol. Sci., B-1,* 119 (1967).

[10] K. K. R. Mocherla and J. P. Bell, *J. Polym. Sci. Polym. Phys. Ed., 11,* 1779 (1973).

[11] R. E. Mehta and J. P. Bell, *J. Polym. Sci. Polym. Phys. Ed., 11,* (1973).

[12] W. O. Statton, J. L. Koenig, and M. J. Hannon, *J. Appl. Phys., 41,* 4290 (1970).

[13] J. B. Park, Ph.D. Dissertation, Univ. of Utah (1972).

[14] P. F. Dismore and W. O. Statton, *J. Polym. Sci. B, 2,* 1113 (1964); in *Small Angle Scattering from Fibrous and Partially Ordered Systems (J. Polym. Sci. C, 13),* R. H. Marchessault, Ed., Interscience, New York, 1966, p. 133.

[15] D. C. Prevorsek, G. A. Tirpak, P. J. Harget, and A. C. Reimschuessel, *J. Macromol. Sci., B-9,* 733 (1974).

THE EFFECT OF TETRACHLOROETHYLENE, 1,2-DIBROMOETHANE, AND THEIR MIXTURES ON THE MORPHOLOGY OF POLY(ETHYLENE TEREPHTHALATE)

ILAN KATZ and MENACHEM LEWIN

Israel Fiber Institute, Jerusalem, Israel

SYNOPSIS

Partially crystalline commercial poly(ethylene terephthalate) fibers were treated in perchloroethylene (PER, $\delta = 9.3$), 1,2-dibromoethane (DBE, $\delta = 10.4$), and their mixtures under no stress and at a temperature range of 60–127°C. While the average crystallite size remained nearly constant, the XRD crystallinity index increased and the orientation of the fibers decreased. DBE was found to be a more effective crystallizing and disorientating agent especially in the temperature range 95–120°C as compared to PER.

A minimum was observed in the crystallinity index isotherms obtained upon changing the composition of the PER–DBE mixtures, indicating a partial decrystallization of the fibers or a dissolution of some crystalline regions occurring simultaneously with the crystallization. Crystallization rates and apparent energies of activation were computed for PET in both solvents.

INTRODUCTION

Amorphous poly(ethylene terephthalate) (PET) has been shown to crystallize following dry heat treatments [1–3] and as a result of interactions with certain organic solvents [4–8]. While heat treatments bring about crystallization at temperatures above the glass transition temperature (T_g) of the polymer, solvent crystallization takes place at temperatures well below T_g, and in many cases at room temperature. Although the interactions of organic solvents with semi-crystalline oriented PET brings about less dramatic changes in structure and properties of the polymer compared to those observed in the amorphous polymer, they are of primary importance for solvent application to textile wet-processing. Little information has been published concerning the effect of solvents on the morphology of PET commercial fibers. It is also yet unclear whether solvent-treated PET has significantly different properties than heat-treated PET. Some changes in mechanical properties as well as a depression of T_g were observed while testing PET in a wide range of organic solvents [9, 10]. These results indicate that morphological modifications, as in the degree of crystallinity and orientation or in the size and number of crystallites, are likely to occur.

The interactions of solvent mixtures with either the amorphous or the partially

Journal of Applied Polymer Science: Applied Polymer Symposium 31, 193–200 (1977)
© 1977 by John Wiley & Sons, Inc.

crystalline polymer have been investigated only to a very limited extent. Lawton and Cates [7] and Kashmiri and Sheldon [11] have shown that mixtures of carbon tetrachloride and ethanol, i.e., of a nonpolar and polar solvent, induce crystallinity in PET whereas the pure solvents alone, at the same temperatures, do not. Measurements of the shrinkage force of PET fibers immersed in organic solvent mixtures of varying compositions revealed that liquid–polymer interactions depended on liquid–liquid interactions [12].

In the present paper, some results of the influence of two nonpolar solvents, with solubility parameters close to that of PET, i.e., perchloroethylene and 1,2-dibromoethane and their mixtures on the crystallinity and orientation of PET fibers, are presented.

EXPERIMENTAL

The fibers used were in the form of a plain weave fabric made of poly(ethylene terephthalate) fibers (Dacron T54 W type, DuPont) by Testfabrics, U.S.A. The fabric was not heat-set. It was scoured with 1.0 g/l. Lissapol D (ICI) and 1.0 g/l. Na_2CO_3 at 50°C for 30 min and then rinsed with distilled water containing 0.1% acetic acid.

The solvents used were perchloroethylene (PER) and 1,2-dibromoethane (DBE), both of an analytical grade.

Solvent mixtures were made on a weight–weight basis, and the fibers were treated in closed vessels for 30 min at several temperatures, followed by air-drying and vacuum-drying at 50°C until a constant weight was attained.

Samples were then finely chopped and randomly placed in an aluminum holder. A GE Model XRD-6 diffractometer was used employing CuK_α radiation with accelerating voltage 40 KV and anodic current 16 mA. Diffractograms were made by continuous scanning over the range of diffraction angles from 9° to 33°. The overall crystallinity index was calculated by using Wlochowicz and Jeziorny's method [13] as follows.

The crystallinity index X was calculated from the equation $X = (1 - X)Fk/h$, where F is the area between the diffraction curve (A) and the background curve (B) (Fig. 1), h is the background diffuse scattering, and k is a constant. The points P_1, P_2, P_3, and P_4 are drawn in order to obtain the background curve. The crystallite lateral dimension L was calculated from the reflection of the (010 + 0$\overline{1}$1) planes using the Scherrer equation [14]:

$$L = K\lambda/\beta_0 \cos \theta$$

where K is 0.89, $\lambda = 1.54$ Å (CuK_α), β_0 is the angular half-intensity breadth of the crystalline reflection, and θ is Bragg's angle.

The orientation of the PET samples was measured by means of birefringence using the Becke line method. The results were expressed in values of ΔN.

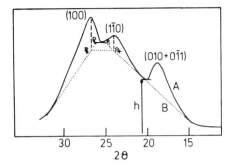

FIG. 1. Typical XRD diffractogram of PET fibers: (A) diffraction curve; (B) background (ref. [13]).

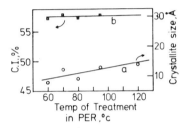

FIG. 2. Effect of treatment temperature of PET in PER on (a) the XRD crystallinity index (C. I.) and on (b) the crystallite size.

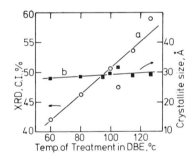

FIG. 3. Effect of treatment temperature of PET and DBE on (a) the XRD crystallinity index (C. I.) and on (b) the crystallite size.

RESULTS AND DISCUSSION

The results in Figures 2 and 3 indicate that both solvents bring about an increase in crystallinity of the PET. The ultimate effect reached by DBE under the experimental conditions applied in this study is much higher than by PER. An XRD crystallinity index of 59% is obtained after 30 min in DBE at 120°C, while the highest value obtained for PER is 49.5 at 120°C. The crystallinity index increases with temperature for both solvents, the increase for DBE being considerably steeper. DBE is a more effective crystallizing agent in the temperature range 95–130°C, while the influence of PER on the crystallization appears to be more pronounced at the lower temperature range.

The higher crystallizing power of DBE seems to be linked with its solubility

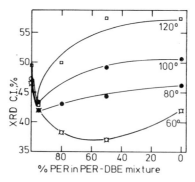

FIG. 4. Effect of composition of the PER–DBE mixture on the XRD C. I. of PET at several temperatures.

parameter (10.4) being closer to the solubility parameter of PET (10.7) than that of PER (9.3). Being a "better" solvent, DBE exhibits stronger penetration power and enables a higher segmental mobility; thus, crystallization is facilitated.

No significant change in the lateral crystallite dimension is observed following treatment in either solvent (see plots b in Figs. 2 and 3) which suggests that both solvents induce crystallization predominantly by creating new crystalline regions possibly of similar lateral dimensions.

Treating PET in various mixtures of the solvents at several temperatures brings about some unexpected changes as compared to the treatments in the pure solvents (Fig. 4). An initial decrease in crystallinity index is observed in the isothermal curves at 5% DBE content. The crystallinity index increases with a further increase in DBE in a nonlinear manner.

The minimum crystallinity points in the curves of Figure 4 appears to depend on the temperature and shifts to higher DBE contents at lower temperatures. At 60°C, the minimum appears to be at 40% DBE in the mixture.

The birefringence values obtained for the treated PET samples, summarized in Table I and Figure 5, are seen to decrease upon treatment with both solvents. The birefringence of the PER-treated samples decreases with increase in temperature of treatment up to 80°C. No further change is observed at higher temperatures. In DBE, the birefringence continues to drop with increasing temperature up to the boiling point. The disorientation effect of DBE is seen to be significantly stronger than of PER, which is parallel to the effect on the crystallization occurring at the same time.

In Table II, approximate kinetic first-order constants calculated from the values of the crystallinity indices of Figures 2, 3, and 4 are presented. The constants were calculated from the expression $\ln(a/a-x) = kt$, where a is the percent of the amorphous regions in the control PET fibers, i.e., $100 -$ percent crystallinity, and "$a-x$" is the % amorphous regions remaining after time t. A first-order kinetics was observed before upon the crystallization of amorphous PET in the presence of benzene [4], and rates of crystallization of a similar order of magnitude to the rates of Table II were found. From Arrhenius plots of $\ln k$ vs. $1/T$, energies of activations for the crystallization of PET in PER and DBE (Figs.

TABLE I
Birefringence (ΔN) of PER- and DBE-Treated PET Fibers[a]

Temp °C	PER	% DBE in PER – DBE Mixture			DBE
		5	20	50	
45	0.154				0.148
60	0.150	0.148	0.146	0.143	0.144
80	0.144		0.144	0.144	0.142
110	0.144				0.140
120	0.144		0.144	0.144	
127				0.144	0.138

[a] Untreated control: $\Delta N = 0.168$.

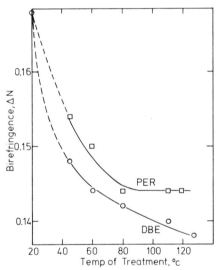

FIG. 5. Change in the birefringence of PET fibers upon treatment with PER and DBE at various temperatures.

6 and 7) and in several of their mixtures (Fig. 8) were computed. While the E_{act} for the DBE-induced crystallization, 8.8 kcal/mole in the temperature range 80–120°C, is close to the value of 10.5 kcal/mole found by Sheldon for the benzene-induced crystallization, the value of 1.68 kcal/mole found for PER is low. It is of interest to note the maximum in the curve of E_{act} versus percent of DBE in the PER–DBE mixture (Fig. 8). The maximum disappears if one takes the higher value of E_{act} for DBE obtained at the temperature range of 60–80°C (Fig. 7).

The results in the present study indicate that the phenomena occurring in the PET are twofold: decrystallization and crystallization. The penetration of the solvents into the fiber is facilitated by the fact that the T_g in both solvents is below 50°C [15]. Upon wetting of the fiber with the solvents, the chains are solvated and the cohesion between the chains in the amorphous regions is interrupted

TABLE II
Rates[a] of Crystallization of PET in PER and DBE

Temp. °C	PER	% of DBE in PER-DBE Mixture			DBE
		5	20	50	
60	5.36	0.78			0.78
70	7.45				
80	6.55	0.78	1.79	1.79	5.0
95	8.05				8.6
100		2.3	3.06	7.78	9.9
115					13.3
120	8.45	2.3	8.95	18.10	20.1

[a] Rates in sec^{-1} × 10^5.

[9, 10]. A solvation complex between the chain segments and the solvent molecules is formed. The mobility of the chains is increased and their orientation decreases. The action of the solvent may be strong enough to open up some small crystalline regions, bring about their dissolution, and thus cause a decrystallization effect. The extent and rate of the decrystallization depends on the nature of the solvent and on the mode of its interaction with the polymer segments. It is influenced by temperature and its activation energy is different for various solvents. The dissolution of the small crystallites removes further stress from the chains in the amorphous regions and facilitates further the disorientation. The decrystallization is however masked partly or fully by the opposing phenomenon: the crystallization which occurs upon the decomposition of the solvent–polymer complex. The rate of crystallization is different for various solvents, depending on the energy needed to decompose the solvation complex. The activation energy of the crystallization could be different from that of the decrystallization, and this difference may again vary from solvent to solvent.

The kinetic constants shown in Table II and the apparent energies of activation in Figures 6, 7, and 8 should therefore be considered as representing the sum of the above two opposing reactions. The relatively rapid and pronounced decrease in orientation for both solvents already at 60°C (see Fig. 5) indicates an efficient penetration of the solvents into the polymer [10]. In the case of DBE, however, the decrease in orientation is more pronounced, showing that DBE is a much more effective agent in breaking the cohesion between the chains. It is reasonable to assume that its decrystallizing effect is also much more pronounced than that of PER. In the low temperature range (60–95°C), the observed rate of crystallization is lower for DBE, indicating the larger contribution of the decrystallization to the overall effect. At higher temperatures, the higher activation energy of the crystallization observed for DBE brings about a reversal in the ratio of crystallization rates for both solvents, and the crystallinity increases much more rapidly for DBE than for PER. This reversal is responsible for the minimum in the plots of crystallinity versus DBE content in the solvent mixture (Fig. 4). The apparent activation energies observed for the solvent mixtures (Fig.

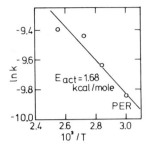

FIG. 6. Arrhenius plot of ln k vs. $1/T$ for PET crystallization in PER.

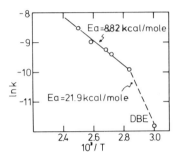

FIG. 7. Arrhenius plot of ln k vs. $1/T$ for PET crystallization in DBE.

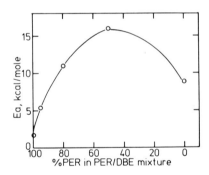

FIG. 8. Plot of activation energy vs. percentage PER in PER–DBE mixtures.

8) are accordingly due to several simultaneously occurring phenomena and depend primarily on the sum of the interactions of the two solvent fractions with the polymer segments and consequently on the decrystallization and on the crystallization reactions.

REFERENCES

[1] H. G. Zachmann, *Makromol. Chem., 74,* 29 (1964).
[2] P. R. Blakey and R. P. Sheldon, *J. Polym. Sci. A, 2,* 1043 (1964).
[3] E. Wunsch and E. Schuller, *Faserforsch. Textiltech., 15,* 391 (1969).
[4] R. P. Sheldon, *Polymer, 3,* 27 (1962).

[5] R. P. Sheldon, *Polymer, 4*, 213 (1963).

[6] W. R. Moore and R. P. Sheldon, *Polymer, 2*, 315 (1961).

[7] E. L. Lawton and D. M. Cates, *J. Appl. Polym. Sci., 13*, 899 (1969).

[8] E. Wiesener, *Faserforsch. Textiltech., 15*, 392 (1964).

[9] A. S. Ribnick, H.-D. Weigmann, *Text. Res. J., 42*, 720 (1972); *ibid., 43*, 176 (1973).

[10] A. S. Ribnick, H.-D. Weigmann, and L. Rebenfeld, *Text. Res. J., 43*, 316 (1973).

[11] M. I. Kashmiri and R. P. Sheldon, *J. Polym. Sci. B, 6*, 48 (1968).

[12] J. H. Lemons, S. K. Kakar, and D. M. Cates, *Amer. Dyest. Rep., 50*, 320 (1961).

[13] A. Wlochowicz and A. Jeziorny, *J. Polym. Sci. A, 2(10)*, 1407 (1972).

[14] W. O. Statton and G. M. Godard, *J. Appl. Phys., 28*, 1111 (1957).

[15] I. Katz and M. Lewin, to be published.

CHEMICAL MODIFICATION OF FIBERS AND FABRICS WITH HIGH-ENERGY RADIATION

V. STANNETT, W. K. WALSH, and E. BITTENCOURT
North Carolina State University, Raleigh, North Carolina 27607

R. LIEPINS and J. R. SURLES
Research Triangle Institute, Research Triangle Park, North Carolina 27709

SYNOPSIS

Some fundamental considerations related to the radiation modification of fibers and fabrics are discussed. Experiments are described on the radiation "grafting" of various phosphorus- and bromine-containing vinyl monomers to polyester, cotton, and their blends to impart flame resistance. It was found that the flame retardancy was more efficient when the grafted polymer was located inside the fiber. The efficiency of the bromine containing polymers was found to be related to the bromine/aliphatic hydrogen ratio and to the thermal stability of the polymers.

Experiments are also described illustrating the successful use of radiation processing with a number of vinyl monomers and oligomers to impart water sorbency, for the bonding of nonwoven fabrics for fabric coating, and for the binding of pigment prints.

INTRODUCTION

The radiation processing of fibers and fiber assemblies, whether they be textile- or paper-based, has had an unusual history. In the mid-fifties, an era of pioneering work in the field of the radiation chemistry of polymers, the DuPont Company embarked on an extensive program with the treatment of textiles using the General Electric Company's newly developed resonance transformer electron accelerator. This work was very imaginative and far-reaching in nature and was subsequently described in an extensive United States patent [1]. It embodies many of the ideas which are still being worked on today. Between then and the mid-sixties, there was considerable activity in the field both in university and industrial laboratories but mainly using cobalt-60 isotope sources. In 1966, the Deering Milliken Company announced the successful development of an electron beam process for imparting both permanent press and soil release properties to polyester and cotton–polyester blends [2]. This was followed by a resurgence in industrial interest in radiation processing in most of the major textile finishing

Journal of Applied Polymer Science: Applied Polymer Symposium 31, 201–227 (1977)

companies and to a lesser extent in some pulp and paper companies. This interest subsided after about 3–5 years, only to be revitalized again in the last few years. The resurgence has been mainly prompted by the growing availability of very high-speed energy delivery industrial accelerators, including models which are self-shielding in nature. The research aims have also shifted somewhat in recent years. Earlier work concentrated on changing the properties of fibers themselves and on imparting crease resistance. More recently, the emphasis has been on fire retardancy and a number of other textile finishing operations such as coatings, including thin back coatings for yarn stabilization, flock adhesives, pressure-sensitive adhesive backings, e.g., for wall coverings, binders for non-woven fabrics, and pigment printing binders.

Parallel with these studies, the more established areas of research such as permanent press, crease resistance in general, antistatic treatments, soil release, and other property changes continue to be investigated. There appears to be a subtle change taking place also in the acceptance by industry of radiation processing equipment. The increased ease of control, lessened floor space requirements, and the improved energy requirements are increasingly attractive to textile finishing plants. The capital investment required, long a much touted negative feature of radiation, has now been shown to be no longer true. In some progressive corporations, the pressure to look at radiation is now coming from the production people rather than from research, as was the case previously. The radiation processing of textiles has been reviewed by Gilbert and Stannett [3], Stannett and Hoffman [4], and by Hoffman [5]. In this paper, some new work will be reported.

SOME FUNDAMENTAL CONSIDERATIONS

The first results of treating polymers, oligomers, and monomers with high-energy radiation is the production of free radicals. Almost all the ionic species which are produced are destroyed by the immediate recombination of the ejected electrons with the parent positive molecule ions. Those ionic species, which do manage to separate for a long enough period of time to initiate chemical reactions, are usually destroyed by the water or other reactive species which are normally present in practical systems.

The free radicals which are produced may induce degradation and/or crosslinking in the case of polymers. With monomers, polymerization takes place. Monomers and reactive oligomers which contain two or more double bonds are rapidly crosslinked, preceded by extensive branching. When the radicals produced by the irradiation of polymers have suitable monomers available to them, graft copolymerization takes place. If the monomers are multifunctional, a combination of grafting and crosslinking occurs. Let us now examine how these various processes have been used with respect to fiber and fabric modifications.

The degradation, i.e., the reduction in molecular weight, which takes place on irradiating many polymers, including cellulose, has not been used in modifying fibers *per se*. However, it has been proposed for the rapid and controlled molecular weight reduction of polymers such as the "ripening" of viscose solutions. In addition, a knowledge of such degradation processes is valuable, since all radiation processing does include any deleterious effects of the radiation as well as the benefits imparted to the fibers. Any crosslinking which takes place is not normally harmful and could, in principle, be used for imparting high-temperature resistance or crease resistance and so on. In fact, no such direct use of radiation for crosslinking has been used, although the crosslinking of polyolefin films and cable covering is the most highly advanced area of radiation processing of high polymers.

The use of radiation to polymerize monomers, particularly when multifunctional in nature or mixed with multifunctional oligomers or the oligomers alone, is of growing importance in textile technology. Briefly, such reactions lead to the direct formation of polymer films and adhesives, rapidly and without solvents. Examples of the use of these reactions will be described in this paper. These reactions are governed by the rather straightforward kinetics of polymerization and gelation described in detail, for example, by Flory [6]. Since the practical use of radiation in radiation processing necessarily involves the use of high dose rate electron accelerators, the "ideal" kinetics are modified by the participation of the primary radicals in the chain termination process. The consequences of this participation on the rates and molecular weights have been discussed in detail by Chapiro [7] for the case of the simple vinyl monomers. The overall effect of using the high dose rates necessary for practical treatments are rather low yields of low molecular weight polymers per unit of dose. The results of such a study with both styrene and methyl methacrylate have been presented [8, 9]. However, with multifunctional vinyl monomers having more than two vinyl groups per molecule, the situation is quite different: the gel effect becomes operative at very low conversions and the rate of polymerization and the molecular weights increase rapidly [10]. Residual monomer can still be a problem in these cases, however, and for this reason high molecular weight oligomers containing several double bonds per molecule are coming into increasing use. When gelation takes place, there is also extensive postpolymerization after the material has left the electron beam.

The grafting of vinyl monomers to fibers has been extensively studied since the early nineteen-fifties. As with polymerization, the yields are rather low, in principle, with electron beam radiation. Again, however, since a great deal of the polymerization is in the swollen fiber, the gel effect is very important and good yields of grafting can be obtained. It is clear that some of this takes place by posteffects. To enhance this effect and to minimize homopolymer and residual monomer, it is highly advantageous to add a small proportion of a multifunctional vinyl monomer. By the judicious control of the swelling, time of contact between the fibers and the monomer, and the reaction time, grafting can be either close to the surface, i.e., "ring grafting," or throughout the fiber. By conducting the grafting in the presence of an inhibitor, air in some cases, the surface can be essentially nongrafted and so-called "core grafting" can be induced. The advantages and disadvantages of these different locations of the grafted side chains will be clear to textile technologists. In general, grafting to fibers superimposes the properties of the grafted polymer onto the fiber without changing the main properties of the fiber such as its mechanical strength. However, the grafting of massive amounts can produce new types of fiber where the properties of the side chains predominate. Good examples are the formation of highly elastic fibers [11–13] or super-absorbent fibers [14]. Such properties can be imparted at more modest degrees of grafting if the structural order of the fiber, such as crystallinity, is destroyed and reformed at a lower level after grafting [12, 13, 15, 16]. Examples of the principles outlined above which have highly practical applications will now be presented.

SOME APPLICATIONS OF RADIATION TO FIBERS AND FABRICS

A number of applications of radiation processing to fibers and fabrics are being actively investigated in the authors' laboratories. The major area at the present time is the grafting of various monomers and monomer combinations to impart flame retardancy. This work will be discussed in some detail. Other areas being studied are moisture absorbency, the bonding of nonwoven fabrics, fabric coatings and the bonding of pigment prints. Progress being made in these fields will be discussed somewhat more briefly.

Fixation of Fire Retardants

Little has been done with grafting of flame retardants to polyester. For this reason, the initial studies were concentrated on working out the methodology for radiation fixation of flame retardants to polyester. Following this, the techniques were applied to 50/50 cotton/polyester blend fabric.

The work with the pure polyester system was divided into two essentially separate phases. In the first, efforts were made to determine the effect of distribution of the flame retardant within the grafted polyester filament upon its flame-retardance efficiency. This was done with two model monomers, one based on phosphorus, and the other on bromine. Diethylvinylphosphonate was selected for the phosphorus studies, while vinylbromide was used for the bromine investigations.

In the second, screening of various phosphorus- and bromine-containing compounds for their flame-retardance efficiency was conducted. Following this, the best candidate systems were scaled up and used on a 50/50 cotton/polyester blend fabric. The initial studies were conducted with γ-radiation, while the scale-up studies used the electron beam and electron curtain radiation sources.

Experimental Procedures

Details of some of the materials and procedures are described in the text itself for convenience.

Substrates Grafted. These included American Enka Co. 150/96 S.D. polyester filament yarn, 0.35% TiO_2, ~0.2% water-soluble finish; cotton/polyester blend—poplin made with 50% Celanese Fortrel #310 and 50% carded cotton, count 96/42 threads to the inch, weight 2.77 yd/lb.

Bromine Compounds Grafted. Among these compounds were vinyl bromide (VBr), vinylidene bromide (VBr_2), 2,3-dibromopropyl acrylate (DBPA), 2,3-dibromopropyl methacrylate (DBPM), 2,4-tribromophenyl methacrylate (TBPM), 2,4,6-tribromophenyl acrylate (TBPA), tribromoneopentyl acrylate (TNPA), 2,2,2-tribromoethyl acrylate (TBEA), 2(2,4,6-tribromophenoxy)ethyl acrylate (TBPOEA), and the bisacrylate of 2-hydroxyethyl ether of tetrabromobisphenol A, BABA-50.

Phosphorus Compounds Grafted. The compounds used were dimethylvinylphosphonate (DMeVP), diethylvinylphosphonate (DEVP), dimethyl phosphonomethyl acrylate (DPA), dimethyl allylphosphonate (DMAP), bis(2-chloroethyl) vinyl phosphonate (Fyrol BB), condensate of Fyrol BB (Fyrol 76), dimethyl 1-acetoxyvinyl phosphonate (DAVP), dimethyl 1-methoxyvinyl phosphonate (DMVP), N-(dimethylphosphonomethyl) acrylamide (NDPA), and bis(2,3-dibromopropyl) phosphoryl-2-oxyethyl methacrylate (BDPOM).

VBr and VBr_2 were courtesy of Ethyl Corporation; DBPA, DBPM, BABA-50 and TBPOEA were courtesy of Great Lakes Chemical Corp.; TBPM, TBPA, DMeVP, DPA, DMAP, DAVP, DMVP, NDPA were courtesy of Hooker Chemical Corp.; and BDPOM courtesy of White Chemical Corp.

Grafting Techniques

γ-Radiation. A series of initial investigations showed no discernible differences in the percent weight gain obtained with scoured and unscoured yarns; thus, all subsequent work was carried out on unscoured samples. The samples were degassed by means of three freeze–thaw cycles to at least 10^{-5} torr. In most cases, the grafting was performed using a mutual irradiation technique in a small glass ampule at dose rates of from 0.01 to 0.1 Mrad/hr. Following irradiation, the fiber samples were extracted with solvent for the homo- or copolymers, first at room temperature and then at elevated temperature. Fibers were vacuum-dried at constant weight and stored in a desiccator for subsequent evaluation. The data for weight gain achieved during grafting was obtained following this drying procedure.

Electron Beam. A High Voltage Engineering Corporation electron accelerator with a maximum beam current of 20 mA and operated at 500,000 V was used in this work. This equipment utilizes a horizontal beam scanned to 48 in. by approximately 6 in. The samples were hung vertically on a conveyor, which carried them in front of the beam twice in each pass through the equipment, so that the samples received half of their total dose from each side. All irradiations were carried out in nitrogen-filled Ziploc polyethylene bags. The beam current and conveyor speed were varied from 0 to 20 mA and 12 to 96 ft/min, respectively. The dose obtained with these variations was a linear function of the ratio of beam current to conveyor speed as previously established. The operating current–speed curve was used to calculate delivered dose for all experiments.

The sample (6 × 6 in.) preparation consisted of drying the fabric in a vacuum oven at 70°C for ∼16 hr and then weighing. Following this, the samples were preswollen in methanol (1–2 hr at 40–50°C) and then in a 50% ethylene dichloride solution (1–2 hr at 50–65°C) of the various monomers. The monomer solution contained also 2–4% of morpholine (acid trap) and 1–5% of bisacrylate of 2-hydroxyethyl ether of tetrabromobis-phenol A (crosslinking agent). Following the swelling procedure, the samples were removed from the solution, squeezed free of excess liquid, and then placed in polyethylene bags under nitrogen, weighed, and irradiated. The work-up of the irradiated samples was similar to that of the fibers described before.

The Electrocurtain Processor*

In the Electrocurtain Processor (Energy Sciences, Inc., Bedford, Mass. 01703), the electrons are generated uniformly along a thin wire filament when heated above its emission temperature. Within the vacuum chamber, the electrons are then accelerated across a potential difference of 150,000–200,000 V, the actual value dictated by the density and thickness of the material being processed. Heavier or thicker fabrics require higher voltages for adequate

TABLE I

Electron Curtain Samples

Formulations

NDPA - 33% H_2O solution

NDPA/DBPA - 50% H_2O emulsion, 4% Triton X-100
  40   60

NDPA/DBPA/TBPM - 43% H_2O/DCE emulsion, 3% Triton X-100
  40   30   30   3.3 1

NDPA/DBPA/Fyrol BB - 50% H_2O emulsion, 4% Triton X-100
  40   30   30

Sample Size - 18 in. x 72 in.

Wet add on - 120 - 160%

Predrying - 94°C (200°F), 7 mins., forced air

    NDPA/DBPA

    NDPA/DBPA/TBPM

    NDPA/DBPA/Fyrol BB

Total Dose - 1 Mrad and 5 Mrads

Atmosphere - essentially ambient air

* The Electrocurtain Processor for our work was used through the courtesy of United Merchants and Manufacturers, Langley, S.C.

electron penetration. After acceleration, the electrons pass through a thin foil window and deposit their energy in the fabric below. The window is used to separate the vacuum chamber from normal atmospheric pressure. This allows fabric to be irradiated directly without complex vacuum interlocks. The amount of radiation delivered to a product can be changed by either varying the speed of the material passing underneath the processor or altering the amount of current being transmitted. This is a processor that has been especially engineered for the industrial environment. For example, for textile mill operations, the speed of the processor would be that of the padding and drying line and the current of the processor could be easily adjusted to the required value. Electrocurtain's capable of continuous processing of 48 in. wide product in either web or sheet form have been designed.

The sample preparation and irradiation conditions are detailed in Table I.

FR Localization Techniques

Surface Coating. Here we used approximately 10% solution of the compound in THF; after immersing the fibers in the solution, they were dried in a vacuum oven at 50°C for 16 hr.

Uniform Grafting. The fibers were preswollen in ethylene dichloride at 70°F for 1/2 hr, then placed in the flame-retardant solution for about 2 hr and following that irradiated.

Surface Grafting. Fibers were placed in the flame-retardant solution and immediately irradiated.

Core VBr Grafting. Fibers were preswollen in ethylene dichloride at 70°F for 1/2 hr, then placed in vinyl bromide for about 2 hr. The excess VBr was decanted and the swollen fiber frozen to −78°C, after which the tube was open to the atmosphere and then was placed in the γ-radiation source.

Oxygen Index

Since the samples in the initial phase of this work were in the form of individual yarns or fibers, it was necessary to develop a modified technique for measuring the oxygen index (OI) of these materials. The development of the sample holder and the procedure for measuring the OI has been described [17]. The procedure permitted not only the determination of the OI values, but also an estimation of the char yield as the material burned in the tester. This procedure was also applied to the evaluation of the various grafted fabric samples.

Scanning Electron Microscopy and X-Ray Microprobe Analysis

The distribution of grafts across the PET fiber cross section was determined in a semiquantitative fashion on selected systems using an x-ray microprobe attachment to the scanning electron microscope.

Thin cross sections of the grafted PET fibers were prepared by embedding in an epoxy or poly(methyl methacrylate) resin, and sectioning with a Reichert

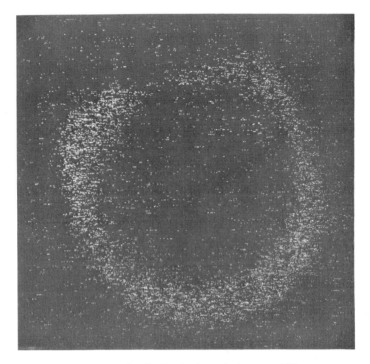

FIG. 1. Bromine distribution in a solution-coated fiber.

"Omu2" ultramicrotome using a glass knife. The epoxy yielded better sections; the poly(methyl methacrylate) was used because of hydrolysis of the vinyl phosphonate grafts by the catalyst used to cure the resin and because of the presence of chlorine in the resin which interferred with some of the analyses. The sections were floated off onto water during the sectioning and then transferred to a drop of water on a carbon stub using an eyelash. The water droplet on the stub was then evaporated and the section examined in the scanning electron microscope. No coating was required.

In some cases, SEM photographs were obtained simply by packing fibers into a hole in a 1-cm deep cylindrical stub and slicing them off in liquid nitrogen at the top of the stub with a razor blade. Such fibers were then coated in an evaporator to prevent charging.

Figures 1–7 depict the various flame-retardant localizations. Figure 1 is a map of the bromine distribution of a solution coated fiber. It shows clearly a thick poly(vinyl bromide) coating which has been pulled away from one section of the filament in the microtoming operation. Figure 2 shows a uniform bromine distribution throughout the filament cross section. Figure 3 shows a surface bromine distribution. Figure 4 is a schematic representation of the bromine distribution in the core grafted sample. Because of the low add-on (1.2%), extremely long exposure times would be needed to collect sufficiently high x-ray dose to make the picture. For this reason only, an x-ray count was made and the results were expressed as the percent bromine distribution in the indicated locations. Figure 5 depicts the phosphorus distribution in a solution-coated fiber.

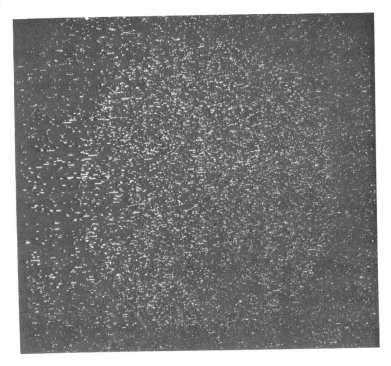

FIG. 2. Uniform bromine distribution.

Figure 6 shows a uniform distribution of grafted phosphorus, and Figure 7 shows a surface distribution of phosphorus. The phosphorus microphotographs are much less clear than the heavier bromine atom counterparts.

OI versus FR Location

Data on the OI versus location of the two model flame retardants are given in Figure 8. The correlations seem quite clear for the two model systems. The placement of the poly(vinyl bromide) in the interior of the fiber enhanced its efficiency of raising the oxygen index the most, followed by uniform grafting and then surface grafting. In the case of DEVP, again the uniform grafting led to a more efficient flame-retardance action, followed by the surface grafting. Thus, both model compounds indicated that better flame-retardance results might be obtained if penetration of the polyester filament by the flame retardant is achieved. The less thermally stable the flame retardant (e.g., PVBr and PDEVP), the more efficient it becomes, and the more it is incorporated in the core of the filament. Thus, it would seem that factors which will delay the volatilization of the bromine and phosphorus from the flame front will increase their efficiency.

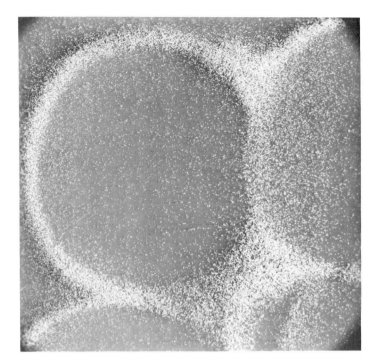

FIG. 3. Surface bromine distribution.

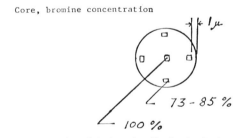

FIG. 4. Schematic representation of the bromine distribution in the core-grafted sample.

Thermal Stability versus Flame Retardance

Analysis of the TGA data of the bromine graft series shows that the apparent thermal stability of a graft and its flame-retardance efficiency may both be related to the structural parameter of the ratio of alpha aliphatic hydrogen to bromine atoms (see Table II). The lower this ratio is, the higher the efficiency. Similarly, the most efficient flame retardants appear to be those with no low decomposition point. The functionality of the phosphorus atom (as a phosphonate) was not changed, the only change being in the hydrocarbon substituents. As a result, the efficiency of the various phosphorus grafts seemed to be related simply to the percentage of phosphorus in the compound; the higher the phosphorus content, the more efficient the compound was. Thus, the following order was indicated: Fyrol 76 $\cong$ DMeVP > DEVP > DPA.

FIG. 5. Phosphorus distribution in a solution-coated fiber.

Electron Beam Grafting

In an effort to determine the feasibility of applying systems of this type in commercial production equipment, various brominated vinyl and vinyl phosphorus compounds were irradiated in the electron beam apparatus. The electron beam grafting results are given in Tables III, IV, and V. The following monomers grafted with an exceptional ease: NDPA, BDPOM, DBPA, and BABA-50. Incorporation of NDPA in a formulation as a comonomer facilitated the grafting of monomers otherwise difficult to graft. The three sets of data showed that grafting of these formulations in an electron beam could be conducted either from an aqueous emulsion, or dichloroethylene solution, or as neat monomer mixtures. The OI's for the NDPA combinations were all quite high, indicating promise in terms of fire retardance.

Electrocurtain Grafting

Electrocurtain is a new generation of radiation processing units of commercial production type. As a further step in our efforts to determine the feasibility of applying our systems in such equipment, we investigated the curing of four flame-retardant compositions on 50/50 PET/cotton blend. The grafting results are given in Table VI. Two major conclusions can be made from these data: (1) grafting of a combination of an acrylamide, acrylate, and a methacrylate is feasible in the Electrocurtain, and 2) the percent add-on is total dose-dependent.

FIG. 6. Uniform phosphorus distribution.

However, much more work is needed to determine the various process variables as well as the durability and effectiveness of the various formulations cured by this technique.

In addition to these somewhat more fundamental studies, a more applied investigation was undertaken on the "grafting" of a multifunctional vinyl phosphonate oligomer to cotton, polyester, and their blends. In using radiation to polymerize a flame retardant in a cellulosic fabric or on the cellulosic component of a blend, water solubility is a chemical requirement if penetration of the cellulosic fibers is to be effected. A multifunctional vinyl phosphonate oligomer (Stauffer Chemical's Fyrol 76) has been found to meet all these requirements. It is presently used commercially as a cellulosic flame retardant in conjunction with N-methylol acrylamide (NAM) (e.g., 20 parts Fyrol 76 to 13.2 parts NAM), using a persulfate catalyst and a thermal cure. This combination can also be cured at room temperature with electron beam, as can the Fyrol 76 alone. Table VII summarizes the results of a number of determinations of the minimum amount of each flame retardant necessary for a 7 in. char length in DOC FF-3-71, for sleepwear. The add-ons were measured before washing.

In the 100% cotton (item 1), the add-on required for both conditions is not very great and the fabric is not excessively stiffened. The heavy weight blend in item 2, however, requires much more, and the hand is only marginally acceptable at 30% add-on of flame retardant. The light weight blend (item 3) is even more difficult, since 20% or more add-on produces large increases in stiffness. In the same type fabric, however, substitution of the normal polyester

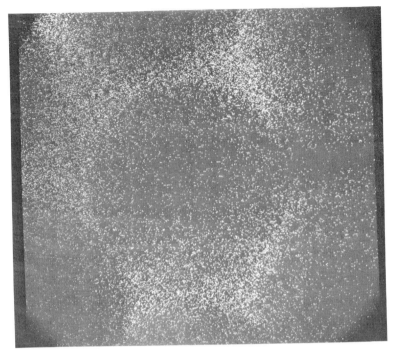

FIG. 7. Surface phosphorus distribution.

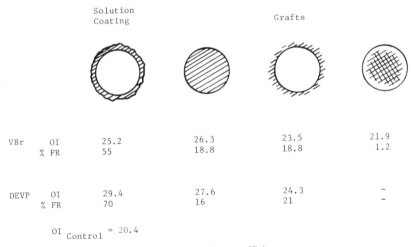

		Solution Coating		Grafts	
VBr	OI	25.2	26.3	23.5	21.9
	% FR	55	18.8	18.8	1.2
DEVP	OI	29.4	27.6	24.3	–
	% FR	70	16	21	–

OI $_{Control}$ = 20.4

FIG. 8. FR location vs. efficiency.

for a flame-resistant one reduces the requirements to a level at which the test requirements can be met without significant hand changes. This combination, therefore, appears quite promising, and studies of the radiation-induced polymerization of Fyrol 76 and similar high-phosphorous retardants are becoming increasingly important.

The polymerization rate under electron beams (1–5 Mrad/sec) is primarily

TABLE II
Alpha Aliphatic Hydrogen/Bromine Ratio versus Thermal Stability versus FR Efficiency

Compound	H/Br	Dec. pt.	ΔOI/1%Br
$-\overset{\overset{O}{\|\|}}{C}-O-\underset{Br}{\overset{Br}{\bigcirc}}-Br$	4	303°C	1.5
$-\overset{\overset{O}{\|\|}}{C}-O-CH_2CH_2O-\underset{Br}{\overset{Br}{\bigcirc}}-Br$	0	—	1.2
$-\overset{\overset{O}{\|\|}}{C}OCH_2-\underset{\underset{Br}{\overset{\|}{CH_2}}}{\overset{\overset{CH_2}{\|}}{\underset{\|}{C}}}-CH_2-Br$	0	300°C	—
$-\overset{\overset{O}{\|\|}}{C}OCH_2-\underset{\underset{Br}{\|}}{\overset{\overset{Br}{\|}}{C}}-Br$	2/3	220°C	1.1
$-CH_2-\underset{\underset{Br}{\|}}{\overset{\overset{Br}{\|}}{C}}-$	1	204°C	1.0
$-CH_2-\underset{\underset{Br}{\|}}{CH}-$	2	150°C	0.5
$-CH_2-\underset{\underset{Br}{\|}}{CH}-\underset{\underset{Br}{\|}}{CH_2}$	2.5	132°C	(0.2)
$-CH_2-\underset{\underset{Br}{\|}}{CH}-\underset{\underset{Br}{\|}}{CH}-CH_2-$	3	127°C	—
$-CH_2-\underset{\underset{Br}{\|}}{\overset{\overset{CH_3}{\|}}{C}}-\underset{\underset{Br}{\|}}{CH}-CH_2$	4	122°C	—

dependent on whether the flame retardant is inside or outside the cellulosic fibers and on the ambient atmosphere. Figure 9 shows the case where the Fyrol 76 is located largely inside the fibers, where atmospheric changes in water and oxygen content do not affect the reaction kinetics. The other extreme can be observed by padding the flame retardant on 100% polyester fabric. In this case, all the Fyrol 76 (which is still liquid even when completely freed of water) is on the fiber's surfaces and its polymerization rate is greatly affected by oxygen and moisture, as Figure 10 demonstrates. In comparing Figures 9 and 10, it is of interest to note that, in nitrogen, the polymerization rate is higher outside the fiber than inside, where restriction of monomer diffusion is likely (the molecular weight of Fyrol 76 is about 1000).

TABLE III
Electron Beam Grafting Results[a]

MONOMER	TOTAL DOSE, MRADS	% ADD ON	OI
–	–	–	18.5
Fyrol BB	2	6	24.0
	4	6	24.3
	8	7	25.2
	16	7	24.3
DAVP/ZnCl$_2$	2	5	20.4
	4	5	20.8
	8	6	21.4
VBr$_2$	2	4	22.7
	4	5	22.7
	8	6	23.3
	16	5	22.7
BABA-50	0.25	19	21.9
	0.50	24	24.0
	1.0	21	24.3
DBPM	2	6	22.2
	4	7	23.0
	8	10	23.5
DBPA	2	49	27.1
	4	51	27.9
	8	66	29.0
BDPOM	0.5	8	24.3
	1	13	26.0
	2	33	30.5
	4	61	33.1
NDPA	0.5	59	32.8
	1	71	33.7
	2	78	34.4

[a] Conditions: 50% monomer, 2–4% morpholine, 0–3% BABA-50, 50% DCE, 50/50 PET/cotton, N_2; preswell in methanol at 50°C, then monomer solution at 60°C.

Scanning electron micrographs have shown that the cotton fibers can hold about 20% of Fyrol 76 and that additions of more than this amount appear on the surface after drying. Polymerization rates of various amounts of Fyrol 76 on fabrics of various blend levels, therefore, fall between the rates shown in Figures 9 and 10.

The fractional conversions presented here are those measured after washing and thus are really measurements of insoluble gel formed during the polymerization. They are affected, therefore, by variations in the double bonds per molecule in the flame retardant as well as by addition of small amount of crosslinking additives. Application of 40% Fyrol 76 and 1–2% of other additives to 50/50 cotton/polyester print cloth gives conversion curves shown in Figure 11. All of the additives increase the rate, including the monofunctional sodium acrylate, which probably acts by increasing the copolymerization rate as well as forming some gel by chain branching.

Water Absorbency

The water absorbency of pure cotton terrycloth toweling was studied as a function of the percent grafting of acrylic acid. The latter was simply padded onto the cloth in aqueous solutions of various concentrations. The irradiations

TABLE IV
Electron Beam Grafting Results[a]

COMPOSITION	TOTAL DOSE, MRADS	% ADD ON	OI
NDPA/TBPM[*]	0.5	23	28.6
	1.0	43	30.5
	2.0	48	31.4
NDPA/BABA-50[*]	0.5	40	26.0
	1.0	44	29.4
	2.0	47	30.9
NDPA/Fyrol BB	0.5	49	30.5
	1.0	63	32.1
	2.0	(51)	32.5
NDPA/BDPOM	0.5	37	32.1
	1.0	56	34.1
	2.0	66	34.6
NDPA/DBPA	0.5	68	32.8
	1.0	79	35.0
	2.0	93	36.6

[a] Conditions: 50/50 PET/cotton, neat monomer mixtures 1:1 by wt, N_2.
[*] Also contains 30% dichloroethane.

TABLE V
Electron Beam Grafting Results[a]

COMPOSITION	TOTAL DOSE, MRADS	% ADD ON	OI
NDPA/DBPA	0.25	17.20	27.6
	0.50	28.63	29.4
	1.00	21.21	30.1
NDPA/BDPOM	0.25	11.86	26.3
	0.50	18.03	26.7
	1.00	16.66	26.7
NDPA/BB	0.25	19.40	28.3
	0.50	22.33	28.6
	1.00	15.51	28.3
NDPA/TBPM	0.25	18.40	26.7
	0.50	29.19	28.3
	1.00	32.38	29.4

[a] Conditions: 50/50 PET/cotton, 50% aqueous emulsions, 6% Triton X-100, 1:1 by wt monomer mixtures, predried at 160°C/min.

were by an electron accelerator (High Voltage Engineering Co., ICT 500) using the procedures outlined in the previous section. Figure 12 shows the percent grafting (inextractable polymer) as a function of the radiation dose at different monomer concentrations together with the moisture regain. About 25% has been

TABLE VI
Electron Curtain Grafting Results

Sample -total dose	% Add on	OI After One Home Wash Cycle
Ungrafted - 1	-	18.5
- 5	-	19.2
NDPA - 1	6	24.7
- 5	10	25.5
NDPA/DBPA - 1	11	25.5
- 5	17	26.3
NDPA/DBPA/TBPM - 1	8	24.3
- 5	11	25.5
NDPA/DBPA/BB - 1	12	26.3
- 5	19	28.6

TABLE VII
Minimum Add-On to Achieve 7-in. Char Length

Fabric	Treatment	% Add on	
		1 wash	50 washes
1. 100% cotton print cloth 3.1 oz./yd^2	Fyrol MAM/Fyrol	10-14% 15-19%	--------- ---------
2. 60/40 cotton/ polyester double knit 9.4 oz./yd ^{2}	Fyrol	21-26%	29%
3. 50/50 cotton/ polyester printcloth 4 oz./ yd^2	Fyrol MAM/Fyrol	23-27% 40%	49% ---------
4. 50/50 cotton/ T-900 F Dacron[a] printcloth 4.3 oz/ yd^2	Fyrol MAM/Fyrol	11.6% 25-27%	15% ---------

[a] DuPont's bromine-containing polyester.

judged in other work as the degree of grafting necessary to give cotton a superior and adequate moisture absorption. This, it can be seen, can be achieved at about 3 Mrad. A dose of 3 Mrad can be achieved at about 300 ft/min under optimum engineering conditions. A full report of these investigations will be presented elsewhere.

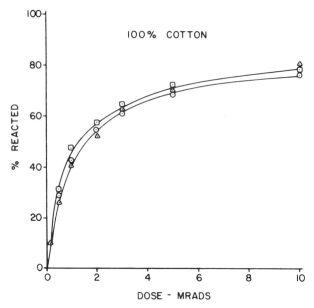

FIG. 9. Conversion of Fyrol 76 padded to 10% on 100% cotton print cloth: (⊙) nitrogen; (△) air;
(□) 60% R.H. in air.

Bonding of Nonwoven Fabrics

Adhesive bonded nonwoven fabrics are normally made by spraying a web of random or partially oriented fibers with latex and calendering to effect penetration into the fabric followed by drying and curing. Application of radiation-curable oligomer–mixtures, without water, and curing with radiation offer the advantage of energy savings in both drying and curing of the latex. The main problem with this approach is getting the viscous binders evenly distributed in the fabric before curing, especially when bonding thicker heavier webs such as needle-punched nonwovens. One solution to this problem is to use the needling process to distribute the binder.

The results of such a study, where a radiation-curable binder was sprayed on one face of a lofty web of randomly oriented polyester fibers, are presented in Figures 13–16. Two such webs were then calendered together (at 3.2 psi), with the uncured adhesive in the center, and the composite was then needled. As the barbed needles passed back and forth through the composite fabric, the usual fiber entanglement occurred, along with movement of the adhesive away from the center of the fabric. Both operations increase the fabric strength, as shown in Figure 13. These fabrics were sprayed with a mixture of 70% Actomer X-70)union Carbide's acrylated oil oligomer, with 2.3 double bonds per molecule) and 30% isodecyl acrylate (IDA). After calendering and needling, the fabrics were cured in the electron beam with 7.5 Mrad in nitrogen. The effect of increasing binder levels on strength is strong and obvious from Figure 13. Not as evident is the effect of increased needling shown by the two central curves. Tripling the needling intensity smoothed and flattened the fabric, decreasing its thickness by about one-third, so the more needled fabrics are actually stronger,

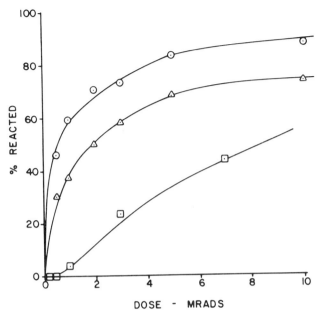

FIG. 10. Conversion of 40% Fyrol 76 on 100% polyester sheeting: (⊙) dry nitrogen; (▲) dry air; (▣) 60% R.H. in air.

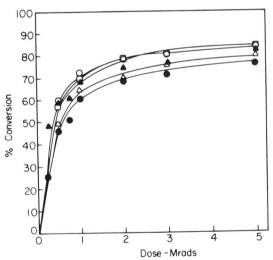

FIG. 11. Conversion of 40% Fyrol 76 on 50/50 polyester/cotton print cloth (●) alone or with (○) 1.5% methylene bisacrylamide (MBA), (△) 1% tetraethylene glycol (TEDA), or (□) 2% acrylic acid (neutralized), irradiated in nitrogen.

when the strengths are corrected for fabric weight. Much more striking, however, is the smoother feel of the fabric due to the more uniform binder distribution. Small grains of cured adhesive droplets could be felt in the fabrics needled at 122 psi but were undetectable in the more heavily needled fabrics which were, however, about three times as stiff, as measured by the bending modulus.

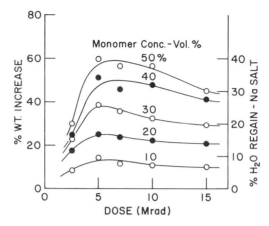

FIG. 12. Electron beam grafting of aqueous acrylic acid onto cotton terrycloth and the resulting moisture regain.

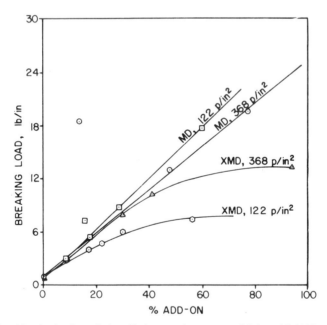

FIG. 13. Breaking loads of needled, radiation cured, nonwoven fabrics with X-70 and isodecyl acrylate in the machine (MD) and cross-machine (XMD) directions. Numbers indicate needling intensity in punches per square inch. Binder expressed as percent weight increase of fabric.

Figure 14 illustrates the result of increased crosslinking in the cured binder by substituting tetraethylene glycol diacrylate for isodecyl acylate. The curves marked "LP" were calendered at 3.2 psi before needling, as were those in Figure 1; therefore, they can be compared with the central (368 psi) curves in that figure. The increased crosslinking had only a slight effect on strength in the cross-machine (XMD) direction, in which the linear density of needle punches is lower, but roughly doubled the strength for fabric breaks in the more strongly needled

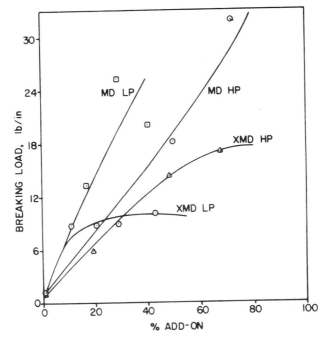

FIG. 14. Breaking load of needled (368 psi), radiation-cured fabrics with X-70 and TEDA. LP and HP indicate calendering at 3.2 and 9.0 psi, respectively, before needling.

machine direction. Another tripling of the bending modulus was also effected by the increased crosslinking, as would be expected. The central curves are for the same system calendered at higher pressure (9.0 psi) before needling and curing. This produced fabrics that were smoother and more isotropic, but did not affect the overall strength level as did the increased needling. It is probable that calendering produces a more uniform binder distribution than needling, because the latter should concentrate the binder in the fiber entanglements around each punch, increasing the effectiveness of the punch in giving strength to the fabric.

The extreme effect of binder distribution can be seen in Figure 15 which compares the average of the outer two curves of Figure 13 with data from an earlier study [B], where the binder (pure X-70) was applied from solvent (perchloroethylene) after needling the fabric. After evaporation of the solvent, the fabric was cured as before, by 7.5 Mrad in nitrogen. It can be seen from Figure 13 that the solvent-applied binder had almost no effect up to about 30% (43% add-on), after which the strength increased rapidly. The binder applied without solvent, however, was effective at the lowest add-ons, and showed a nearly linear relationship. Presumably the solvent-applied binder was evenly distributed and much of it was wasted in adhesion to single fibers "floating" between cross-over points, while the needled-in binder was concentrated at the most favorable locations in the fabric.

The question of grafting and its possible affect on adhesion always arises in radiation-cured composite systems. It is difficult to obtain sufficient evidence

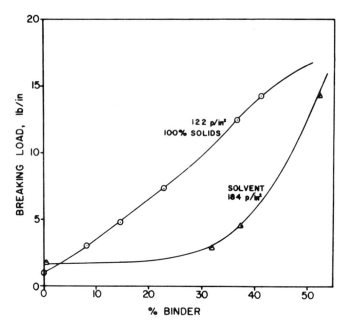

FIG. 15. Breaking loads of fabrics with binder applied neat or from solvent. Top curve is average of breaks in machine and cross-machine directions. Bottom curve includes breaks in 45° direction. Abscissa expressed as percent binder on total fabric weight.

to make a clear statement about this, but scanning electron micrographs of bonded polyester fibers after failure indicate strongly that it does not occur. Interfacial failure was always observed, whether the binders were applied from solvent or not, and with several oligomer–monomer systems. Figure 16 shows an example of an X-70 isodecyl acrylate bond after failure. Clean, interfacial separation of binder and fiber can be seen clearly. Obviously, swelling agents, heat, etc., might change this situation, but these would complicate the system and counter-balance the energy savings obtainable by the more simple application.

Fabric Coating

Fabrics are coated with either thin, yarn-conforming films for yarn stabilization against seam-slippage, or smoother, coherent coatings to give leather-like products. Latexes are used predominantly for the former type of fabric, which normally goes into upholstery with the coating on the back of the fabric. The latter types of coating are often vinyl plastisols or solvent-applied urethane, which go into apparel, upholstery, and industrial applications and are usually applied to the fabric face.

The former type of coating has been applied using a 100% reactive, radiation-curable composition and cured with ultraviolet to give fabrics comparable to those coated with latex. Of more interest, however, are the leather-like coatings. These often use chemicals comparable in cost to the radiation-curable

FIG. 16. Scanning electron micrograph of broken bond between polyester fiber (17 μ) and cured binder of X-70 and isodecyl acrylate.

compositions, and which suffer considerably in comparison with latex costs. In the study described here, a solution of an acrylo-urethane oligomer (two acrylic end groups) with a monofunctional reactive diluent from Hughson Chemical Company was transfer-coated on nylon jersey (lock-knit) fabric in three ways. Direct transfer coating (1) involved placing a thin (1–6 mils) coating of resin on Mylar film (with wire-wound rods); calendering the fabric to the coated film; curing the composite; and removing the Mylar, which acts as a release paper. The pre-cure method (2) is the same, but a small radiation dose (1/4 Mrad) in air is given the coated Mylar prior to applying the fabric. In the tie-coat method (3), the coated Mylar is fully cured as in method (1). All these methods prevent the collapse of the coating around the yarns and fibers and leave a coating with the surface characteristics of the release paper which, in this case, is smooth.

Figure 17 shows the effect of several of these variables on the adhesion of the coating to the fabric. An uncured, coated Mylar film was calendered to the fabric after receiving various pre-cure doses in air, and the composite was given 5 Mrad in nitrogen. The effect of the first dose is to increase the viscosity of the resin and prevent penetration of the fabric into the coating. At high pre-cure doses, the coating approaches full cure and fabric penetration and adhesion are low. As the pre-cure dose decreases the surface tackiness and penetrability both increase, giving better adhesion. At zero, the system reduces to method one, and

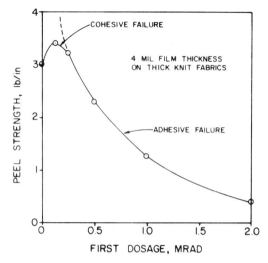

FIG. 17. Effect of pre-cure dose (in air) on peel adhesion of transfer-coated fabrics. All samples were cut in the warp (low extensibility) direction.

the intermingling of the fabric and coating is such that normal adhesive failure cannot occur and the coating actually tears away from the fabric during the peel test.

This process strongly affects the stiffness of the final coated fabric, as would be supposed. Figure 18 shows that the flexural rigidity is fairly constant with pre-cure dose except at low doses (1/4 Mrad), where it increases sharply due to severe interpenetration of the coating and fabric. At pre-cure doses above this,

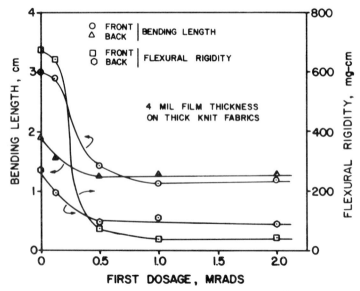

FIG. 18. Effect of pre-cure dose on cantilever bending stiffness of 5 oz/yd^2 nylon jersey (lock-knit) coated as in Fig. 5. "Front" means coated side up during test.

the bulk of the coating is cured enough to prevent penetration, and further ir-radiation only reduces surface tackiness.

In coherent face coating such as described here, the physical properties of the cured film are of more critical importance than in back coating. Checking, cracking, and adhesion failure in the fabric due to repeated flexing are common modes of failure and the following comparison presented in Table VIII shows how the film properties affect the flex life. Coated fabrics were made like those in Figure 17 with pre-cure doses of 1/4 and 1/2 Mrad, using different oli-gomer–monomer combinations to give different film properties.

Radiation-curable compositions generally, however, usually give films with properties less than desirable for fabric coating applications, particularly tear strength, toughness, and elongation to break. This is presumably due to their highly branched and crosslinked nature. Development activities currently under way in numerous laboratories will certainly minimize these deficiencies, since this application has so many obvious advantages for radiation processing.

Bonding of Pigment Prints

Pigment printing of fabrics usually involves applying a paste of water, latex binder, thickener, pigment, water, and other auxiliaries to the area to be printed, followed by drying and curing. The concept of printing with a pigment dispersion in a radiation-curable resin and curing offers the advantage of a simpler, "dry" system with energy savings. Conventional equipment for printing fabric is not suited for such a fully reactive paste because it places too much resin in the printed area, which stiffens the fabric and costs too much. Printing methods and equipment used on paper and smooth substrates apply a reasonable amount of paste, but do not give good coverage on hairy, porous substrates such as fab-ric.

It has been found, however, that such techniques will work on fabrics if higher pressure is used in the transfer step. This compresses the fabric and squeezes the pigment dispersion into all parts of the fabric surface, eliminating the many small, uncoated islands in the colored area. The pigment paste was coated on Mylar as described in the previous section, and this coating was transferred to a rubber sheet which was pressed onto the fabric in a Carver press at 555 psi.

Tables IX–XI illustrate the effect of electron beam radiation dose and binder

TABLE VIII
Effect of Free Film Properties on Flex Fatigue of Coated Fabrics

| | Free Film | | | Coated Fabric |
	Breaking Strength psi	Young's Modulus psi	Elongation at break	Bally Flex-cycles to failure
Film A	700	1000	140%	4-800
Film B	1000	500	400%	>220,000

TABLE IX

Properties of 50/50 Polyester/Cotton with 75% Acrylo-Urethane Oligomer (MW 1087), 22.5%
Vinylpyrrolidone, and 2.5% Dye

Dose Mrads	% Add-on	Flexural Rigidity (g. cm)	Crock-fastness (wet)	Crock-fastness (dry)	Washfastness
0	3.8	.13	--	4	1
1	4.95	.24	2-3	3-4	2
2	4.32	.40	2-3	3-4	2-3
4	5.50	.74	2-3	3-4	3
7.5	7.87	1.36	2-3	3-4	2

TABLE X

Properties of 50/50 Polyester/Cotton Fabrics Printed with 75% Acrylo-Urethane Oligomer
(MW 1953), 22.5% Vinylpyrrolidone, and 2.5% Dye

Dose Mrads	% Add-on	Flexural Rigidity (g. cm)	Crock-fastness (wet)	Crock-fastness (dry)	Washfastness
0	4.0	--	--	1-2	1
1	4.0	.13	2-3	4	1-2
2	5.0	.17	2-3	4	3-4
4	5.1	.23	3	3-4	2
7.5	7.9	.27	3	3-4	3

TABLE XI

Properties of 50/50 Polyester/Cotton Fabrics Printed with 75% Acrylo-Urethane Oligomer
(MW 3000), 22.5% Vinylpyrrolidone, and 2.5% Dye

Dose Mrads	% Add-on	Flexural Rigidity (g. cm)	Crock-fastness (wet)	Crock-fastness (dry)	Washfastness
0	4.7	.10	2-3	1-2	1
1	3.7	.13	2-3	2-3	2
2	3.81	.17	2-3	4	2-3
4	3.61	.20	2-3	4	3
7.5	4.1	.23	2-3	4	3-4

properties on printed fabric properties. Acrylo-urethane oligomers (from Thiokol
corporation) were mixed with a paste of phthalocyanine pigment (Graphtol Blue
6825-0, Sandoz) and vinylpyrrolidone (G.A.F.). The linear oligomers were made
from poly(ethylene adipate) and toluene diisocyanate (TDI) and were end-

capped with hydroxylethyl acrylate. As oligomer molecular weight increased, the separation between the double bonds on each end of the molecules increased, decreasing the crosslinking and giving more flexible films. The glass transition of the cured films also decreased (due to the lower fraction of TDI) also contributing to flexibility. This is reflected in the lower fabric stiffness and better wash fastness as shown in progressing from Table IX to Table XI. Properties change only gradually with dose above 2 Mrad, as the films are fully cured at this point (1% extractable). The wash test was AATCC 61-1972 3a and the crock test was AATCC 8-1969.

It has been shown that radiation curing of fully reactive pigment prints can be done in the laboratory, but it is still not clear whether this process, which is similar to gravure-offset printing, with high-pressure transfer, can be duplicated on production equipment and much more development will be required to test this point.

The Fyrol 76 treatments were done with the assistance of Mr. Jim Ennis, nonwoven fabric bonding with Mr. Ennis and Mr. Houng Chu, pigment print bonding with Mr. Ashok Makati, and fabric coating with Mr. Khen Hemachandra. The support of the National Science Foundation (RANN) for one of the authors (W.K.W.) (Grant GI 43105) is gratefully acknowledged. The sponsors do not necessarily endorse or concur with the findings and conclusions presented here.

The authors are also indebted to Drs. Robert H. Barker, R. S. Gregorian, and N. Morosoff for many stimulating discussions. Part of this work was supported by National Bureau of Standards ETIP Contract No. 4-35963.

REFERENCES

[1] E. E. Magat and D. Tanner, U.S.Pat. 3,188,228 (June 8, 1965).

[2] *Chem. Eng. News, 44,* 23 (1966); *Wall Street Journal,* June 1, 1966.

[3] R. D. Gilbert and V. Stannett, *Isotop. Radiat. Technol., 4,* 403 (1967).

[4] V. Stannett and A. S. Hoffman, *Amer. Dyest. Rep., 57,* 998 (1968).

[5] A. S. Hoffman, *Isotop. Radiat. Technol., 8,* 84 (1970).

[6] P. J. Flory, *Principles of Polymer Chemistry,* Cornell U. Press, New York, 1953.

[7] A. Chapiro, *Radiation Chemistry of Polymeric Systems,* Interscience, New York, 1962.

[8] D. R. Squire, J. A. Cleaveland, T. M. H. Hossain, W. Oraby, E. P. Stahel, and V. Stannett, *J. Appl. Polym. Sci., 16,* 645 (1972).

[9] C. C. Allen, W. Oraby, T. M. A. Hossain, E. P. Stahel, D. R. Squire, and V. Stannett, *J. Appl. Polym. Sci., 18,* 709 (1974).

[10] C. C. Allen, W. Oraby, D. R. Squire, E. P. Stahel, and V. Stannett, *J. Macromol. Sci.-Chem., A8(5),* 965 (1974).

[11] J. L. Williams and V. Stannett, *Text. Res. J., 38,* 1065 (1968).

[12] J. L. Williams, D. K. Woods, V. Stannett, S. B. Sello, and V. Stannett, *Int. J. Appl. Radiat. Isotope, 26,* 159 (1975).

[13] J. L. Williams, V. Stannett, L. G. Roldan, S. B. Sello, and C. V. Stevens, *Int. J. Appl. Radiat. Isotope, 26,* 169 (1975).

[14] P. K. Chatterjee and R. F. Schwenker, U.S. Pat. 3,889,678 (June 17, 1975).

[15] J. L. Williams and V. Stannett, *J. Polym. Sci. B, 10,* 665 (1972).

[16] J. L. Williams and V. Stannett, U.S. Pat. 3,814,676 (June 4, 1974).

[17] R. Liepins, *J. Fire Flammability, 6,* 326 (1975).

[18] W. K. Walsh, et al., High Energy Radiation for Textiles: Assessment of a New Technology. 3rd Semiannual Report to NSF, RANN on Grant GI 43105, November, 1975.

[19] J. W. Vogler, W. K. Walsh, and M. H. Mohamed, *Tappi, 58*(9), 125 (1975).

FUNCTIONAL FINISHES FOR NATURAL AND SYNTHETIC FIBERS

STEPHEN B. SELLO

*J. P. Stevens & Co., Inc., Technical Center, Garfield,
New Jersey 07026*

SYNOPSIS

Diverse bulk and surface properties are imparted to natural and synthetic fibers through chemical finishing. Durable finishes are attained by several routes: chemical modification of the fiber surface; covalent bonding with the reactive groups of the fiber molecules; *in situ* polymerization; insolubilization of preformed polymers; deposition of chemicals of low surface tension; insolubilization of appropriate chemicals in thermoplastic fibers by thermosol treatment. While crosslinking of cellulose imparts dimensional stabilization and increased resilience, the formation of hydrolytically stable crosslinks in wool results in permanent setting, but modification of the surface scale structure is required for shrinkage control. Phosphorus compounds acting in the condensed phase are the most efficient flame retardants on 100% cellulosic substrates, but flame retardants possessing both condensed and vapor-phase activity are the most suitable for polyester/cellulose blends. Water repellency is attained with long-chain hydrocarbons, fluorochemicals, and silicones. Surface treatment of hydrophobic fibers with hydrophilic finishes results in improved soil release and redeposition properties, and treatment with hydrophilic electroconductive polymers leads to the dissipation of accumulated electrostatic charges. The chemistry and mechanism of these bulk and surface modifications are reviewed.

INTRODUCTION

In the last three decades, extensive work has been carried out for the purpose of developing new textile products using new fibers and fiber blends as well as treating textiles with chemical finishes to impart specific functional properties. Many investigations have been reported which were designed to lead to a better understanding of the mechanism by which the various finishes act, and progress has been made to replace the empirical approaches with those developed by systematic scientific studies. It is beyond the scope of this paper to give detailed chemical finishing procedures, but it is an objective to consider the various possibilities of attaining certain functional properties through the selection of appropriate finishing chemicals and procedures.

The functional properties included in this report can be categorized as arising from bulk and surface characteristics. While the crosslinking and flame-retardant treatment of cellulosic substrates should be considered as bulk modifica-

Journal of Applied Polymer Science: Applied Polymer Symposium 31, 229–249 (1977)
© 1977 by John Wiley & Sons, Inc.

tions, water- and oil-repellent finishes and stain release and antistatic treatments of hydrophobic fibers are surface modifications.

Only those chemical finishing procedures are being discussed which are capable of assuring improved functional properties to the textile for a prolonged period of time, which includes multiple launderings and dry cleanings. Such durable finishes can be attained by several routes, such as modification of the fiber surface (e.g., chlorination or bromination of wool to minimize felting and laundering shrinkage); insolubilization of appropriate chemicals via covalent bonding with the reactive groups of the fiber molecules (e.g., crosslinking of cellulose to attain dimensional stabilization and crease resistance or reacting cellulose with organophosphorus compounds to reduce its flammability); insolubilization of monomers or oligomers by *in situ* polymerization (e.g., flame-retarding treatment of cellulosic substrates by the free-radical polymerization of organophosphorus oligomers or by the condensation polymerization of appropriate phosphorus- and nitrogen-containing monomers); insolubilization of preformed polymers (e.g., wool shrinkproofing with preformed polymers or the antistatic treatment of hydrophobic fibers via the crosslinking of preformed water-soluble electroconductive polymers); deposition of specific insoluble polymers (e.g., water- and oil-repellent treatments with flourochemicals).

Since many of these durable textile finishes have undesirable side effects (e.g., crosslinking of cellulose reduces the extensibility of the fiber which contributes to its strength, or the high level of modification required to render cellulose substrates self-extinguishing alters the physical and aesthetic properties of the textiles), it is of great importance to select the appropriate chemicals and processes which impart the permanent improvement in the desired functional properties without substantial impairment of the other aesthetic and wear characteristics.

MODIFICATION OF BULK PROPERTIES

Cellulose Crosslinking

Permanent press cellulosic or cellulose-containing fabrics are produced by setting the fabric in the desired configuration via a crosslinking reaction between hydroxyl groups on adjacent cellulose chains. Such a crosslinking imposes restraints on the movement of the chains and thus imparts to the fabric a tendency to recover from any subsequent deformation and to return to the state in which the crosslinks were introduced. The extent of changes in fiber and fabric properties resulting from the crosslinking reaction depends on the degree of crosslinking and also on the condition of the fiber at the time the crosslinks were introduced. Low levels of crosslinking are capable of imparting dimensional stability, while higher degrees of modification are necessary to attain wash/wear properties. The retention of set creases or pleats requires the highest level of crosslinking [1].

Only those crosslinking reactions can be considered suitable for textile applications which proceed with satisfactory rate under relatively mild conditions.

Crosslinking agents penetrate the fiber but do not enter the crystalline portion of the cellulose; thus, the crosslinking takes place exclusively in the amorphous region. The crosslinking can be carried out in the collapsed state of the fiber by heat-curing and by vapor-phase reactions or in its swollen state by wet-curing techniques. Cellulose fabrics crosslinked in the collapsed state exhibit high dry crease recovery, low moisture regain, and low water imbibition. Those crosslinked in the swollen state possess high wet and low dry crease recovery, high moisture regain, and high water imbibition [2].

Dry Cure

The most widely used method is the dry cure process, which is the most economical, imparts the highest level of dry crease recovery and crease retention, but causes the greatest losses in strength and abrasion properties. Both pre- and postcure processes have gained commercial acceptance, and acid-catalyzed reactions are utilized almost exclusively. The structure and reactivity of the customary crosslinking agents are shown in Table I [3].

TABLE I
Structural Formulas of Customary Reactants for Dry Cure Process

CUSTOMARY REACTANTS	REACTIVITY	DURABILITY TO			LIGHTFASTNESS OF REACTIVE AND DIRECT DYESTUFFS
		LAUNDERING	CHLORINE	HYDROLYSIS	
X-HN-CO-NH-X	↑ INCREASING REACTIVITY	+(+)	—	+	++
X-N / \ N-X (O)		+++	+	+	+
X-N / \ N-X (O)		+++	+++	+	+
ROCONX$_2$		+++	++(+)	++(+)	+++
X-N / \ N-X, HO OH (O)		+++	+	+++	+++
X-N / \ N-X, R¹ R¹ (O)		+++	++(+)	++(+)	+++
NX$_2$ / N N \ X$_2$N-N-NX$_2$		+++	+(+)	++	+(+)

Note: X = CH$_2$OR, R = H, alkyl, R¹ = alkyl, hydroalkyl, alkoxyalkyl; R = alkyl; +++ very good, ++ good, + fair, − bad.

Reprinted in part from *Amer. Dyest. Reptr., 63,* 48 (1973). Used with permission.

If high acid stability is required, crosslinking agents of low reactivity should be employed. Bis(hydroxymethyl) or bis(alkoxymethyl) derivatives of carbamates, 4,5-dihydroxy-(or dialkoxy-)ethylene ureas, and 4-hydroxy-(or alkoxy-)5,5-dialkylpropylene ureas are the most important reactants in this group.

Interestingly, there is a correlation between the hydrolytic stability of the reactant and its effect on the lightfastness of direct and reactive dyeings. The crosslinking agents possessing high hydrolytic stability do not impair the lightfastness, while those of low hydrolytic stability have an adverse effect on the lightfastness of direct and reactive dyeings [4].

N,N'-Dimethylol-4,5-dihydroxyethylene urea (DMDHEU) is the most widely used crosslinking agent. It exhibits high acid hydrolytic stability and relatively low reactivity, while N,N'-dimethylolethylene urea (DMEU) possesses high reactivity and only limited hydrolytic stability. It is generally accepted that all the four reactive groups of DMDHEU react with cellulose [eq. (1)], but the 4,5-dihydroxyethylene ureas undergo also conversion to hydantoins [eq. (2)] which are responsible for the increased chlorine retention of the crosslinked cellulose since the free N-C groups combine readily with chlorine. The dimethylol derivative of 4,5-dialkoxyethylene urea yields a less chlorine-retentive cellulose ether than the 4,5-dihydroxy derivative [4]. In addition, DMDHEU yields some colored products under strongly acidic curing conditions in the presence of zinc nitrate catalyst, but by buffering the impregnating solution, the discoloration can be minimized [3].

The reaction of cellulose with formaldehyde has been well documented in the literature [2, 5, 6]. Formaldehyde condenses with cellulose to form inter- and intramolecular crosslinks with the loss of water. The reaction can be carried out in substantially anhydrous solvent media, in the wet swollen state, by dry-cure process or in the vapor phase. The latter process, employing SO_2 as catalyst, has gained some commercial acceptance for the treatment of garments made of cotton or cotton blends [7].

The alkali-catalyzed elimination of water or alcohol from bis(2-hydroxyethlyl) or bis(2-alkoxyethyl) sulfones at elevated temperatures under anhydrous conditions is also an effective way to crosslink cellulose [eq. (3)] [8].

where R = H or lower alkyl groups. Alkali-catalyzed reactions have not gained commercial acceptance, however, because of discoloration of cellulose in alkaline heat-curing. It should also be considered that the cellulose ethers formed have higher acid, but lower alkaline hydrolytic stability than those formed with N-methylol crosslinkers under acidic conditions.

Wet Cure

By crosslinking cellulose with the wet-cure method in the swollen and with the moist-cure technique in the partially swollen state, high wet crease recovery can be attained without undue reduction in strength properties. By adjusting the residual moisture content, the ratio of wet and dry crease recovery can be varied.

Since the crosslinking with N-methylol crosslinkers takes place under strongly acidic conditions, reactants of high acid hydrolytic stability should be used which resist the low pH for a prolonged period of time. The yield of crosslinking reaction is generally lower under these conditions than in the dry-cure process. It should be considered that some hydrolysis of the reactant takes place in the acidic medium, leading to the release of formaldehyde and to the subsequent cross-linking of cellulose through methylene ether bonds and to the formation of free NH groups which are responsible for higher chlorine retention [3].

Epihalohydrin, dihalopropanol, and the saturated derivatives of divinyl sulfone and of vinylsulfonium compounds are suitable for alkaline wet-curing at ambient temperature. Ionic derivatives of divinyl sulfone [9] and bis(vinylsulfonyl) compounds [10] containing various leaving groups (pyridinium, sulfate, thio-sulfate, acetate) have been successfully utilized for cellulose crosslinking. The rate of elimination depends on the structure of the leaving group, but the addition of the activated vinyl groups to the cellulosic hydroxyl groups is the rate-determining step [eqs. (4) and (5)] [2].

Elimination:

$$SO_2 \underset{CH_2CH_2Y}{\overset{CH_2CH_2Y}{<}} + \ 2 \ NaOH \longrightarrow SO_2 \underset{CH=CH_2}{\overset{CH=CH_2}{<}} + \ 2 \ NaY + H_2O \qquad (4)$$

where Y = $(-N^+C_5H_5)Cl^-$, $-OSO_3Na$, $-SSO_3Na$, $-OOCCH_3$

Addition:

$$SO_2 \underset{CH=CH_2}{\overset{CH=CH_2}{<}} + \ 2 \ HO-Cell \overset{OH^-}{\longrightarrow} SO_2 \underset{CH_2CH_2O-Cell}{\overset{CH_2CH_2O-Cell}{<}} \qquad (5)$$

The activating effect of the sulfonium group is similar to that of the sulfone group, but that of the sulfoxide group is insufficient for reaction with cellulose under mild conditions. The disodium tris(2-sulfatoethyl) sulfonium inner salt has been used for cellulose crosslinking in Europe [11].

$$NaO_3SOCH_2CH_2 \overset{\oplus}{>} S \overset{\ominus}{-} CH_2CH_2OSO_3$$
$$NaO_3SOCH_2CH_2$$

$$CH_2=CHCON \underset{CH_2-CH_2}{\overset{CH_2-CH_2}{\diagdown \diagup}} NCOCH=CH_2$$

Lower reactivity activated vinyl compounds (e.g., acrylamides) and their ionic saturated derivatives can be employed by use of an alkaline steaming technique. A good example of this type of reactant is 1,4-diacrylic piperazine [12]. The lower the reactivity, the higher is the alkaline hydrolytic stability of the cross-linked cellulose ether formed.

The base-catalyzed reaction of epihalohydrin gives a crosslinked product, but many side reactions lead to the formation of polymeric crosslinks and side chains [13]. The crosslinking reaction of dihalopropanol with cellulose [14] proceeds through an epihalohydrin intermediate [15] and leads to the formation of similar complex products.

The wet-curing techniques in alkaline medium have failed to gain any commercial acceptance.

Unsymmetrical Crosslinking Agents

The characterization of crosslinked cellulose by direct chemical methods is extremely difficult because the crosslinked products—especially when covalent crosslinks are introduced in the nonswollen fibers—cannot be penetrated by chemical reagents. It should also be considered that many side reactions (e.g., single-ended reaction with cellulose, polymer formation) occur when poly-functional crosslinking agents are employed. The use of unsymmetrical agents which have two groups capable of reacting with cellulosic hydroxyl groups under widely different conditions has been the most feasible approach to establish the extent of the formation of true crosslinks. Examples of such reagents are N-methylolacrylamide (NMA) and 2-methoxyethylvinyl sulfone (MVS). Both of these chemicals can be reacted with cellulose stepwise which makes it possible to characterize the single-ended product prior to the formation of crosslinks in the second reaction step.

In the case of NMA, while the single-ended reaction takes place by acid-catalyzed heat-curing, the crosslinking occurs when the activated vinyl group reacts with cellulosic hydroxyl groups at room temperature in the presence of an alkaline catalyst [Eqs. (6) and (7)] [16].

In the case of MVS, the formation of single-ended reaction product takes place

$$CH_2=CHCONHCH_2OH + HO-Cell \xrightarrow[\Delta]{H^+} Cell-OCH_2NHCOCH=CH_2 \quad (I.) \qquad (6)$$

$$I. + HO-Cell \xrightarrow{OH^-} Cell-OCH_2NHCOCH_2CH_2O-Cell \qquad (7)$$

$$CH_3OCH_2CH_2SO_2CH=CH_2 + HO-Cell \xrightarrow{OH^-} Cell-OCH_2CH_2SO_2CH_2CH_2OCH_3 \qquad (8)$$
$$(II.)$$

$$II. + HO-Cell \xrightarrow[\Delta]{OH^-} Cell-OCH_2CH_2SO_2CH_2CH_2O-Cell \qquad (9)$$

at room temperature in the presence of an alkaline catalyst, while the crosslinking step can be effectuated by alkaline heat-curing [eqs. (8) and (9)] [17].

It has been reported earlier [18] that in cotton, the maximum crease recovery is attained at D.S. = 0.04–0.08 which is equivalent to 1 crosslink for 2–4 anhydroglucose units in the accessible region (estimated at about 20%).

Location of Crosslinks

The great increase in wrinkle recovery and the severe losses in strength properties of cotton crosslinked with formaldehyde are attributed to the formation of short methylene crosslinks. It has been postulated that crosslinking cellulose with longer-chain, more flexible, difunctional reactants might lead to a more favorable ratio between increases in crease recovery and losses in tensile strength [19], but the improvement has been found to be only moderate. It has also been established that reagents of functionality greater than 2 are not necessarily more effective than difunctional ones [12].

It has been attempted to minimize the losses in strength properties by controlling the location and distribution of the crosslinks. Skin crosslinking has been obtained by applying the reagent and catalyst from a nonswelling solvent such as ethanol [20]. Core crosslinking is achieved by applying the reagent and catalyst uniformly from aqueous medium and after drying, employing a nonswelling solvent to apply a catalyst poison [20] or to remove the reagent and catalyst from the surface [21]. The improvement in relationship between fabric crease recovery and strength has been only moderate.

In recent years, it has been reported that by using a controlled minimum wet pickup (MA process) for the treatment of cellulose or cellulose-containing substrates, a more uniform distribution of the crosslinking agents can be attained due to a lower degree of migration in drying. Better relationships between permanent press performance and losses in abrasion resistance are attributed to this improved crosslink distribution [22].

Wool Modification

Dimensional Stabilization

In the case of cellulosic substrates, both the dimensional stabilization and setting can be attained by crosslinking; however, in the case of protein fibers, which contain disulfide crosslinks, even the introduction of nonreversible crosslinks does not aid in reducing the shrinkage in laundering. Felting shrinkage is caused by an irreversible, unidirectional movement of the fibers during the laundering of the wool fabric [23] and is attributed to the scale structure and shape of the wool fiber. It can be reduced by chemical procedures such as oxidation with halogens, hypohalites, or peroxy compounds, which degrade the scales or otherwise modify the surface of the fiber to reduce the directional friction effect. These oxidative procedures result in some loss of mechanical properties.

The elimination or reduction of the felting shrinkage can be obtained by depositing polymers on the fiber surfaces. Such a procedure does not cause weight or strength losses, but might lead to stiffening or alteration of the hand of the wool fabric. Therefore, it is of great importance to use such polymers which are efficient in low concentration and are capable, for instance of masking the scales with a thin layer of polymer coating or of spot-welding the fibers at only a small number of places [24].

The formation of polyamides from diamines in aqueous solution and dicarboxylic acid chlorides in a water-immiscible organic solvent or polyureas from diamines and aromatic diisocyanates *in situ* the fiber have been promising approaches to eliminate felting shrinkage in laundering [25, 26]. Later, the formation of polyamide was suggested by a single-step emulsion process using a mixture of diamine and activated dicarboxylic acid ester [27]. The two monomers do not react at low temperatures, but do form polyamide at ambient temperature or above. Similarly, polyurea can be formed *in situ* the fiber by a single-step process using long-chain aliphatic diisocyanates which possess low reactivity and can be employed from aqueous emulsion in combination with appropriate polyamines. The polyurea is formed at elevated temperature, preferably in steam-curing. The aliphatic diisocyanates can be replaced with hindered aromatic ones, but their activation requires very severe heat-curing conditions. By using a combination of aliphatic diisocyanates and polyols in the presence of appropriate catalysts, polyurethane can similarly be formed *in situ* the fiber [28].

An alternative approach is the use of preformed polymers. The early procedures with hard resins, such as urea–formaldehyde or melamine–formaldehyde precondensates [29], required high concentrations of the polymers to give a reasonable shrinkage control. Among the softer polymers, the most interesting are the linear *N*-alkoxymethyl polyamides [30], polyethylene-imines [31], water-dispersible self-crosslinking polyacrylates [32], isocyanate- [33, 34], aziridine-, and epoxy-terminated [35, 36], and thiolated polymers [37]. The polymers must spread over the fiber surface and adhere to ensure durability in wear, laundering, and dry cleaning. The spreading of the polymers can be greatly improved by preliminary modification of the wool, such as halogenation, to increase the critical surface tension of the fiber [38].

The polymers formed or deposited within the wool fiber can effectively yield the desired fiber properties, and greater changes in properties can be attained with smaller amount of polymers if it is anchored to the fiber as a graft polymer [39]. The dimensional stabilization of wool with water soluble basic polyaziridines, specifically with alkylene imine-terminated poly(ethylene oxide) [40] and with amidoaziridine- or epoxy-terminated polyurethanes derived from poly(propoxylated glycerol) [35] are such approaches. The functional groups of these reagents possess reactivity with the nucleophilic centers in the wool fiber. The reagents are also prone to polymerization as new reactive sites are formed in ring opening. While the aziridine-terminated polymers are effective shrinkproofing agents alone, the epoxy-terminated ones are effective only in the presence of coreactants such as polyamines.

$$M = \quad CH_2OM \atop CHOM \atop CH_2OM \qquad \quad \left[CH{-}CH_2O \atop \underset{CH_3}{|} \right]_n \qquad CHCH_2OCONH \atop \underset{CH_3}{|} \left\langle \quad \right\rangle CH_3 \qquad NHCON \atop CH-CH_3 \atop CH_2$$

The stiffening effect of the polymers with poly(propylene oxide) backbone is less than that of the polymers with poly(ethylene oxide) structure. Polymer treatments do not impair the abrasion resistance of wool and may even improve it.

Wool Setting

Processes for setting wool fabrics or garments with creases or pleats stable to wear, wetting, or dry cleaning depend primarily on hydrogen bond and disulfide bond rearrangements which are inherently reversible [41] and thus not stable to washing or hot water exposure.

An effective approach to permanently set wool is by combining the cleavage of the disulfide bonds with the formation of hydrolytically stable thioether crosslinks [42]. It can be achieved in a single-step steampressing operation using alkali reactive difunctional crosslinking agents, e.g., activated vinyl or cleavable onium compounds. Among the activated vinyl compounds, 1,4-diacrylic piperazine [12] and among the cleavable onium compounds, xylylene-bis(dimethylanilinium bromide) [43] are preferred because of their stability in aqueous alkaline solutions and high reactivity in alkaline steaming.

The cleavable onium compounds must of course be applied at higher concentrations than the activated vinyl compounds due to the high molecular weight of the group eliminated in the course of reaction. The alkali concentration must also be higher because stoichiometric amounts of alkali are required for the elimination [eq. (10)].

Inhibition of shrinkage and felting is essential if the set fabrics are to withstand vigorous washing. When curing of the shrinkproofing polymer takes place simultaneously with or subsequent to the setting, stabilization of set is superior compared with that obtainable when the polymer treatment preceeds the setting. The improved effect is due to fiber encapsulation and bonding [44]. The crease retention performance is improved by moistening the fabric prior to steam setting. This wetting step may be avoided by adding hygroscopic agents to the sensitizing solution, but satisfactory conditions for this approach have not yet been established.

$$H_3C{\overset{\underset{CH_3}{|}}{\overset{\oplus}{N}}}{-}CH_2\left\langle \quad \right\rangle CH_2{-}{\overset{\underset{CH_3}{|}}{\overset{\oplus}{N}}}{-}CH_3 \cdot 2Br^{\ominus} + 2\ H{-}S{-}Wool \xrightarrow{\ K_2CO_3\ }$$

$$\tag{10}$$

$$Wool{-}S{-}CH_2\left\langle \quad \right\rangle CH_2{-}S{-}Wool + 2 \left\langle \quad \right\rangle N(CH_3)_2 + 2\ KBr$$

Flame-Retardant Finishes

While textile treatments such as permanent press, water repellency, stain release, etc., are attained at <5% weight increase, most of the durable flame-retarding finishes are effective only at 15–25% weight gain. Therefore, the durable flame-retarding treatments are costly and it is a special problem to attain self-extinguishing properties without significant alteration of the mechanical and aesthetic characteristics.

Nonmelting Fibers

Cellulosics. Cellulosic fibers can be rendered flame-resistant by phosphorus chemicals which generally act in the substrate by altering the route of thermal decomposition in such a manner that the proportion of flammable volatile products to the amount of carbon and water formed is reduced. They also divert the oxidation of carbon residue from CO_2 to CO, and thereby produce insufficient energy to sustain the afterglow.

The esterification of cellulose with phosphoric or sulfamic acid or the ammonium salts of these acids are known methods for rendering cellulose flame-resistant [45, 46]. The severe afterglow of cellulose sulfate and the ion exchange of cellulose phosphate with alkaline earth salts, which destroy its flame resistance, has inhibited the commercial acceptance of these procedures. A single-step process based on simultaneous sulfation and phosphorylation overcomes these drawbacks. The combined process involves the use of ammonium sulfamate and phosphoric amide [$O=P(NH_2)_3$] [47, 48]. The utilization of phosphoric amide as phosphorylating agent is related to its limited hydrolytic stability.

The introduction of substituents on the amido groups of phosphoric amide increases the hydrolytic stability of the P—N bond. By coreacting N,N',N'',-trialkylphosphoric amide with trimethylol melamine *in situ* the fiber, excellent durable flame resistance is attained whenever the reactants are used in such a proportion that all the amide hydrogens are substituted [49, 50].

Hydrolytically stable phosphonate derivatives of cellulose can be obtained by reacting cellulose with N-methyloldialkylphosphonopropionamide [$(RO)_2$-$P(O)CH_2CH_2CONHCH_2OH$] [51, 52]. It can be applied by an acid-catalyzed heat-curing technique in conjunction with aminoplast resins. The hydrolytic stability of dialkyl phosphonates depends on the alkyl substituents.

Another approach to obtain a phosphonate derivative of cellulose is via the free-radical polymerization of a vinylphosphonate oligomer [53] *in situ* the fiber. It can be homopolymerized or copolymerized with acrylamides.

$$\begin{array}{c} RO \\ \diagdown \\ R' \diagup \end{array} \overset{O}{\underset{\|}{P}}-O-(CH_2CH_2O-\overset{\overset{O}{\|}}{\underset{\underset{CH=CH_2}{|}}{P}}-OCH_2CH_2O-\overset{O}{\underset{\underset{R'}{|}}{\overset{\|}{P}}}-O-)_n R$$

Polymers containing the hydrolytically very stable phosphine oxide structures are obtained by insolubilizing tetrakis(hydroxymethyl)phosphonium chloride (THPC) with amines and amides.

$$(HOCH_2)_4P^+Cl^-$$

The THPC–amide heat cure finish [54, 55] or the coreaction of the THPC/urea precondensate with ammonia vapor [56] as well as the THPOH–amide heat cure [57] or the THPOH–ammonia vapor-phase reaction [eq. (11)] [58] are excellent approaches to obtain flame-resistant cellulosic textiles. Oxidation of the intermediate trivalent phosphorus in the polymer to the pentavalent state is essential to achieve the maximum durability of the polymer. In all these reactions, the THPC and THPOH can be replaced with other tetrakis(hydroxymethyl)phosphonium salts, such as the sulfate, oxalate, phosphate, etc. [59].

In most of the procedures designed to flame-retard cellulosic substrates with organophosphorus chemicals, nitrogen is present, in some cases derived from the nitrogen content of the phosphorus compound, in other cases from the addition of a nitrogen-bearing component to the flame-retardant formulation. It has been established that nitrogen contributes to the flame-retarding effectiveness of organophosphorus compounds [51, 60–62], although its efficiency depends on the source of nitrogen [63]. While nitriles are ineffective, amides, especially in the form of urea and melamine, are the most effective. The nitrogen/phosphorus synergism is demonstrated experimentally, but its mechanism is still unresolved.

Although the flame retardants acting in the condensed phase are the most effective on 100% cellulosic substrates, halogen-containing compounds which function in the vapor phase can also be used as flame-retarding agents. They function mainly by reducing the flammability of the mixture of air and vapor distilled from the heated cellulose. Bromine is more effective than chlorine, and antimony oxide possesses a synergistic effect with the halogen-containing chemicals. The evolution of HCl or HBr from the halogenated compound and its interaction with Sb_2O_3 to yield SbOCl and $SbCl_3$ to form a free-radical trap in the vapor phase explain the enhanced flame retardance [64]. An example of this approach is the coating of cellulosic or cellulose containing fabrics with a composition consisting of decabromodiphenyl oxide, antimony oxide, and an acrylic binder [65].

Protein Fibers. Wool possesses relatively high ignition temperature, low flame temperature, and low heat of combustion. These characteristics are due to the morphological and chemical structure of the wool fiber. The relatively low flammability level of wool is also connected to its high moisture content, and thus under bone-dry conditions—as required in certain flammability standards—wool might fail the flame test.

$$P^+(CH_2OH)_4OH^- \xrightarrow{NH_3} \quad ---\overset{\overset{O}{\parallel}}{\underset{\mid}{P}}-CH_2-\left[NH-CH_2-\overset{\overset{O}{\parallel}}{\underset{\underset{\overset{N}{\diagdown}}{\underset{R\ H}{}}}{\underset{\mid}{P}}}-CH_2-\right]_n-NH-CH_2--- \quad (11)$$

THPOH

$$R = H \text{ or } \left[\quad\right]$$

The flame resistant treatments of wool are based on the exhaustion of negatively charged metal complexes, treatment with halogenated acids and with
phosphorus chemicals.

Negatively charged complexes of titanium and zirconium with fluorides, ci-
trates, or other α-hydroxycarboxylic acids are exhausted on positively charged
wool fiber (pH < 3) at elevated temperatures. Titanium complexes are more
effective than the zirconium ones, probably because of better penetration of the
fiber with the smaller titanium complexes. However, the zirconium compounds
do not affect the color. The F/Ti or F/Zr ratio of the complexes has to be at least
5 but preferably 6 to improve the natural flame resistance of wool effectively.
Only negatively charged penta-, hexa-, or heptafluorozirconate or titanate can
be exhausted onto the positively charged wool. For washable wool, zirconium–
tungsten complexes are preferred [66].

The metal complex in the wool fiber must form an insoluble acid residue that
is volatile or decomposes at temperatures below the ignition temperature of wool.
The acid residue is probably responsible for the increased amount of char, and
for the formation of acid fragments on mild heating which trap the flammable
volatiles so that the flame is extinguished.

Chlorendic acid [67] and tetrabromophthalic acid or anhydride [68] can also
be exhausted onto the wool to improve the flame resistance of the wool fabric.
Organophosphorus compounds are also effective flame retardants for wool fibers;
e.g., the homopolymerization of oligomeric vinyl phosphonate or its copolymerization with acrylic compounds renders wool flame-resistant. Such a
treatment can be combined with appropriate shrinkproofing finishes [28].

Thermoplastic Fibers (Polyester and Nylon)

Thermoplastic fibers are considered less flammable than the cellulosics because they possess a relatively low melting point and the melt drips rather than
remains to propagate the flame when the source of ignition is removed.

The most common approach of flame-retarding polyester is the application
of tris(2,3-dibromopropyl)phosphate (TDBP) $[(BrCH_2CHBrCH_2O)_3P{=}O]$
by a thermosol or exhaustion technique [69]. It should be considered that
polyester fiber cannot retain permanently more than 4–5% TDBP calculated
on the weight of the fiber.

Interestingly, halogens are less effective flame retardants on nylon than on
polyester, and most of the flame retardants effective on cellulosic or polyester
substrates have only insufficient effectiveness on nylon. Thiourea appears to
be the most effective on nylon [70]. Some durability can be imparted by
employing methylolated thiourea or by applying aminoplast resins in conjunction
with thiourea [71]. The effectiveness of thiourea and of certain other chemicals
in flame-retarding nylon is related to their ability to depress the melting point

of the polymer and to increase the ease with which the melt drips away [70].

Blends of Thermoplastic and Nonmelting Fibers

Among the blends of thermoplastic and nonmelting fibers, polyester/cellulosic blends have the greatest commercial importance. It has been shown that the flammability behavior of the blends cannot be predicted from the flammability of the single-fiber structures [72–75]. The treatment of one of the fiber components with a flame-retarding agent specifically effective for that component does not necessarily render the two-component blend flame-resistant [72, 74] unless the treated component comprises at least 85% of the blend [73]. An effective flame-retarding treatment must reduce the flammability of each of the components. It has also been established that it suffices to flame-retard only one fiber component while the second one can be completely free of finish if a flame-retarding agent is used which is effective on both types of fibers [76].

While phosphonates are the more effective flame retardants on 100% cellulosic substrates, phosphine oxides exhibit higher effectiveness on polyester/cellulosic blends. This observation was confirmed with specially designed analogous structures [77].

$$\begin{array}{c} R \\ \diagdown \\ \diagup \\ R \end{array} \overset{O}{\overset{\|}{P}}-CH_2CH_2\overset{O}{\overset{\|}{C}}-NHCH_2OH \qquad \text{where } R = CH_3O \quad \text{(Phosphonate)} \\ \qquad\qquad\qquad\qquad CH_3 \quad \text{(Phosphine Oxide)}$$

Phosphonium compounds show similar effectiveness as the phosphine oxides because they are converted to phosphine oxides in the course of insolubilization.

It should be considered that chemicals employed from aqueous medium to the blend fabric are deposited preferentially onto the hydrophilic cellulosic component, and the adhesion of the polymer is significantly better to the cotton than to the hydrophobic polyester fiber. Thus, the extremely high chemical add-on on the cellulosic component results in significant impairment of the aesthetic and mechanical properties. Therefore, it would be desirable to employ booster flame-retarding chemicals to the polyester component (e.g., brominated compounds which can be thermosoled into the fiber) or to use inherently flame-resistant fiber in order to accommodate a portion of the chemical load on the polyester fiber and to reduce it on the cotton component. It should also be considered that the thermally more stable aromatic bromo compounds seem to be more effective on the blend fabric than the aliphatic counterparts. Antimony enhances the effectiveness of bromine compounds, but its synergistic effect is diminished when phosphorus is present [78].

Although the flame-retardant finishing of polyester/cellulosic blends is not a commercially solved problem, the investigations in the last five years have succeeded in determining the main requirements of a finish which should be anticipated to be effective on the blend fabric.

SURFACE MODIFICATIONS

Water- and Oil-Repellent Finishes

The deposition of hydrophobic films consisting of saturated hydrocarbon or fluorocarbon chains attached to the fiber surface by reactive polar groups is the most usual technique by which water repellency is attained on textiles.

The treatment of synthetic textiles with wax dispersion in conjunction with melamine– or urea–formaldehyde resin results in water repellency with limited durability. Aluminum salts and zirconium and chromium complexes of fatty acids have been combined with these waxes to improve the durability of the treatment [79].

A very significant development has been the introduction of fatty alkyl derivatives containing groups capable of reacting with cellulosic hydroxyl groups [80]. It should be considered that the yield of cellulose etherification drops rapidly with increasing chain length of the agent. The maximum water repellency is attained when the hydrocarbon chain is about 16–20 carbon atoms long. Longer chains tend to coil, and shorter chains exhibit reduced water repellency because of the proximity of the terminal polar groups.

Fatty alkyl derivatives containing Y-CH_2Cl moiety, where Y is a —O—, —S—, —NH—, —CONH— group, are reactive and can be solubilized as the quaternary halides of tertiary amines, particularly pyridine. A successful example of this is the stearamidomethyl pyridinium salt [81].

$$[C_{17}H_{35}CONHCH_2\overset{+}{N}\bigcirc]\ Cl^{\ominus}$$

The extent of the reaction with cellulose is limited to about 1 stearamidomethyl group per 150 anhydroglucose units; the formation of some methylenedistearamide *in situ* the fiber is also a possibility. The durability of the finish in laundering is very good.

The water repellency of silicones [82] is due to the low surface tension of the methyl-on-silicone groups; e.g., the critical surface tension of a polydimethylsiloxane film is approximately 22 dyne/cm. The structure of the siloxanes used for these treatments can be represented as follows:

$$--O--\underset{R^2}{\overset{R^1}{Si}}--O--\underset{R^2}{\overset{R^1}{Si}}--O--\underset{R^2}{\overset{R^1}{Si}}--$$

When R_1 and R_2 are both methyl groups, the silicone is the nonreactive type; when R_1 is methyl and R_2 is hydrogen, the silicone is the reactive type.

The nonreactive polydimethylsiloxanes are added to the fiber with titanium or zirconium compounds which, on curing, hydrolyze to metal oxide films and then coordinate with the siloxane structure. The reactive poly(methyl hydrogen siloxanes), which account for most of the silicone textile water repellents, crosslink on the fiber surface via the reaction of the residual Si—H bonds with water or other sources of hydroxyl to give crosslinked structures. Tin and zinc catalysts are generally used and, interestingly, in this case titanium and zirconium catalysts are ineffective. Such silicone finishes are durable and possess thermal and oxidative resistance because of the inherent stability of the siloxane chain.

Fluorochemicals have revolutionized both oil- and water-repellent finishing. Two basic types of perfluoro compounds are used. These are the perfluorocarboxylic acids and polymers of perfluoroalkyl esters of acrylic acids [83].

$$-\text{\{CH}_2-\text{CH}\}-$$
$$\overset{|}{\underset{|}{\text{CO}}}$$
$$\text{OCH}_2(\text{CF}_2)_n\text{CF}_3$$

Poly(perfluoroalkyl acrylates) exhibit increasing oil repellency as the number of fluorine atoms increases. A fluorochemical possessing satisfactory oil repellency should contain at least one perfluorogroup as large as C_7F_{15}. The water-repellent property is less affected by the size of the perfluoro chain than is the oil repellency. The CF_3 end group present in a fluorochemical film gives the surface a very low critical surface tension (approx. 6 dyne/cm). This low CST explains the fact that selected fluorochemicals are the most effective organic compounds in rendering textiles oil-repellent [84].

It is interesting to note that by combining a fluorochemical with other selected water repellents (e.g., stearamidomethylpyridinium chloride) in well-specified proportions, a less expensive oil- and water-repellent finish with superior durability can be obtained [85].

The surface roughness of the fluorine-containing film influences its surface tension. When the fluorochemical is deposited onto a fabric surface, the fluorochemical is present as an uneven film which is expected to have slightly higher CST than a smooth film. Any mechanical action which roughens the fabric and/or the fluorochemical surface film, reduces the effectiveness of the treatment. This explains why the oil repellency of a cotton fabric treated with a fluorochemical is diminished in laundering but is at least partially restored by hot ironing, which serves to smooth out increased roughness of the thermoplastic fluorochemical film caused by abrasion in laundering [84].

Soil Release Finishes

The oleophilic character of synthetic fibers increases their tendency to retain oily soils and their hydrophobic character impedes the penetration of the detergent solution and the solubilization of the oily stains in the course of laundering.

We must differentiate between three terms: (1) soil repellency is the ability of the textile to resist soiling or staining; (2) soil release is the ability to remove the soil in laundering; (3) soil redeposition is the pickup of soil suspended in wash water and results in greying of the textile [86]. Since all these properties are associated with the surface of the polymer, it is feasible to modify the surface characteristics of the fiber without altering its mechanical properties.

An early method used to improve soil release and prevent soil redeposition was treatment of polyester textiles with polyglycols. Sodium polyglycolate is capable of reacting with poly(ethylene terephthalate), the ester interchange leading to the chemical bonding of the polyglycolate to the fiber surface. Since the ester-interchange reaction is very sensitive to time, temperature, and moisture, this approach has been abandoned [87].

The alternative method for the surface modification of polyester fiber has been the application of a hydrophilic polyester derived from terephthalic acid and poly(ethylene glycol) [88]:

$$\left[OC - \bigcirc - COO(CH_2CH_2O)_n CH_2CH_2 - O \right]_x$$

This is designed to form a surface coating, the adhesion of which is claimed to be enhanced by a "co-crystallization" mechanism. The hydrophilic surface impedes wetting by oily stains and enhances wettability in aqueous launderings, thus preventing soil redeposition and improving soil release properties.

Acrylic polymers employed as stain release finishes to permanent press polyester/cellulosic blends provide the fiber surface with a polar wettable coating to impart hydrophilic, oleophobic character. Copolymers of ethylacrylate and acrylic acid, e.g.,

$$\left[\begin{array}{c} CH_2 - CH \\ COOH \end{array} \right]_x \left[\begin{array}{c} CH_2 - CH \\ COOC_2H_5 \end{array} \right]_y$$

are examples of such finishes [89].

Acrylic acid can also be grafted onto the fiber surface by radiation techniques. The preparation of nylon 6,6–acrylic acid graft copolymer by high-energy electron-beam radiation has been reported [90]. A newer approach which has reached commercial acceptance employs an electrical discharge in argon gas to graft acrylic acid onto a polyester surface and thereby provides improved wettability with enhanced soil release and reduced soil redeposition characteristics [91, 92]. Typical hydrophilic polyester generated by this process has a poly(acrylic acid) add-on of 0.05–0.10%. The grafted product retains its wettability through 50 launderings although there is a gradual loss of graft due to abrasion, about half of it being lost after 20 washings.

Another route to soil release finishing has been based on imparting soil repellency by the use of fluorochemicals [93]. The use of conventional fluorochemical soil-repellent treatments proved to be unsuitable for soil release finishing. While the soil-repellent treatment prevents penetration of oil under some conditions, it impedes the release of any oily stain which has succeeded in reaching the fiber.

"Dual action" fluorochemicals were developed to overcome this deficiency; these are soil repellent, soil releasing copolymers containing alternating hydrophilic and fluorochemical segments [94].

Hydrophilic segments:

$$HS-[-CH_2CH-COO(CH_2CH_2O)_4COCHCH_2-S-]_xH$$
$$CH_3 CH_3$$

Fluorochemical segments:

$$+CH\text{-----}CH_2+$$
$$COOC_2H_4N-SO_2C_8F_{17}$$
$$CH_3$$

These finishes form continuous films which possess low surface energy in air and low interfacial energy in water. It has been postulated that the fluorinated segments lie on the fabric surface in air, and the hydrophilic ones lie on the surface in water.

Antistatic Finishes

Whereas most of the natural fibers are composed of strongly hydrogen-bonding polymers and exhibit hygroscopicity, some synthetic polymers such as polyolefins, polyamides, polyesters, and polyacrylonitriles are comparatively hydrophobic. The presence of water in the polymeric material increases the electrical conductance and therefore, under identical conditions, synthetic fibers are better electrical insulators than the natural fibers [95]. The only chemical mechanism by which accumulation of electrical charge by synthetic fibers can be prevented effectively is by increasing their electrical conductivity to a level comparable to that shown by hydrophilic fibers [96–98].

A textile fiber (other than the metal fibers) must contain water in order to be static-free. Electrolytic conductance requires the presence of a medium of high dielectric constant, and water is the only such medium which can be replenished from the atmosphere when lost. In order to assure the presence of water in or on the fiber, a hygroscopic compound must be associated with the fiber, either as an integral part of the fiber or added to it as a finish. With durable antistatic finishes hygroscopicity, while increasing their efficiency, will tend to reduce durability, since it causes swelling of the finish in laundering. The swollen finish shows reduced resistance to wet abrasion. The ideal durable antistatic finish should thus combine the highest possible moisture regain at medium and low relative humidities with the lowest possible swelling in water.

Since every ionizing group is hygroscopic, antistatic finishes can be prepared without the presence of nonionic hygroscopic groups in the molecule. Quaternary ammonium polymers retain more water at low relative humidity than other polymers. This suggests why these types of polymers are more conductive as a group than other polyelectrolytes [99]. Ionizing antistatic agents are, in general, more efficient antistats than mere nonionic hygroscopic ones. The latter serve as a conductive vehicle for the ions present in the fabric and rely solely on tramp electrolytes, e.g., the impurities present in the last rinse water. Only a finish

composed of an ion-exchange resin of suitable structure can establish its efficiency level independently of the nature of the rinse water.

Effective durable antistatic finishes can be generally obtained by forming
in situ a crosslinked polymeric network containing ionic and hygroscopic groups,
thus providing the fiber surface with an insoluble conductive sheath. Durable
antistatic finishes obtained according to this general principle from crosslinked
polyamine resins were first reported in 1959 [100], and many subsequent developments have been based on related reactions yielding polymers of similar
structure [101–104]. For instance, insoluble polymers were formed on a polyester
fabric surface by reacting *in situ* [eq. (12)] a trifunctional crosslinking agent,
e.g., 1,3,5-tris(3-methoxypropionyl)-*s*-perhydrotriazine (TMPT), with a
polyhydroxy polyamine (PHPA) [105]. The polymer formed contains a varying
number of oxyethylene groups, contributing to hygroscopicity, and also ionizable
amino groups. The tightness of the crosslinked polymeric network can be altered
by varying the polyamine/crosslinker ratio and the average molecular weight
of the PHPA reagent by controlling the reaction conditions used for its preparation.

It has been established that crosslinked polymers formed *in situ* on the surface
of hydrophobic fibers impart more effective antistatic properties to the fabric
if they contain both hydrophilic moieties and ionizable groups than if they
contain only one of these.

CONCLUSIONS

Diverse functional properties can be imparted to natural and synthetic fibers
by chemical finishing procedures. In many cases, a compromise should be made
because certain functional properties can be attained only by impairing other
aesthetic and wear characteristics.

The introduction of new fibers and fiber blends has made chemical modifications, especially the flame-retarding treatments, more complex. For instance
the flammability behavior of the blend fabric and the effectiveness of a flame-
retardant finish on the blend cannot be predicted from the flammability behavior
of the individual fiber structures or from the efficiency of a specific flame retardant finish on the individual fiber component.

It is a special problem to attain a uniform distribution of chemicals on blends
of hydrophilic and hydrophobic fibers. For instance, while in the permanent press

$$\left[\text{N-(CH}_2\text{CH}_2\text{O)}_n\text{CH}_2\text{CH}_2\right]_x + \text{CH}_3\text{OCH}_2\text{CH}_2\text{CON} \quad \text{NCOCH}_2\text{CH}_2\text{OCH}_3$$

PHPA   TMPT   (12)

$$\xrightarrow[\Delta]{\text{OH}^-} \text{Crosslinked Polymer}$$

treatment of polyester/cellulosic blends the crosslinking agents should be deposited exclusively on the hydrophilic cellulosic component, flame-retarding chemicals should be uniformly distributed on the hydrophilic cellulosic and hydrophobic polyester components to avoid the undesirable side effects of an extreme chemical load on one of the fiber components.

For many end uses multiple chemical modifications are required, and it is not practical or possible to subject a textile to a sequence of finishing operations. Thus, some of the chemical finishing and dyeing procedures should be combined in a single-step operation. It is known that water repellents, such as the quaternary pyridinium types, silicones, or fluorochemicals, can be combined with N-methylol resins to attain permanent press and water repellency in a single step. Permanent press resins can also be employed together with stain release finishes, such as the acrylic acid copolymers or the "dual action" fluorochemicals. The combination of a flame-retarding treatment with pigment dyeing and a permanent press finish on cellulosic textiles, or simultaneous wool shrinkproofing and flame retarding are possible approaches to attain multifunctional properties in a single-step operation.

The author expresses his gratitude to Ms. Catherine V. Stevens for the critical review of this manuscript.

REFERENCES

[1] G. C. Tesoro, *J. Amer. Oil Chem. Soc., 45,* 351 (1968).
[2] G. C. Tesoro and J. J. Willard, in *Cellulose and Cellulose Derivatives,* N. Bikales and L. Segal, Eds., Wiley, New York, 1971, pp. 835-875.
[3] W. Ruemens, G. Burkhardt, H. Petersen, and W. Ruettiger, *Amer. Dyest. Reptr., 62,* 43 (1973).
[4] H. Petersen, *Text. Res. J., 38,* 156 (1968).
[5] W. J. Roff, *J. Text. Inst., 49,* T646 (1958).
[6] J. T. Marsh, *J. Soc. Dyers Col., 75,* 244 (1959).
[7] R. Swidler and J. P. Gamarra, *Text. Chem. Col., 3,* 41 (1971).
[8] G. C. Tesoro, *Text. Res. J., 32,* 189 (1962).
[9] G. C. Tesoro, P. Linden, and S. B. Sello, *Text. Res. J., 31,* 283 (1961).
[10] G. C. Tesoro, *J. Appl. Polym. Sci., 5,* 721 (1961).
[11] F. H. Burkitt, *Text. Res. J., 34,* 336 (1964).
[12] G. C. Tesoro and S. B. Sello, *Text. Res. J., 36,* 158 (1966).
[13] J. B. McKelvey, R. R. Benerito, R. J. Berni, and B. G. Burgis, *J. Appl. Polym. Sci., 7,* 1371 (1963).
[14] D. M. Gagarine (to Deering Milliken Research Corp.) U.S. Pat. 2,985,501 (1961).
[15] A. Fairbourne, G. P. Gibson, and D. W. Stephens, *J. Chem. Soc., 1965* (1932).
[16] J. L. Gardon, *J. Appl. Polym. Sci., 5,* 734 (1961).
[17] G. C. Tesoro and A. Oroslan, *Text. Res. J., 33,* 93 (1963).
[18] G. C. Tesoro, *Text. Res. J., 30,* 192 (1960).
[19] E. I. Valko, B. Bitter, and R. S. Perry, *Text. Res. J., 34,* 849 (1964).
[20] C. J. Gogek and E. I. Valko, *Textilveredlung, 2,* 423 (1967).
[21] J. J. Willard, G. C. Tesoro, and E. I. Valko, *Text. Res. J., 39,* 413 (1969).
[22] M. Schwemmer, H. Bors, and A. Goetz, *Textilveredlung, 10,* 15 (1975).
[23] G. F. Flanagan, *Text. Res. J., 36,* 55 (1966).
[24] O. C. Bacon and D. E. Maloney, *Amer. Dyest. Reptr., 56,* 319 (1967).
[25] R. E. Whitfield, L. A. Miller, and W. L. Wasley, *Text. Res. J., 31,* 704 (1961); *ibid., 32,* 743 (1962); *ibid., 33,* 440, 752 (1963).
[26] R. E. Whitfield, *Appl. Polym. Symp. No. 18,* 559 (1971).

[27] H. K. Rouette, H. Zahn, and M. Bahra, *Textilveredlung, 2,* 474 (1967).

[28] S. B. Sello, C. V. Stevens, and R. O. Rau, Paper presented at the 5th International Wool Research Conference, Aachen, West Germany, Sept. 1975.

[29] P. J. Alexander, *J. Soc. Dyers Col., 66,* 349 (1950).

[30] D. L. C. Jackson, *Text. Res. J., 24,* 624 (1954).

[31] C. E. Pardo, Jr., (to the U.S. Dept. of Agriculture), U.S. Pat. 2,817,602 (1957).

[32] H. D. Feldtman and J. R. McPhee, *Text. Res. J., 35,* 150 (1965).

[33] F. Reich, *Mell., 50,* 305 (1969).

[34] Deering Milliken Research Corp., Brit. Pat. 1,181,373 (1970).

[35] S. B. Sello and G. C. Tesoro, *Appl. Polym. Symp. No. 18,* 627 (1971).

[36] S. B. Sello, G. C. Tesoro, and R. Wurster, *Text. Res. J., 41,* 556 (1971).

[37] I. W. S. Nominee Co., Ltd. and Ciba A.G., Ger. Offen. 1,934,678 (1970)

[38] H. D. Feldtman and J. R. McPhee, *Text. Res. J., 34,* 634 (1964).

[39] J. B. Speakman, *J. Amer. Leather Chem. Assoc., 53,* 492 (1958).

[40] G. C. Tesoro and S. B. Sello, *Text. Res. J., 34,* 523 (1964).

[41] J. B. Caldwell, S. J. Leach, J. Meschers, and B. Milligan, *Text. Res. J., 34,* 627 (1964).

[42] S. B. Sello and G. C. Tesoro, *Text. Res. J., 43,* 309 (1973).

[43] H. Rath, *Mell., 38,* 181 (1957); *ibid., 38,* 656, 1406 (1957).

[44] J. R. Cook and J. Delmenico, *J. Text. Inst., 62,* 27, 62, 438 (1971).

[45] M. Sander and E. Steininger, *J. Macromol. Sci.-Rev. Macromol. Chem., 2,* 57 (1969).

[46] E. E. Gilbert, *Sulfonated and Related Reactions,* Interscience, New York, 1965.

[47] P. Isaacs, M. Lewin, C. Stevens, and S. B. Sello, *Textilveredlung, 8,* 158 (1973); *Text. Res. J., 44,* 700 (1974).

[48] M. Lewin and P. Isaacs (to the State of Israel), Ger. Offen. 2,127,188 (1972).

[49] S. B. Sello, I. E. Pensa, and R. Wurster, *Textilveredlung, 10,* 183 (1975).

[50] S. B. Sello (to J. P. Stevens & Co., Inc.), U.S.Pats. 3,632, 297; 3,681,060, (1972).

[51] G. C. Tesoro, S. B. Sello, and J. J. Willard, *Text. Res. J., 39,* 180 (1969).

[52] R. Anishaenslin, C. Guth, P. Hofmann, J. Maeder, and H. Nachbur, *Text. Res. J., 39,* 375 (1969).

[53] E. D. Weil (to Stauffer Chemical Co.) U.S. Pat. 3,695,925 (1972).

[54] W. A. Reeves and J. D. Guthrie, *Text. World, 104,* 101 (1954).

[55] G. L. Drake, Jr., R. M. Perkins, and W. A. Reeves, *J. Fire Flamm., 1,* 78 (1970).

[56] W. A. Reeves and J. D. Guthrie (to the U.S. Dept. of Agriculture), U.S. Pat. 2,772,188 (1956).

[57] J. V. Beninate, E. K. Boylston, G. L. Drake, Jr., and W. A. Reeves, *Text. Res. J., 38,* 267 (1968).

[58] J. V. Beninate, R. M. Perkins, G. L. Drake, Jr., and W. A. Reeves, *Text. Res. J., 39,* 368 (1969).

[59] G. Hooper, in *Proceedings of the 1973 Symposium on Textile Flammability,* E. Greenwich, Ed., R. I. LeBlanc Research Corp., 1973, p. 50.

[60] G. C. Tesoro, *Textilveredlung, 2,* 435 (1967).

[61] G. C. Tesoro, S. B. Sello, and J. J. Willard, *Text. Res. J., 38,* 245 (1968).

[62] S. B. Sello, B. J. Gaj, and C. V. Stevens, *Text. Res. J., 42,* 241 (1972).

[63] W. A. Reeves, R. M. Perkins, B. Piccolo, andG L. Drake, Jr., *Text. Res. J., 40,* 223 (1970).

[64] J. Lyons, *The Chemistry and Uses of Fire Retardants,* Wiley-Interscience, New York, 1970.

[65] V. Mischutin, *Text. Chem. Col., 7,* 40 (1975).

[66] L. Benisek, *J. Text. Inst., 65,* 102, 140 (1974).

[67] M. Friedman, J. F. Ash, and W. Fong, *Text. Res. J., 44,* 555 (1974).

[68] M. Friedman, J. F. Ash, and W. Fong, *Text. Res. J., 44,* 994 (1974).

[69] E. Baer, in *Proceedings of the 1973 Symposium on Textile Flammability,* E. Greenwich, Ed., R. I. LeBlanc Research Corp., 1973, p. 117.

[70] D. Douglas, *J. Soc. Dyers Col., 73,* 258 (1957).

[71] H. Stepnicka, *Ind. Eng. Chem. Prod. Res. Devel., 12,* 29 (1973).

[72] G. C. Tesoro and C. Meiser, Jr., *Text. Res. J., 40,* 430 (1970).

[73] W. Kruse and K. Filipp, *Mell., 49,* 203 (1968).

[74] W. Kruse, *Mell., 50,* 460 (1969).

[75] G. C. Tesoro and J. Rivlin, *Text. Chem. Col., 3,* 156 (1971).

[76] P. Linden, L. G. Roldan, S. B. Sello, and H. S. Skovronek, *Textilveredlung, 6,* 651 (1971).

[77] G. C. Tesoro, *Text. Chem. Col., 5,* 235 (1973).

[78] C. V. Stevens and S. B. Sello, in *Proceedings of the Third Symposium on Textile Flammability,* LeBlanc Research Corp., New York, 1975, p. 186.

[79] J. M. May, *Amer. Dyest. Reptr., 58,* 15 (1969).

[80] H. A. Schuyten, J. D. Reid, J. W. Weaver, and J. G. Frick, Jr., *Text. Res. J., 18,* 396 (1948).

[81] F. V. Davis, *J. Soc. Dyers Col., 63,* 260 (1947).

[82] J. K. Campbell, *Text. Chem. Col., 1,* 370 (1969).

[83] E. J. Grajeck and W. H. Petersen, *Text. Res. J., 32,* 320 (1962).

[84] D. C. M. Dorset, *Text. Mfr., 96,* 112 (1970).

[85] C. G. DeMarco, A. J. McQuade, and S. J. Kennedy, *Mod. Text., 41,* 50 (1960).

[86] G. C. Tesoro and J. J. Willard, *Tenside, 6,* 258 (1969).

[87] D. A. Garrett and P. N. Hartley, *J. Soc. Dyers Col., 82,* 252 (1966).

[88] Imperial Chemical Industries, Ltd., U.S. Pat. 3,416,952 (1968).

[89] Deering Milliken Research Corp., U.S. Pat. 3,377,249 (1968).

[90] E. E. Magat, I. K. Miller, D. Tanner, and J. Zimmerman, *J. Polym. Sci., 8,* 615 (1964).

[91] J. D. Fales, paper presented at the Radiation Processing of Textiles Conference, North Carolina State University, Raleigh, N.C., May 1975.

[92] S. M. Suchecki, *Text. Ind., 3,* 91 (1975).

[93] R. E. Read and G. C. Culling, *Amer. Dyest. Reptr., 57,* 881 (1967).

[94] P. O. Sherman, S. Smith, and B. Johanessen, *Text. Res. J., 39,* 449 (1969).

[95] E. I. Valko and G. C. Tesoro, in *Kirk Othmer Encyclopedia of Chemical Technology, Vol. 2,* 1963, p. 204.

[96] E. I. Valko and G. C. Tesoro, *Mod. Text., 38,* 62 (1957).

[97] J. W. S. Hearle, *J. Text. Inst., 48,* P40 (1957).

[98] G. J. Sprokel, *Text. Res. J., 27,* 501 (1957).

[99] R. J. Dolinsky, and W. R. Dean, *Chem. Technol.,* 1971, 304.

[100] E. I. Valko and G. C. Tesoro, *Text. Res. J., 29,* 21 (1959).

[101] E. I. du Pont de Nemours & Co., U.S. Pat. 3,021,232 (1962).

[102] Onyx Oil and Chemical Co., U.S. Pats. 3,063,870; 3,070,552 (1962).

[103] Imperial Chemical Industries Ltd., Brit. Pat. 985,325 (1965).

[104] Millmaster Onyx Corp., U.S. Pat. 3,553,111 (1971).

[105] G. C. Tesoro, S. B. Sello, and W. K. Lee, *Textilveredlung, 1,* 208 (1966).

RECENT ADVANCES ON CHEMICAL MODIFICATIONS
OF WOOL AND WOOL BLENDS

D. P. VELDSMAN

S. A. Wool & Textile Research Institute of the CSIR,
Port Elizabeth 6000, P.O. Box 1124, South Africa

SYNOPSIS

In this paper, some evidence is advanced on the useful application of chemical technology in achieving dimensional stability in wool and wool/cotton fabrics and also in improving their whiteness and durable press performance. Wrinkling performance could also be improved by the use of elastomeric polymers.

For any polymer to spread evenly over a wool fiber, its critical surface energy should be lower than the critical surface tension of the fiber. Proper adhesion of the polymer to the fiber is also of great importance with regard to durability to washing. Yellowing during dyeing could be reduced significantly by not exceeding a dyeing temperature of 85°C in the presence of specific dyeing auxiliaries. Bleaching of wool/cotton blends could be carried out with hydrogen peroxide by a pad–dwell or pad–steam process. Durable press treatments of wool/cotton blends involve the use of aminoplast resins and a polymer former. The use of silicones which are compatible with aminoplast resins is worth mentioning. Finally, wrinkling performance of wool fabrics could be improved by the use of silicones. The significance of this application depends on whether the fabrics have been aged or de-aged prior to wrinkling.

INTRODUCTION

Throughout the years, the subject of chemistry has found wide application in wool textile technology, especially in the fields of dyeing and wet-finishing. The purpose of this paper is to briefly outline what contributions the South African Wool and Textile Research Institute (SAWTRI) has made recently in this particular field. For this purpose, attention will be paid to shrink–resist treatments, simultaneous dyeing and shrinkproofing, bleaching, durable press treatments, and wrinkling.

SHRINK–RESIST TREATMENTS

For many resin applications to untreated wool, the critical surface tension (CST) of which is in the vicinity of 30 dyne/cm, difficulty is experienced in getting the resin to spread as, for adequate spreading, the CST of wool should be higher than the surface energy of the resin [1]. In this respect, it is worth

Journal of Applied Polymer Science: Applied Polymer Symposium 31, 251–256 (1977)
© 1977 by John Wiley & Sons, Inc.

mentioning that SAWTRI is currently carrying out some interesting work on the surface energies of various polymers. For this purpose, two solvents having widely different nonpolar values for their surface tensions are used for measuring their angles of contact on the resin. If the polar and nonpolar (disperse) components of their surface tensions are known, substitution of these values and their contact angles on a specific resin into eq. (1) leads to a calculation of the surface energy of a specific resin. The interfacial tension between a liquid L and a solid S is given by Wu's equation [2]:

$$\gamma_{LS} = \gamma_L + \gamma_S - \frac{4\gamma_L{}^d\gamma_S{}^d}{\gamma_L{}^d + \gamma_S{}^d} - \frac{4\gamma_L{}^p\gamma_S{}^p}{\gamma_L{}^p + \gamma_S{}^p} \tag{1}$$

where γ_{LS} = interfacial tension between liquid and solid, γ_L = surface tension of liquid, γ_S = surface free energy of solid, and γ^d and γ^p = the dispersion and polar components of the surface tension.

$$\gamma_S = \gamma_L \cos \theta + \gamma_{LS} \tag{2}$$

By substituting eq. (1) into eq. (2):

$$(1 + \cos \theta)\, \gamma_L = 2 \left[\frac{2\gamma_L{}^d\gamma_S{}^d}{\gamma_L{}^d + \gamma_S{}^d} + \frac{2\gamma_L{}^p\gamma_S{}^p}{\gamma_L{}^p + \gamma_S{}^p} \right] \tag{3}$$

Equation (3) contains two unknown quantities, viz., $\gamma_S{}^d$ and $\gamma_S{}^p$. When the contact angles are measured, and if the dispersion and polar components of the testing liquid, $\gamma_L{}^d$ and $\gamma_L{}^p$, are known, then for *two* contact angles of two testing liquids, the two unknowns can be calculated from the two quadratic equations.

Having obtained the CST of the wool fiber and surface energy of the resin to be applied, it has been found in most cases that the CST of the fiber had to be *increased* to exceed the surface energy of the resin. This can easily be achieved by means of a *prechlorination* procedure, mostly applied on a continuous basis. It is, however, the experience that obtaining an *even* chlorination is not an easy matter, especially when a padding system with hypochlorite and sulfuric acid is used. SAWTRI has had great success in this respect by using DCCA (di-chloro-isocyanuric acid) plus a mixture of sulfuric and acetic acids, which lowers the pH to 2.0 or below. At this pH level, padding on 1.5% active chlorine produces a very even and rapid chlorination of the *surface* of the fiber [3], thereby increasing the CST to levels well beyond 50 dyne/cm [1].

Proper spreading, however, is not the only prerequisite for proper shrink-proofing. The polymer layer must *adhere* very strongly to the fiber surface. If, after several launderings, the polymer peels off, shrink-resistance will be lost. It has recently been shown by Cook et al. [4] that by washing a polymer–shrink–resist-treated fabric in a Borate buffer, a satisfactory shrink-resistance is recorded. Once a surface-active agent is added to the buffer, the resin may come off after several washes.

Finally, it is essential that the polymer must not be brittle but flexible. In this respect, the glass transition temperature of the polymer is quite important. SAWTRI found [5], for example, that an acid colloid of Aerotex M3 (an alkylated methylolmelamine, Cyanamid) rendered superior shrinkproofing and satisfactory handle to the fabric.

Continuous shrinkproofing of tops by the well-known IWS chlorine/Hercosett process proceeds satisfactorily on a commercial scale. Recently, SAWTRI published its chlorine/aminoplast continuous shrinkproofing technique [6]. In this case, DCCA is used at a pH of 1.8–2.0 for prechlorination and the Aerotex M3 is *padded on* in colloidal form. Curing takes place at 100°C. Currently, the IWS is using SAWTRI's *prechlorination technique* on an increasing scale because of its efficiency and even treatment. Tops could now be processed at much higher speeds (14 m/min, i.e., almost double) by using this prechlorination technique instead of hypochlorite/sulfuric acid.

Shrinkproofing of fabrics is another important field for the application of resins, both on a continuous or batch basis. Recently, SAWTRI investigated the use of a mixture of a polyamide–epichlorhydrin (Hercosett, Hercules Inc.) resin and various polyacrylates [7]. Unfortunately, difficulties were experienced in obtaining good reproducibility. It was found that some of the Hercosett/polyacrylate dispersions were very unstable and frequently coagulated which, in turn, led to poor shrink-resistance. Sometimes, *overly stable* emulsions also led to poor shrink-resistance. One fact became clear, however, and that was: as the percentage of active chlorine used for prechlorination drops, there is a corresponding increase in the amount of Hercosett, which is required for maintaining a certain level of shrink-resistance. For a specific chlorination level, a certain minimum concentration of Hercosett in the Hercosett/polyacrylate dispersion is required. If the Hercosett concentration drops below this level, shrinkage increases drastically in spite of the addition of relatively large amounts of polyacrylate.

Application of *resins in a solvent* such as perchloroethylene is not new. So, for example, is the application of Synthappret LKF (Bayer) in perchloroethylene solution and, more recently, the application of silicone resins such as DC 109 (Dow Corning). The conversion of the latter type of resin into a state whereby it could be applied in an *aqueous medium* is something which various laboratories are paying attention to. Such a breakthrough has already been made with *cotton* where a silicone resin, viz., Q2-4011B (Dow Corning), can be applied in an aqueous medium [8]. It is also compatible with aminoplast resins and could, therefore, be padded on simultaneously.

Of course, current attempts are mainly concerned with the application of resins *without any pretreatment*. In this respect, the use of the bisulfite adduct of Synthappret LKF (BAS), plus a polyacrylate, in the ratio of 1:2, has given outstanding results when padded on to fabric, followed by drying [9]. This treatment has the advantage of being compatible with reactive dyes and when applied at pH 7 with 30% urea (o.m.f.) and a wetting agent, such as Tergitol TMN (Union Carbide), a satisfactory process for the *simultaneous* dyeing and shrinkproofing of wool has been found. Of course, BAS could be used alone but, in order to make the process more economical, a mixture of BAS and a polyacrylate in the ratio of 1:2 is used. In this regard, the use of a water-soluble, amphoteric derivative of Synthappret LKF, used for shrinkproofing by an exhaust method, is worth noting [10].

There are, of course, other resins or processes being developed for shrink-proofing without any pretreatment. Certain polybutadienes would appear to be satisfactory as well as Bunte salt polyethers [11].

Another aspect which has received little attention is determining the heats of adsorption and desorption of resins on wool. For this purpose, SAWTRI uses a microcalorimeter and wool, prechlorinated as well as untreated, is subjected to treatments with various polymer solutions. Very interesting results have already been obtained.

YELLOWING AND BLEACHING

The yellowing and bleaching of wool are considered as relatively old problems, and yet no major breakthrough has been achieved in this field. SAWTRI considered reductive bleaches, such as Blankit D (BASF), as one of the most efficient bleaching methods, although, as expected, this does not present subsequent yellowing in sunlight [12]. If only an *optical bleaching agent* (OBA) could be found which has a high fastness to light when applied to wool, this would be a major achievement.

The bleaching of wool in *nonaqueous media* has also received attention. Wool was successfully bleached with emulsions of hydrogen peroxide in perchloroethylene [13]. Further improvements can be effected by using the peroxide in conjunction with an optical brightening agent (OBA) on condition that the OBA is reasonably stable in aqueous emulsions of H_2O_2 [14]. Only a few commercially available OBA's qualified in this respect.

During a normal dyeing process at the boil, keratin fibers tend to yellow significantly, especially during prolonged dyeing cycles. In an investigation at SAWTRI [15], it was found that mohair, and for that matter wool also, could be dyed effectively at temperatures below the boil (85°C) using specifically acid milling dyes and 1.5–2.0% of an auxiliary such as Albegal B (Ciba-Geigy) at a pH of approximately 4.

Wool/cotton blends have become increasingly important, not only because of the comfort factor arising from the use of such blends, but also because processing could be carried out on either the worsted or cotton systems of converting such blends into yarns. *Bleaching* is normally one of the processes prior to a durable press treatment. Although research on the bleaching of wool and cotton in separate form has been dealt with extensively, very little has been done about intimate blends of wool and cotton.

SAWTRI's investigations on the bleaching of wool/cotton blends indicated that a satisfactory process for bleaching 50/50 blends involved the application of 1% hydrogen peroxide (100% active, on mass of fabric) by the pad–dwell process [16] using sodium silicate as stabilizer. This, however, is a slow process; consequently, a more rapid process, although not as effective as the previous one in terms of whiteness, involves the use of the pad–steam method with hydrogen peroxide and Stabiliser C (Laporte Chem).

DURABLE PRESS TREATMENTS OF WOOL AND WOOL/ COTTON BLENDS

It is well known by now that wool fabrics could be treated for durable press by application of Synthappret LKF (Bayer) in a perchloroethylene medium followed by steam curing. Even the water-soluble derivatives of this polyurethane resin can be applied for the same purpose should a solvent process not be acceptable.

Wool/cotton blends have received considerable attention at SAWTRI, and various resins have been applied to achieve a durable press performance. In blends comprising at least 40% cotton, it was found unnecessary to shrinkproof the wool. Aminoplast resins (crosslinking agents), or mixtures of these resins and polymer formers, were used [17].

An alkylated methylol melamine such as Aerotex M3 (Cyanamid) applied at the 15% level (o.m.f.) to a 55/45 wool/cotton blend (210 g/m^2) seemed to produce reasonably satisfactory DP ratings, although the flex abrasion resistance was lowered [18] by at least 40%. For a 70/30 cotton/wool blend, it has been found that the level of resin treatment required was largely dependent on the fabric structure and mass: lighter-weight fabrics required less aminoplast resin for similar crease recovery angles [19]. In this blend, the use of 5% Aerotex M3 (Cyanamid) plus 5% Fixapret CPN (BASF) with 1% Mystolube S (Catomance) is recommended for the heavier type of fabric (200 g/m^2) and half this amount for lighter-weight fabrics (140 g/m^2) to achieve a DP rating of about 4 after washing.

Recently, a 70/30 cotton/wool blend was treated by padding with a mixture of Fixapret CPN (BASF) and a silicone resin (Q2-4011B, Dow Corning) with very promising results [20].

WRINKLING

Although the wrinkling of wool and wool blends is largely dependent on fabric mass and structure, it is also true to state that the use of elastomeric polymers can improve the wrinkling performance of wool fabrics in a considerable manner. SAWTRI [21] recently showed that a silicone polymer (DC 109, Dow Corning), when applied at the 10% level (o.m.f.), can improve wrinkling significantly depending on whether the wool fabric had been aged or de-aged* prior to the application of the resin. Firstly, those fabrics which had been *de-aged* prior to wrinkling performed the worst. Secondly, once the fabrics had been *aged* prior to wrinkling, the difference in wrinkling performance between untreated and treated was not all that great. This infers that if *ageing* could be made more permanent, especially to wet treatments, considerable improvements could be brought about in the wrinkling propensity of a fabric.

* *Aged* by storing at 20°C and 65% RH for 3 months after finishing; *de-aged* by wetting out, hydro-extracted, steam-pressed while still damp. (More recently, a final drying operation at 45°C was also applied.)

REFERENCES

[1] E. Weideman, SAWTRI Techn. Rep. No. 178 (1972).
[2] S. Wu and K. J. Brzozowski, *J. Colloid Interface Sci., 37* (4), 686 (1971).
[3] E. C. Hanekom and F. A. Barkhuysen, *SAWTRI Bull., 8*, 19 (1974).
[4] J. R. Cook, H. D. Feldtman, and B. E. Fleischfresser, paper presented at the International Wool Textile Conference, Aachen, Sept. 1975.
[5] E. C. Hanekom, G. W. P. de Mattos, and Denise Fryer, SAWTRI Techn. Rep. No. 154 (1971).
[6] E. C. Hanekom, F. A. Barkhuysen, and N. J. J. van Rensburg, SAWTRI Techn. Rep. No. 259 (1975).
[7] N. J. J. van Rensburg, SAWTRI Techn. Rep. No. 257 (1975).
[8] N. J. J. van Rensburg, SAWTRI Techn. Rep. No. 283 (1976).
[9] H. M. Silver, N. van Heerden, and Pam Schouten, SAWTRI Techn. Rep. No. 238 (1974).
[10] J. A. Rippon; *J.S.D.C., 1975,* 406.
[11] D. M. Lewis and V. A. Bell, IWS P.D. Report No. 166 (1975).
[12] N. J. J. van Rensburg, SAWTRI Techn. Rep. No. 162 (1972).
[13] N. J. J. van Rensburg, SAWTRI Techn. Rep. No. 143 (1970).
[14] N. J. J. van Rensburg, SAWTRI Techn. Rep. No. 197 (1973).
[15] M. A. Strydom, SAWTRI Techn. Rep. No. 246 (1975).
[16] R. A. Leigh, SAWTRI Techn. Rep. No. 282 (1976).
[17] E. C. Hanekom, Miriam Shiloh, and R. I. Slinger, SAWTRI Techn. Rep. No. 173 (1972).
[18] Miriam Shiloh and E. C. Hanekom, SAWTRI Techn. Rep. No. 161 (1972).
[19] Miriam Shiloh and E. C. Hanekom, SAWTRI Techn. Rep. No. 195 (1973).
[20] N. J. J. van Rensburg, to be published.
[21] I. W. Kelly and L. Hunter, *SAWTRI Bull., 9*, 22 (1975).

SOME STRUCTURE–PROPERTY RELATIONSHIPS IN POLYMER FLAMMABILITY: STUDIES ON POLY(ETHYLENE TEREPHTHALATE)

A B. DESHPANDE and E. M. PEARCE
Polytechnic Institute of New York, Brooklyn, New York 11201

H. S. YOON
Korean Institute of Technology, Seoul, Korea

R. LIEPINS
Camille Dreyfus Laboratory, Research Triangle Institute, Research Triangle Park, North Carolina

SYNOPSIS

Attempts to correlate variations in poly(ethylene terephthalate) structure such as end-group concentration, polymer molecular weight, and diethylene glycol content led to the relationship that the log of Oxygen Index is proportional to the log of $1/[COOH]$. Incorporation of the tetrabromobisphenol-A moiety into PET via copolymerization as compared to its use as an additive indicates an advantage in Oxygen Index flammability. This would appear to be related to the opportunity for the additive diffusing more rapidly to the flame front. In the case of phosphorus containing nonreactive additives based on triphenylphosphine oxide, the opposite appears to occur. In the latter, triphenylphosphine oxide appears to operate in the vapor phase, whereas copolymer or reactive additive moieties operate in the solid phase, thus accounting for these differences.

INTRODUCTION

Several studies have reviewed the thermal and oxidative degradation of poly(ethylene terephthalate) (PET) [1–3]. No attempts have been made to relate structural features of the polymers and their subsequent effects on degradation reactions to polymer flammability measurements for PET, although previous studies of this nature have been reported by us for nylon 6 [4].

Flame retardation of PET has been recently reviewed extensively by Lawton and Setzer [5]. The generalized effects obtained from halogen- and phosphorus-containing flame retardants are discussed in this review and related to oxygen index [6] flammability measurements. In addition, our previous studies have reported on the effect of flame retardants in PET on some other fiber properties such as color and transparency [7].

Journal of Applied Polymer Science: Applied Polymer Symposium 31, 257–268 (1977)

Our present studies have attempted to provide insights into two primary questions: (1) What are the effects of various PET structural parameters, such as end-group concentration, diethylene glycol concentration, and initial polymer molecular weight, on the thermal degradation and flammability as measured by the oxygen index? (2) What differences, if any, occur in flame retardation effectiveness and mechanism for bromine- or phosphorus-containing flame retardants when these structures are incorporated into the polymer additives or are copolymerized into the polymer backbone?

EXPERIMENTAL

Thermooxidative Degradation Studies

A Mettler Thermoanalyzer TA-1, as well as DuPont 950 or 900, were employed for the thermooxidative degradation studies. The samples were heated in the presence of circulating air and weight loss was measured as a function of sample temperature.

Determination of Transition Points

Simultaneous differential thermal analysis and thermogravimetric analysis were carried out during some of the thermooxidative degradation on the Mettler Thermoanalyzer TA-1. Transition points of selected samples were also confirmed with the Mettler TA-2000 and the DuPont 950 Differential Scanning Calorimeter.

Oxygen Index Flammability Measurements

Oxygen index measurements were performed on the Oxygen Index Flammability Gauge (General Electric Co. Model No. 280 FM11B). Polyester powder samples were supported in an inverted 2-cm diameter porcelain crucible lid as the sample holder. The crucible lid was supported on the metal bar inside the Pyrex glass chimney. This test significantly differs from other reported Oxygen Index experiments on PET [5]. In this procedure, since the sample was not held in a vertical position, thermoplastic melt-flow viscosity characteristics as a significant factor in influencing Oxygen Index values were eliminated [4].

The sample of polyester sample was ignited in the presence of an appropriate oxygen/nitrogen mixture. The flow rate was in the range of 200–300 ml/sec. The Oxygen Index was determined at that point when the nitrogen gas pressure was increased or oxygen gas pressure was decreased and the flame was essentially extinguished. An average Oxygen Index was obtained from three to five observations.

Materials

A series of specially prepared polyesters were obtained from American Enka. These samples varied in molecular weight, carboxyl end-group concentration, and diethylene glycol concentration (Table I).

The bromine-containing compounds used were tetrabromobisphenol A and the bishydroxyethyl ether of this compound, obtained from Great Lakes Chemicals.

Bis(2-hydroxyethyl)terephthalate was obtained from American Enka and was recrystallized from water and dried under vacuum at 50°C; mp 108–109°C [10].

The phosphorus-containing monomers were synthesized as follows.

Synthesis of Phenylphosphonyldi(p-methyloxybenzoate)

$$H_3COOC - \underset{}{\bigcirc} - O - \overset{\overset{O}{\parallel}}{\underset{\underset{C_6H_5}{|}}{P}} - O - \underset{}{\bigcirc} - COOCH_3$$

The preparation of this monomer was based on the reaction of phenylphosphonyldichloride and p-methylhydroxybenzoate [8], but the detailed procedure was not available. An ethereal solution of phenylphosphonyl dichloride was added dropwise to an ethereal solution of p-methylhydroxybenzoate containing

TABLE I
Poly(ethylene Terephthalate) Samples[a]

Polymer Number	Intrinsic Viscosity dl/g	COOH Meq/kg	DEG %	COOCH₃ Meq/kg	Mn ppm	Sb ppm
Z-422-1P	.47	8	1.00	<3	86.5	274
Z-423-1P	.64	11	0.80	<3	82.1	263
Z-424-1P	.54	12	2.06	<3	82.1	250
Z-425-2P	.74	16	0.94	<3	58.9	296
T-443-1P	.57	42	1.63	<3	73.0	307
T-444-1P	.63	12	1.63	<3	80.0	278
T-451-1P	.64	28	1.75	<3	77.7	258
T-453-1P	.56	30	1.75	<3	87.4	278
T-461-1P	.71	16	1.96	<3	82.0	254
T-461-1P-D	.61	70	1.96	<3	82.0	254
T-461-2P	.64	23	0.93	<3	77.8	268

[a] Samples were prepared and analyzed by Dr. A. Meierhofer of the American Enka Corp.

pyridine at 25°C. The white product and pyridine hydrochloride precipitated together. Pyridine hydrochloride was removed by dissolving in water. A 50% yield was obtained after drying under vacuum at 50°C. The recrystallized product from a benzene–heptane mixture had a melting point of 145°C. The structure of the monomer was confirmed by infrared and NMR spectra.

Synthesis of Phenylphosphonyldi(p-methyl Benzoate)

$$H_3COOC-\underset{}{\bigcirc}-\overset{\overset{O}{\underset{}{\parallel}}}{\underset{\underset{C_6H_5}{|}}{P}}-\bigcirc-COOCH_3$$

This monomer was prepared by following the procedure of Morgan and Herr [9]. The product was obtained by a three-step procedure. p-Toluene magnesium bromide was reacted with phenylphosphonyldichloride and phenylphosphonylditoluene was obtained. This was oxidized with potassium permanganate in the presence of pyridine and the acid so obtained was further esterified. The recrystallized white product from benzene–heptane mixture had a melting point of 162°C (lit. [7] 165°C).

Copolymerization

Bis(2-hydroxyethyl)terephthalate and the phosphorus-containing monomers were copolymerized by standard procedures for the bulk polymerization of polyesters [11]. Elimination of methanol occurred during the copolymerization of the appropriate combination of monomers. The bromine-containing polyesters were previously prepared and characterized by Yoon [12].

Viscosity

A polymer solution of 0.1% concentration was prepared in 30:50 phenol–tetrachloroethane mixture [13]. The viscosities were measured at 30°C using a Ubbelohde viscometer. The intrinsic viscosity was calculated by using standard equations.

Melt-Blending of Poly(ethylene Terephthalate)

The PET sample and an additive were dry-mixed, melted and stirred at 270°C under nitrogen. As soon as the sample completely melted (~5 min), the heat source was removed and the melt-blended PET was cooled rapidly to 25°C for further evaluation.

TABLE II
Thermogravimetric Weight Loss and Melting Point of Polyesters

Polymer Number	Weight Loss (at 500° C)[a]	Melting Point[c] °C
Z-422-1P	82[b]	270
Z-423-1P	80[b]	260
Z-424-1P	76[b]	270
Z-425-2P	93	255
T-443-1P	90	250–270
T-444-1P	89	--
T-451-1P	90	252
T-453-1P	91	255
T-461-1P	89	260
T-461-1P-D	89	252
T-461-2P	90	249

[a] Run on DuPont 950 TGA in a boat-shaped crucible.
[b] Run on Mettler Thermoanalyzer TA-1 in "deep" dish crucible.
[c] DTA on Mettler Thermoanalyzer TA 2000 at a heating rate of 15°C/min.

RESULTS AND DISCUSSION

Relationship of Polyester Structural Parameters with Thermal Degradation and Flammability

The polyester samples used in this study were carefully characterized as to viscosity, end groups, ethylene glycol content, and residual catalysts (Table I).

TGA analysis of the polyester samples studied on Mettler Thermoanalyzer TA-1 No. 13 using a deep dish-shaped aluminum crucible showed weight losses of 60–80% at 500°C (Table II). In order to verify the results, one sample was analyzed in a nitrogen atmosphere and the rate of heating was changed from 6°C/min to 15°C/min. However, there was no change in the observed weight loss at 500°C. Subsequently, the other samples were analyzed on the DuPont 950 thermal analyzer using an open boat-shaped crucible under the same conditions. The weight losses were considerably higher and generally around 90%, which is the expected range for the polyester samples [5, 7].

The anomalous results using the deep-dish crucible suggested that char formation initially occurred on the surface of the polyester under these conditions. In any case, diffusion-controlled processes are important and they relate either to the char formed or the particle geometry of the sample holder. The effect of diffusion rate can also be seen from the TGA measurements at different heating rates: 6°C/min to 50°C/min (Fig. 1).

As a measure of flammability, the series of polyester samples was subjected

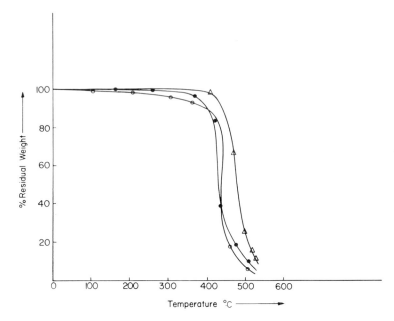

FIG. 1. Relationship of residual weight at 500°C with the heating rate of polyester sample T-443-1P: (⊙) 20°C/min; (●) 6°C/min; (△) 50°C/min; air atmosphere.

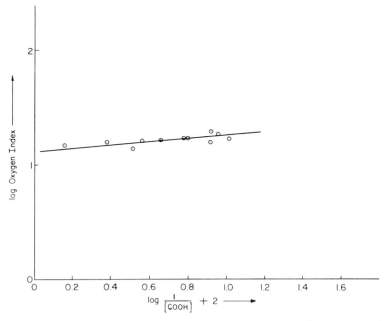

FIG. 2. Relationship of Oxygen Index with the carboxy group content of polyester samples.

to an Oxygen Index (OI) evaluation in powder form. The OI values obtained were low and in the range of 16–21 (Table III). Correlations of OI with [COOH], DEG content, intrinsic viscosity, metal content, and weight loss were attempted. At this point, only [COOH] was found to have some relationship to OI. The log OI was observed to be linearly proportional to the reciprocal of log [COOH] (Fig. 2). A log–log plot was necessary to remove scatter in the data. This discrepancy on [COOH] would be consistent with carboxyl-catalyzed degradation reactions such as acid-catalyzed ester hydrolysis reactions which lead to low molecular weight fragments. It has been noted previously that hydrolysis occurs much more rapidly than ester pyrolysis, and can be autocatalytic because of increasing [COOH] content [14]. The concentration of carboxyl groups tends to increase during the process of poly(ethylene terephthalate) thermolysis [1]. Thus, the effect of [COOH] on flammability should become increasingly important as the degradation continues. The rate of ester pyrolysis should be relatively unaffected by the structural variations under study.

Bromine-Containing PET

A series of polyester copolymers and homopolymer mixtures were prepared and OI's were determined. These polymer systems contained brominated flame retardants at a level sufficient to give incorporation of 4–8% bromine on a molar basis. The PET copolymers contained the bishydroxyethyl ether of tetrabromobisphenol A, whereas the homopolymer mixtures contained PET and tetrabromobisphenol-A as an additive. When these samples were analyzed thermogravimetrically up to 500°C in an air atmosphere (15°C/min heating rate), the weight loss at 500°C of the copolymers was usually found to decrease as the bromine content increased. These results are shown in Table IV. Whether the decreased weight loss at 500°C was really related to bromine content or was a contribution due to the aromatic nature of the bisphenol-A structure cannot be decided at present.

These samples were also evaluated in terms of Oxygen Index (OI). The OI generally increased with increasing bromine content. The values of the OI as measured as powdered samples in a dish were higher for the additive + homopolymer mixture than for the copolymers (Table IV).

Table IV also shows results for the OI measured vertically and indicates the significant effect that the sample test specimen geometry has on the measured results. However, these results also indicate a small but consistent higher OI for the additive mixtures as compared to the copolymers.

The results suggest that the additives in the mixture are capable of diffusing more rapidly than the analogous structures in the copolymers to the flame front, thereby accounting for these differences.

Phosphorus-Containing PET

The 1:1 molar copolymer was prepared by the reaction of bis(2-hydroxyethyl)terephthalate and phenylphosphonyldi(p-methyloxybenzoate) by heating at 160°C for 8 hr at 0.1 mm pressure. It contained 4.5 mole % phosphorus (theory for 1:1 copolymer, 4.8%). The copolymer contained 4.5 mole % phosphorus (calcd. 4.8%), was pale brown, and had a 0.78 inherent viscosity at 30°C.

TABLE III
Oxygen Index of Poly(ethylene Terephthalate)

Polymer Number	Oxygen Index, %			
	I	II	III	Average
Z-422-1P	18.1	17.9	18.4	18.1
Z-423-1P	18.4	20.2	20.2	19.6
Z-424-1P	22.42	21.3	21.4	21.7
Z-425-2P	19.8	18.1	18.1	18.6
Z-443-1P	16.6	17.2	16.6	16.8
Z-444-1P	16.6	16.6	17.2	16.8
T-451-1P	16.4	16.6	16.6	16.5
T-453-1P	16.6	16.6	17.3	16.4
T-461-1P	16.6	16.6	17.2	16.8
T-461-1P-D	16.4	16.3	16.3	16.3
T-461-2P	16.6	17.3	17.3	17.0

TABLE IV
Thermogravimetric Weight-Loss Analysis and Oxygen Index of Copolymers and Homopolymer Mixtures

Sample No.	Sample Type	Bromine Content %	Weight loss[a] % (500°C)	OI[b] %	OI[c] %
C-4	Copolymer	4.03	81.3	15.7	27.2
C-6	Copolymer	6.28	79.1	17.3	29.5
C-8	Copolymer	7.62	73.7	18.8	31.0
M-4	Mixture	4.0	79.1	17.2	28.2
M-6	Mixture	6.0	74.0	22.2	30.5
M-8	Mixture	8.0	77.2	23.9	31.8
PET	--	None	90.0	15.6	20.0

[a] Heating rate, 15°C/min.
[b] In dish.
[c] Vertical.

This sample could be spun manually and had a T_g of 42°C and a melting point of 191°C. The sample was evaluated by OI and weight loss was evaluated by TGA. The OI value was about 30.8 and the weight loss was around 84% at 500°C (Fig. 3).

The 1:1 molar copolymer obtained by the reaction of bis(2-hydroxyethyl)-terephthalate and phenylphosphonyldi(p-methyl benzoate) was prepared in a

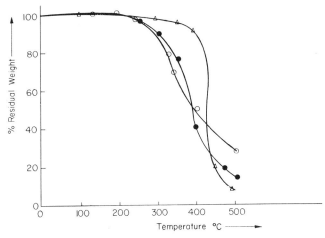

FIG. 3. Thermogravimetric analysis of copolymers obtained from bis(2-hydroxyethyl)terephthalate and (●) phenylphosphonyldi(p-methyloxybenzoate) and (⊙) phenylphosphonyldi(p-methyl benzoate) and comparison with (△) PET. Scanning rate, 15°C/min; air atmosphere.

similar fashion heating for 8 hr at 200°C under 0.1 mm pressure. The sample contained about 5 mole % of phosphorus, was dark brown, and had a $[\eta]_{inh}$ of 1.3. It could be spun manually. The polymer had a T_g of 42°C and a melting point of 172°C. The OI value was 29.0 and the weight loss by TGA was 72% (Fig. 3).

These copolymer samples were compared to two additive systems. The phenylphosphonyldi(p-methyl benzoate) monomer and triphenylphosphine oxide, respectively, were added to PET as additives. The concentration of phosphorus was 5 mole % in order to compare the flammability effects directly with the phosphorus-containing copolymers. Triphenylphosphine oxide had previously been shown to act in the vapor phase while reducing the flammability of PET [15].

The monomer and triphenylphosphine oxide (TPPO) samples were first studied alone by TGA to understand their potential behavior as additives. Figure 4 indicates that triphenylphosphine oxide essentially is degraded and volatilized 100% at 500°C. Some oxidation of TPPO occurs and is indicated by an initial early small gain in weight, whereas phenylphosphonyldi(p-methyl benzoate) degraded from the beginning and showed a weight loss of about 90% at 500°C.

PET samples were melt blended with phenylphosphonyldi(p-methyl benzoate) monomer and TPPO, respectively, and then were studied by TGA and the Oxygen Index. As seen in Figure 5, PET melt-blended with phenylphosphonyldi(p-methyl benzoate) degraded above 220°C and decomposed above 290°C. The weight loss was 75% at 500°C and gave an OI of 29. Ester-interchange reactions are rapid and thus could account for the very similar results with the copolymer. PET melt-blended with TPPO showed an early weight loss, with a final weight loss of 96% at 500°C as compared to PET itself which has a weight loss of 90% at 500°C. The OI was 24.5, which is higher than the 17.0 value for

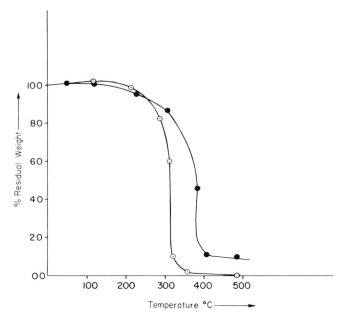

FIG. 4. Thermogravimetric analysis of (⊙) triphenylphosphine oxide and (●) phenylphosphon-yldi(*p*-methyl benzoate). Scanning rate, 15°C/min; air atmosphere.

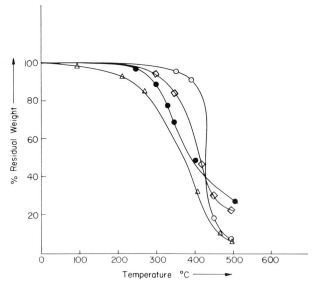

FIG. 5. Thermogravimetric analysis of (⊙) PET, (●) PET copolymer with phenylphosphonyl-di(*p*-methyl benzoate), (△) PET melt-blended with triphenylphosphine oxide, and (◊) PET melt-blended with phenylphosphonyldi(*p*-methyl benzoate). Scanning rate, 15°C/min; atmosphere, air (5.72 l./hr).

the unmodified polyester sample. These results suggest that TPPO apparently acts in the vapor phase [15] while phenylphosphonyldi(*p*-methyl benzoate)

TABLE V
Phosphorus-Containing Additives and/or Comonomers in PET ($\sim$5 mole % P)

Structure	Type	%Weight loss (500°C)	OI
PET	homopolymer	90	17.0
PET + H_3COOC—◯—O–P(=O)(φ)–O—◯—$COOCH_3$	copolymer	84	30.8
PET + H_3COOC—◯—P(=O)(φ)—◯—$COOCH_3$	copolymer	72	29
	additive	75	29
PET+ $\phi_3P=O$	additive	96	24.5

appears to operate primarily in the condensed phase. A simple powder blend of PET and the phenylphosphonyldi(p-methyl benzoate) also had an OI of 26.5.

A summary of the weight loss and Oxygen Index data is shown in Table V. The contrast between the vapor-phase action of triphenylphosphine oxide and the condensed-phase action of phenylphosphonyldi(p-methyloxybenzoate) and phenylphosphonyldi(p-methyl benzoate) was notable. It would also appear that a nonreactive phosphorus additive (TPPO) which operates in the vapor phase was a less efficient flame retardant for PET than those which are reactive or can be copolymerized and also act in the condensed phase.

The authors acknowledge the support of this research under a contract from the National Bureau of Standards ETIP Contract No. 4-35963 to Clemson University. They are grateful to Dr. R. H. Barker for his encouragement and discussions; Dr. A. Meierhofer and American Enka Corp. for the polyester samples; the Mettler and DuPont companies for occasional use of their thermoanalytical equipment; the General Electric Co. for the Oxygen Index equipment; Stauffer Chemical Corp. and Dr. E. Weil for samples of phenylphosphonyldichloride; and Mr. K. Nangrani and J. Shah for their laboratory assistance.

REFERENCES

[1] L. H. Buxbaum, *Angew. Chem. Int. Ed.*, 7(3), 182 (1968).
[2] P. D. Richie, *Soc. Chem. Ind.* (London) Monograph No. *13*, 107 (1961).
[3] I. Marshall and A. Todd, *Trans. Faraday Soc.*, *49*, 67 (1953).
[4] H. K. Reimschuessel, S. W. Shalaby, and E. M. Pearce, *J. Fire Flamm.*, *4*, 299 (1973).
[5] E. L. Lawton and C. J. Setzer, in *Flame Retardant Polymeric Materials*, M. Lewin, S. M. Atlas, and E. M. Pearce, Eds., Plenum Press, New York, 1975, p. 193.

[6] C. P. Fenimore, in *Flame Retardant Polymeric Materials,* M. Lewin, S. M. Atlas, and E. M. Pearce, Eds., Plenum Press, New York, 1975, p. 371.

[7] P. J. Koch, E. M. Pearce, J. A. Lapham, and S. Shalaby, *J. Appl. Polym. Sci., 19,* 227 (1975).

[8] Stauffer Chemical Co., *Phenylphosphorous Dichloride and Derivatives,* 1972, p. 11.

[9] P. W. Morgan and B. C. Herr, *J. Amer. Chem. Soc., 74,* 4526 (1952).

[10] W. R. Sorenson and T. W. Campbell, *Preparative Methods of Polymer Chemistry,* Interscience, New York, 1968, p. 152.

[11] F. B. Cramer, *Macromol. Synth., 1,* 17 (1963).

[12] H. S. Yoon, unpublished results.

[13] M. L. Wallach, *Makromol. Chem., 103,* 19 (1967).

[14] L. A. Wall, *Soc. Chem. Ind.* (London) Monograph No. *13,* 146 (1961).

[15] R. H. Barker, National Bureau of Standards Special Publication 411, *Fire Safety Research* (Proceedings of a Symposium held at NBS, Gaithersburg, Md., Aug. 22, 1973), p. 37.

FLAME RESISTANCE OF CHEMICAL FIBERS

D. W. VAN KREVELEN

Akzo Research & Engineering b.v., Arnhem, The Netherlands

SYNOPSIS

Flammability and flame stability are relative rather than absolute concepts. The principles underlying these phenomena are discussed on the basis of stability diagrams and combustion mechanisms. There is a distinct correlation between the oxygen index as a measure of the flammability and the elementary composition of the polymer. Another correlation is that between the oxygen index and the pyrolysis residue of a polymer. Since pyrolysis is an essential step in the combustion process, this correlation is not unexpected.

From these relationships, conclusions can be drawn with respect to the flame-resistant polymer structures. The influence of the concentration of flame-retardant additives on the oxygen index can be approximated by a linear function with two constants, characteristic of the polymer flame-retardant combination and the nature of the flame-retardant element, respectively. A short survey is given of the flame-retardant modifications of established fibers and of new polymers with intrinsic flame resistance.

INTRODUCTION

The flammability hazard of textile materials is an old problem. The first patent for flameproofing cotton, with a mixture of alum, ferrous sulfate, and borax, was issued in England in 1735. The famous French chemist Gay Lussac has been active in the field and recommended (in 1821) a mixture of ammonium phosphate, ammonium chloride, and borax as a flame-retardant for linen and jute.

Yet the problem of flammability has never been of such interest as it is today since the introduction (or imminent introduction) of stringent legislation.

People have always used flammable materials for clothing and also for building and auxiliary materials. The problem at hand has become acute because of the present-day phenomenon of the high concentration of a great number of people in confined spaces of considerable height (office buildings, department stores, hospitals) and because of the special fire hazards in big aircraft (jumbo jets).

In this paper, we shall discuss the relationship between flammability and chemical structures of fiber polymers.

Journal of Applied Polymer Science: Applied Polymer Symposium 31, 269–292 (1977)
© 1977 by John Wiley & Sons, Inc.

BASIC CONCEPTS

Flammability, Ignition, Combustion, and Extinguishment

Flammability is a relative rather than an absolute concept. There are no absolutely flameproof materials; there are only widely different degrees of fire resistance and flammability. Flammability is a property of a *system* comprising the basic elements: combustible material, air, and heat.

A fire does not originate without heat. After ignition, the flame itself produces the heat required for the constant production of combustible gases.

Frank-Kamenetzky [1] introduced a useful diagram which shows the influence of the nature of a reactive (flammable) material and its environment. It is a graph in which heat generation and heat loss are plotted as a function of temperature (see Fig. 1A).

The generation of heat depends on the rate of combustion and, like any chemical reaction, increases exponentially with temperature until a maximum is reached which corresponds to complete conversion. The loss of heat is roughly proportional to the difference in temperature between the combustion zone and the environment and can therefore be represented by an approximately straight line.

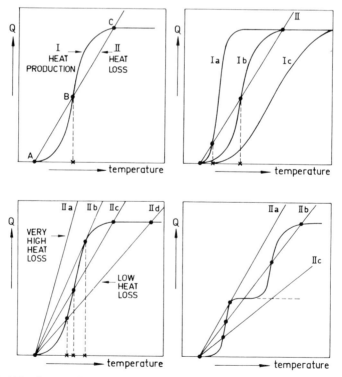

FIG. 1. Stability diagram of combustion: (A) flame stability diagram; (B) three degrees of flammability; (C) four different surroundings; (D) two-stage combustion.

Equilibrium between heat generation and heat loss is realized at the points of intersection of I and II, so in points A, B and C, A (ambient temperature) and C (stationary temperature) are stable and B is unstable. To the left of B, the heat loss exceeds the heat generation; to the right of B this is just the reverse. Therefore the temperature corresponding to B is the ignition temperature. During a fire, the material must be heated to such an extent by heat fed back from the flame that at least this ignition temperature is reached. The temperature at B is also the temperature of self-extinguishment; at lower temperatures, heat loss exceeds heat generation.

Figure 1B shows three materials with different degrees of flammability but in the same environment. The first is highly flammable, the second moderately flammable, while the third is flameproof under these conditions.

Figure 1C represents a material in different environments. An increased heat loss may be caused by a higher rate of flow of the air, less insulation, etc. It is evident that a flammable material may be barely flammable or even nonflammable under different conditions. It depends on whether there *is* a point B or not. Sometimes the combustion (or the preceding decomposition) takes place in two steps. This situation is shown in Figure 1D, again with new possibilities.

Oxygen Index as a Criterion of Flammability and Extinguishment

In 1966, Fenimore and Martin [2] introduced the so-called Oxygen Index (OI) as a measure of the flammability of a material, i.e., its tendency to continue burning after ignition. This OI is the numerical value of the quantity

$$n = \phi(O_2)/[\phi(O_2) + \phi(N_2)] \qquad (1)$$

TABLE 1
Oxygen Indices of Polymers

POLYMER	OI	POLYMER	OI
POLYFORMALDEHYDE	0.15	WOOL	0.25
POLYETHYLENE OXIDE	0.15	POLYCARBONATE	0.27
POLYMETHYL METHACRYLATE	0.17	NOMEX®	0.285
POLYACROLONITRILE	0.18	PPO®	0.29
POLYETHYLENE	0.18	POLYSULPHONE	0.30
POLYPROPYLENE	0.18	PHENOL-FORMALDEHYDE RESIN	0.35
POLYISOPRENE	0.185	NEOPRENE®	0.40
POLYBUTADIENE	0.185	POLYBENZIMIDAZOLE	0.415
POLYSTYRENE	0.185	POLYVINYLCHLORIDE	0.42
CELLULOSE	0.19	POLYVINYLIDENE FLUORIDE	0.44
POLYETHYLENE TEREPHTHALATE	0.21	POLYVINYLIDENE CHLORIDE	0.60
POLYVINYLALCOHOL	0.22	CARBON	0.60
NYLON 66	0.23	POLYTETRA FLUOROETHYLENE	0.95
PENTON®	0.23	(TEFLON®)	

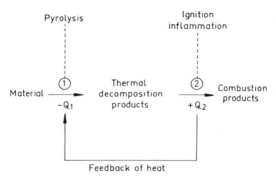

FIG. 2. Consecutive reactions during burning.

i.e., the minimum fraction ϕ of oxygen in an N_2–O_2 mixture that is just sufficient to maintain combustion (after ignition).

The test is made under standardized conditions. Table I gives OI values of a series of polymers. A material is to be considered flammable when the OI value is lower than 0.26. Investigations have shown that the OI value is dependent on the temperature, the weight and the construction of the specimen, the moisture content, etc., but when carried out under standard conditions the test is nevertheless a precision test which is exactly reproducible and can be applied to a wide variety of materials (plastics, films, textiles, foams, rubbers, etc.). It may even be said that, at the moment, the OI test is the most valuable criterion in flammability research, although it cannot replace all the existing criteria. Phenomena like afterglow and dripping are not assessed by means of the OI.

The main advantages of the OI test are that numerical data are obtained and the the the OI value usually shows a linear relationship with the concentration of flame retardants added.

Basic Scheme of the Combustion of Solid Substances

Figure 2 gives a broad outline of what happens during the combustion of a solid [3]. Actually there are two successive chemical processes, viz., decomposition and combustion, which are linked together via thermal feedback. An ignition mechanism is essential.

Primarily the material decomposes (pyrolysis), which requires heat. The decomposition products are combusted (after ignition), which generates heat. This heat is again (partly) utilized for the pyrolysis. The process is heterogeneous, so that the contact area plays an important role.

From this basic mechanism, some interesting conclusions can be drawn. First of all, the thermal effects $-Q_1$ and $+Q_2$ and the contact area (A) are of special importance. A high Q_1 value, a low Q_2 value, and a low A value favor the flame resistance.

Furthermore, it has long been known that the hydrogen content of the material has a great influence on the resistance to decomposition and on flammability. The lower the hydrogen content, the more flame-resistant the material. On the

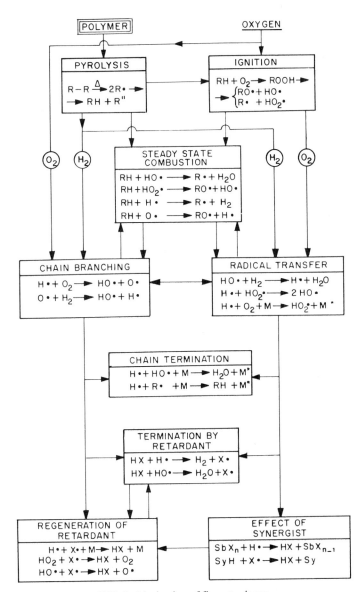

FIG. 3. Mechanism of fire retardancy.

one hand the pyrolysis residue becomes larger, while on the other the amount of combustible gases, and hence Q_2, becomes smaller.

The combustion of the gases generated by pyrolysis is a chain reaction.

Molecular Model of Combustion

A molecular model of the complicated process of combustion is given in Figure 3. Again, the model shows the essential processes of pyrolysis and ignition from which the combustion emerges. Steady-state combustion is possible due to a dynamic equilibrium of chain branching and chain termination.

The role of flame retardants may be twofold: they can influence the pyrolysis (viz., by diminishing the formation of "fuel") or influence the chain reaction (viz., by increasing the chain termination). In the first case, their action is mainly confined to the solid phase; in the second, it takes place in the gas phase.

Sometimes the action of flame retardants is enhanced by so-called synergists. Their main role seems to be the regeneration of the active retardant species.

Conclusions from the Foregoing

From the foregoing, the following conclusions may be drawn:

(1) Flammability is a property of a system. Not only the combustible material but also the environmental conditions are decisive.

(2) Under specific conditions, every combustible substance has a very definite tendency to ignite and extinguish.

(3) The oxygen index is a suitable criterion for describing this tendency.

(4) The tendency towards ignition (or extinguishment) can be influenced by certain additives, the flame retardants. They can act either in the solid phase (influencing the pyrolysis and hence diminishing the generation of fuel) or in the gas phase (influencing the chain reactions by increasing the chain termination).

The resistance to inflammation of a material may therefore be described by the following expression:

$$OI = (OI)_0 + f(FR) \tag{2}$$

where $(OI)_0$ is the oxygen index of the pure material, $f(FR)$ is the functional influence of the flame retardant, and OI is the oxygen index of the composition (material + flame retardant).

We shall now first discuss the factors that influence $(OI)_0$ and then the factors determining $f(FR)$.

FACTORS INFLUENCING THE INTRINSIC FLAMMABILITY $[(OI)_0]$

Flammability of Pure Polymers

In Table I, the oxygen indices of the most important "pure" polymers are given. The question arises whether there exists a direct relationship between oxygen index and polymer composition. To a certain extent this is indeed so. Since H and OH radicals play an essential role in the combustion mechanism, it is only natural to expect a correlation between the hydrogen content of a substance (the source of H and OH) and its oxygen index. This correlation is considerably influenced, however, when the polymer contains halogen, since halogen causes chain termination in the gas phase on decomposition.

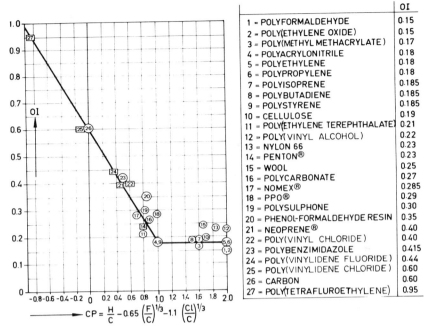

	OI
1 = POLYFORMALDEHYDE	0.15
2 = POLY(ETHYLENE OXIDE)	0.15
3 = POLY(METHYL METHACRYLATE)	0.17
4 = POLYACRYLONITRILE	0.18
5 = POLYETHYLENE	0.18
6 = POLYPROPYLENE	0.18
7 = POLYISOPRENE	0.185
8 = POLYBUTADIENE	0.185
9 = POLYSTYRENE	0.185
10 = CELLULOSE	0.19
11 = POLY(ETHYLENE TEREPHTHALATE)	0.21
12 = POLY(VINYL ALCOHOL)	0.22
13 = NYLON 66	0.23
14 = PENTON®	0.23
15 = WOOL	0.25
16 = POLYCARBONATE	0.27
17 = NOMEX®	0.285
18 = PPO®	0.29
19 = POLYSULFONE	0.30
20 = PHENOL-FORMALDEHYDE RESIN	0.35
21 = NEOPRENE®	0.40
22 = POLY(VINYL CHLORIDE)	0.40
23 = POLYBENZIMIDAZOLE	0.415
24 = POLY(VINYLIDENE FLUORIDE)	0.44
25 = POLY(VINYLIDENE CHLORIDE)	0.60
26 = CARBON	0.60
27 = POLY(TETRAFLUROETHYLENE)	0.95

$$CP = \frac{H}{C} - 0.65\left(\frac{F}{C}\right)^{1/3} - 1.1\left(\frac{Cl}{C}\right)^{1/3}$$

FIG. 4. Correlation between oxygen index and elementary composition: (O) halogen-free polymers; (□) halogen-containing polymers.

We have been able [4] to find a parameter, the composition parameter (CP), which describes the combined influence of hydrogen and halogen content:

$$CP = (H/C) - 0.65(F/C)^{1/3} - 1.1(Cl/C)^{1/3} - x(Br/C)^{1/3} \qquad (3)$$

where (H/C), etc., are the atomic ratios of the respective elements in the polymer composition. The coefficient x is still uncertain due to lack of sufficient data, but probably is of the order of 1.6.

Figure 4 shows the relationship between CP and OI. Although the data scatter considerably, there is an obvious correlation. It can be described by the formula:

$$\begin{aligned} OI &\approx 0.175 && \text{if } CP \geqslant 1 \\ OI &\approx 0.60 - 0.425\, CP && \text{if } CP \leqslant 1 \end{aligned} \qquad (4)$$

For polymers not containing halogen the parameter CP results in the H/C, which is a rough measure of the "aromaticity" of a substance. All aromatic polymers have a CP value <1. This already indicates that polymers with intrinsic flame resistance have to be aromatic in character.

Flammability and Pyrolysis

We have already stressed that pyrolysis is the essential first step in the combustion of a polymer. In fires, the pyrolysis is attended with oxidative degradation which accelerates the pyrolytic decomposition. For obtaining scientifically in-

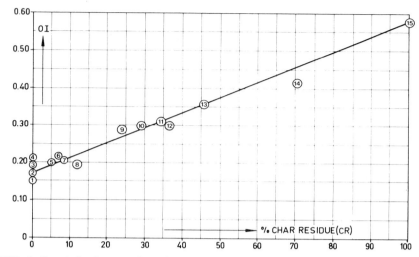

FIG. 5. Correlation between OI and char residue: (1) polyformaldehyde, (2) polyethylene + polypropylene, (3) polystyrene + polyisoprene, (4) nylon, (5) cellulose, (6) poly(vinyl alcohol), (7) PETP, (8) polyacrylonitrile, (9) PPO, (10) polycarbonate, (11) Nomex, (12) polysulfone, (13) Kynol, (14) polyimide, (15) carbon.

terpretable and reproducible results, it is necessary to study the pyrolysis in oxygen-free atmospheres, e.g., in nitrogen or argon.

A significant correlation [5] has been found between the char residue and the oxygen index of polymers, as is shown in Figure 5. This linear relationship can be represented by the formula:

$$OI \times 100 = 17.5 + 0.4CR \qquad (5)$$

where CR is the char residue as a percentage by weight at 850°C.

In view of this interesting correlation, it seemed useful to conduct an elaborate investigation into the relation between residue of pyrolysis and the constitution of polymers. This study has recently been published [6, 7], so that we will confine ourselves to the conclusions.

The results of pyrolysis experiments with nearly 100 polymers made it possible to quantify a *char-forming tendency* (*CFT*). This quantity is defined as the amount of residue (at 850°C) per structural unit divided by 12, i.e., the amount of C-equivalent per structural polymer unit. It was found that, in principle, each functional group in the polymer contributes in its own characteristic way to the final pyrolysis residue, i.e., each functional group has its own char-forming tendency. Aliphatic groups as such have no CFT, and linked to aromatic systems they even have a negative CFT because they provide hydrogen for the dispro-portionation to gaseous fuels. Aromatic groups have a more or less strong CFT.

Table II gives the CFT values for the various structural groups. It should be emphasized that the char-forming tendency is a *statistical concept*. The fact that a phenyl group with 6 C-atoms has a CFT value of 1 C-equivalent means

TABLE II
Group Contributions to Char Formation

GROUP	CFT in C-equiv.	GROUP	CFT in C-equiv.
ALIPHATIC GROUPS		**HETEROCYCLIC GROUPS**	
—CHOH—	$1/3$	(N—N / —C—O—C— ring)	1
ALL OTHER *)	0	(—C=C—, N—NH, CH ring)	$3\frac{1}{2}$
AROMATIC GROUPS		(HC—S, C—N ring)	$3\frac{1}{2}$
(phenyl)	1	(HC=CH, C—N, N— ring)	$3\frac{1}{2}$
(ortho-disubst. phenyl)	2	(HN, C—N benzo ring)	7
(meta-disubst. phenyl)	3	(—C, O, N benzo ring)	7
(para-disubst. phenyl)	4	(—C=N, HC=N benzo ring)	9
(trisubst. phenyl)	6	(O=C—N, N—C=O ring)	11
(biphenyl)	6	(H N, N H double benzo ring)	10
(two para-linked phenyls)	10	(—N—C(=O)...C(=O)—N— imide)	12
(anthracene-type)	14	(—N=C, N=C ... C=N, C=N— ring)	10
—O—C(=O)—⟨phenyl⟩—C(=O)—O—	$1\frac{1}{4}$	(—C=N—⟨⟩—N=C—, N, N ring)	15
CORRECTIONS DUE TO DISPROPORTIONING (H-SHIFT): GROUPS DIRECTLY CONNECTED TO AROMATIC NUCLEUS			
$>CH_2$ and $>CH-CH_2-$	−1		
$-CH_3$	$-1\frac{1}{2}$		
$>C(CH_3)_2$	−3		
$-CH(CH_3)_2$	−4		

* No halogen groups included.
Note: System is not valid for halogen-containing polymers.

that, on the average, only one out of six phenyl groups in the polymer goes into the residue and five out of six go into tar and gas. If the aromatic ring contains four side groups (i.e., if 4 H-atoms are substituted), all the rings land in the residue; the CFT value is 6, etc.

By means of these group contributions, the amount of pyrolysis residue (CR) of a polymer can be estimated very rapidly, viz., with eq. (6):

$$CR = \frac{\sum_{i} (CFT)_i \times 12}{M_{unit}} \times 100 \tag{6}$$

Here CR is the pyrolysis residue expressed as a percentage. Figure 6 compares the calculated values with the measured values. By means of Figure 5 and with

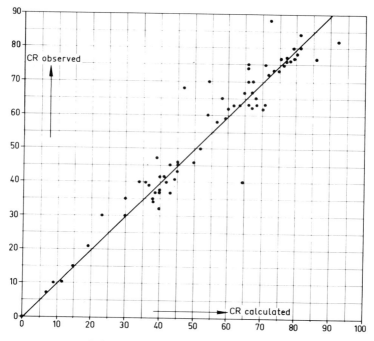

FIG. 6. Calculated vs. observed CR values.

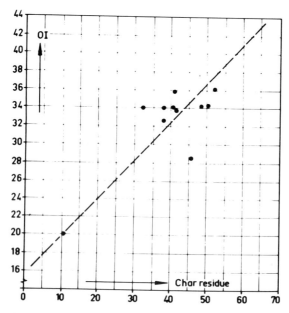

FIG. 7. Relationship between OI and CR for cellulose treated with flame retardants. From Einsele [8].

the aid of eq. (5), the oxygen index can be calculated as soon as the pyrolysis residue is known. The relationship found between OI and CR *holds only for polymers not containing halogen*, for halogen acts in a different way, viz., by influencing the mechanism of combustion.

On the other hand, eq. (5) has a wider application and also holds for the combination of polymer with additive, *provided* that the main effect of the flame retardancy is based on influencing the decomposition and increasing the pyrolysis residue. This applies to many flame-retardant additives that are used for cellulose. They have mainly a dehydrating effect.

An extensive investigation into this matter has been carried out by Einsele [8]. The results of his experiments are represented in Figure 7 and fit with the correlation found.

Conclusions

The main conclusions are as follows.

(1) The oxygen index of pure polymers can be estimated by means of a composition parameter which is a function of the atomic ratios of hydrogen and halogen to carbon.

(2) For intrinsic flame resistance, an aromatic structure seems to be necessary, unless a considerable amount of halogen is present in the polymer structure.

(3) A significant correlation exists between the oxygen index and the residue of pyrolysis. The correlation, however, is not valid for halogen-containing polymers.

(4) The residue of pyrolysis can be estimated by means of an additive quantity: the char formation tendency. This probably is a basic quantity for flame resistance.

ACTION OF FLAME RETARDANTS [f(FR)]

Flame-Retardant Additives

Flame-retardant additives may have three effects: (1) shielding of the contact area between "fuel" and oxygen; (2) influence on the pyrolysis; and (3) influence on the mechanism of combustion. The first effect is mainly physical; the second and third effects are clearly chemical in nature.

Hoke [9] has made a very conveniently arranged classification of flame retardants, which is shown in Figure 8. The elements that are primarily responsible for flame retardance are phosphorus and the halogens, often in conjunction with each other. Phosphorus compounds strongly affect pyrolysis and char formation, while halogens considerably influence polymer breakdown and the combustion. With respect to these primary flame retardants, other elements have a synergistic effect: nitrogen in combination with phosphorus and antimony in combination with the halogens.

The most important phosphorus compounds used as flame retardants are given

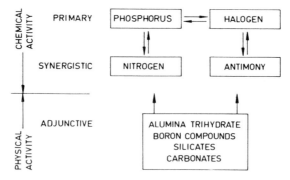

FIG. 8. Classification of flame retardants.

in Table III. Their effect is based mainly on the formation of P_2O_5 and its hydrates, which strongly enhance the formation of pyrolysis residues.

The halogen compounds most commonly used are listed in Table IV. The

TABLE III
Structures of P-Containing Flame Retardants

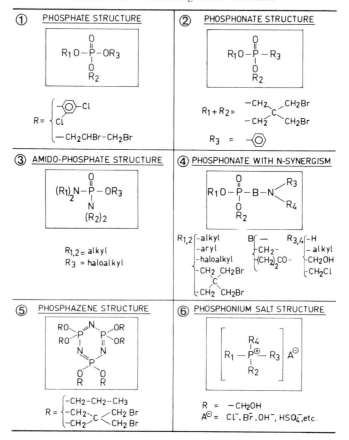

TABLE IV
Structures of Halogen-Containing Flame Retardants

BASIC STRUCTURE	CHLORINATED	BROMINATED
PARAFFINS	CHLOROWAX®	—
AROMATIC STRUCTURES	PERCHORODIPHENYL PERCHLOROTRIPHENYL (AROCLOR®)	HEXABROMOBENZENE PENTABROMOTOLUENE OCTABROMO–DIPHENYL DECABROMO–DIPHENYL PENTABROMO–PHENYL ALLYL ETHER PENTABROMO–DIPHENYL ETHER OCTABROMO–DIPHENYL ETHER TRIBROMOPHENOL TETRABROMO–PHTHALIC ACID TETRABROMO–BISPHENOL A TETRABROMO–BIS(HYDROXY PROPYL) BISPHENOL A
ALICYCLIC STRUCTURES	HEXACHLOROCYCLO– PENTADIENE CHLORENDIC ACID 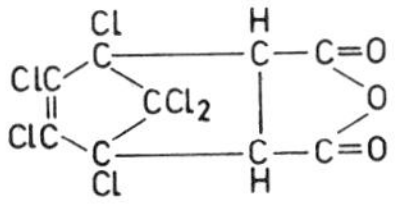 : PERCHLORINATED DICYCLO– PENTADIENYL-CYCLO– OCTADIENE 1,5 (DECHLORANE 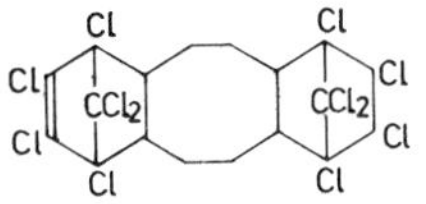 ®)	PENTABROMO–CHLORO– CYCLOHEXANE HEXABROMOCYCLO– DODECANE BICYCLOHEXANYL– ETHENE HEXABROMIDE
HALOGENATED PHOSPHATE ESTERS		TRIS 2,3 DIBROMOPROPYL PHOSPHATE

action of synergists is probably the generation of halogen hydrides which are the essential chain terminators.

The mechanism of the synergism between halogens and antimony has been studied by Pitts and co-workers [10]. Their interpretation is presented in Scheme I. It is seen that SbOCI acts as a "reservoir" for the separable $SbCl_3$ and is

$$SbOCl \xrightarrow[245-280°C]{(SbCl_3)} Sb_4O_5Cl_2 \xrightarrow[410-475°C]{(SbCl_3)} Sb_3O_4Cl \xrightarrow[475-565°C]{(SbCl_3)} Sb_2O_3$$

Scheme I

therefore the actual flame retardant. The strong endotherm effect at 245°C reduces the rate of decomposition.

TABLE V
Average Requirements for Fire-Retardant Elements to Render Common Polymers Self-Extinguishing*

POLYMER	%P	%Cl	%Br	%P+%Cl	%P+%Br	%Sb$_4$O$_6$+%Cl	%Sb$_4$O$_6$+%Br
CELLULOSE	2.5–3.5	>24	–	–	1+9	12-15+9-12	–
POLYOLEFINS	5	40	20	2.5+9	0.5+7	5+8	3+6
POLYVINYL CHLORIDE	2–4	40	–	NA	–	5-15% Sb$_4$O$_6$	–
ACRYLATES	5	20	16	2+4	1+3	–	7+5
POLYACRYLONITRILE	5	10–15	10–12	1-2+10-12	1-2+5-10	2+8	2+6
STYRENE	–	10–15	4–5	0.5+5	0.2+3	7+7-8	7+7-8
ACRYLONITRILE–BUTADIENE-STYRENE	–	23	3	–	–	5+7	–
URETHANE	1.5	18–20	12–14	1+10-15	0.5+4-7	4+4	2.5+2.5
POLYESTER	5	25	12–15	1+15-20	2+6	2+16-18	2+8-9
NYLON	3.5	3.5–7	–	–	–	10+6	–
EPOXIES	5–6	26–30	13–15	2+6	2+5	–	3+5
PHENOLICS	6	16	–	–	–	–	–

*From Lyons [11].

In general, it may be said that flame retardants, by their effect on the substrate, *lower* the temperature at which thermal decomposition begins. Together with their endotherm decomposition, this leads to a lower rate of pyrolysis. Moreover, the gaseous pyrolysis products may affect the chain reactions in the flame.

As a rule, flame retardants can be incorporated into the polymer in three ways: (a) by aftertreatment of a semifinished product (e.g., a staple fiber, a woven fabric, or a textile end product), the so-called "topical treatment"; (b) by admixture with the polymer melt or solution, the so-called "spinning dope"; (c) by reaction as comonomer during the polymerization.

Table V, derived from Lyons [11], gives a survey of the amounts of flame retardants that are required for the various polymer classes (as percentages by weight).

Quantitative Aspects of the Effectivity of Flame-Retardant Additives

My collaborators Magré and Klos [12] have made a systematic study of the effectivity of flame retardants on polystyrene. They used bromine compounds of different structures (aromatic, aliphatic, and cycloaliphatic), and also the combination of bromine compounds with synergists, viz., "dimers" of diisopropyl benzene (obtained by oxidative coupling); their formula is given in Scheme II.

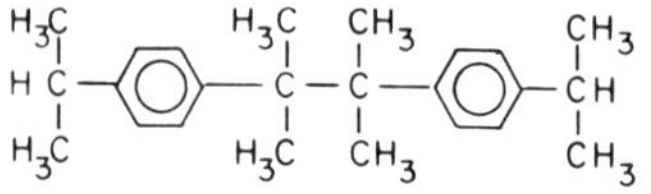

Scheme II

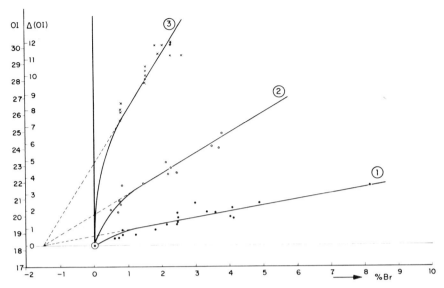

FIG. 9. Flame retardancy in polystyrene: (1) 5 FRS containing bromine in the aromatic part of the molecule; (2) 4 FRS containing bromine in the aliphatic (alicyclic) part of the molecule; (3) hexabromocyclodecane with additions of oligomers of diisopropylbenzene.

The results of their experiments are given in Figure 9, in which the effectivity of the flame retardant, $\Delta(OI)$, i.e., the increase in OI by the flame retardant, is plotted as a function of its concentration. All data can be represented by three curves: the lowest (curve 1) for the aromatic bromine, the middle (curve 2) for the aliphatic and cycloaliphatic bromines, and the upper (curve 3) for the cycloaliphatic bromine + synergist. The three curves have the same shape: they are steepest at low concentrations, then show a bending, and are linear at concentrations $> 1\%$ Br. For concentrations $> 1\%$ they may be represented by the following formula:

Curve 1:
$$OI \times 100 = 18.3 + 0.3 \left[(\%Br)_{ar} + 1.6 \right]$$

Curve 2:
$$OI \times 100 = 18.3 + 1.1 \left[(\%Br)_{al} + 1.6 \right] \qquad (7)$$

Curve 3:
$$OI \times 100 = 18.3 + 3.0 \left[(\%Br)_{act} + 1.6 \right]$$

The general form is:
$$(OI - OI_0) \times 100 = \Delta(OI) \times 100 = a(C + b) \qquad (8)$$

Here C is the concentration of the flame-retardant element (%), a is the efficacy coefficient, and b is a constant characteristic of the flame retardant.

Formula (8) was found to be generally applicable to a series of 125 flame retardants investigated in the Akzo Research Laboratories. Figures 10–12 give

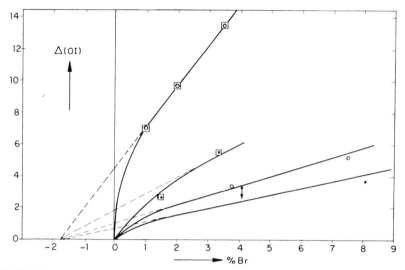

FIG. 10. Flame retardancy in polypropylene: (●) aromatic bromine; (○) aliphatic bromine; (□) with Sb_2O_3 as synergist.

examples of the effectivity of several classes of flame retardants on different polymers. Constant b is larger if the atomic weight of the flame-retardant element is higher; as a first approximation one may use:

$$b = 0.02 \times A$$

where A = atomic weight of fire-retardant element. This means that, in order

TABLE VI
Survey of Different Classes of Flame Retardants[a]

FR ELEMENT			a					
ELEMENT	b	FUNCTION	PETP	PA6	PS	PP	CELL	PAN
P	0.6	GENERAL	1.5	1.1	(1.5)*	1.3	1.5	1.5
		AMIDOPHOSPHATE	2.6	1.1			2.4	
CL	0.7	AROMATIC	0.8	0.1	0.1	0.2		
		ALIPHATIC	0.8	0.1	0.5	0.2		0.1
		ALIPH.,ACTIVATED						
		AROMATIC+ Sb_2O_3						
		ALIPHATIC+Sb_2O_3				1.1		
Br	1.6	AROMATIC	0.5	0.1	0.3	0.45		
		ALIPHATIC	0.7		1.1	0.6		0.3^5
		ALIPH.,ACTIVATED			3.0			
		AROMATIC+ Sb_2O_3				1.0		
		ALIPHATIC+Sb_2O_3				2.6		

[a] General formula: $\Delta(OI) \times 100 = a[C + b]$ where $C > 1\%$; C is the concentration in wt %.
*Phosphines, phosphinoxides.

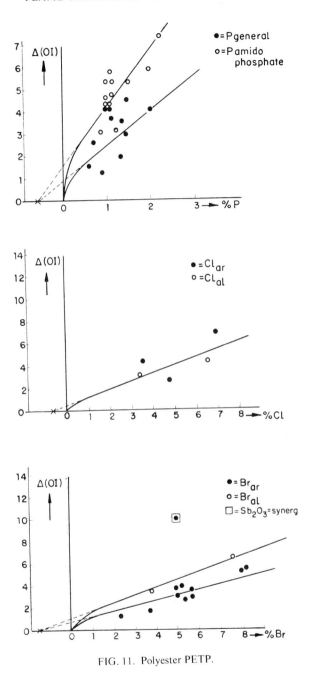

FIG. 11. Polyester PETP.

to obtain universal curves, the concentration of the flame-retardant element
should be plotted as an atomic percentage.

Table VI gives a very interesting survey of the different classes of flame retardants and their activity in five important polymers: polyester, polyamide,

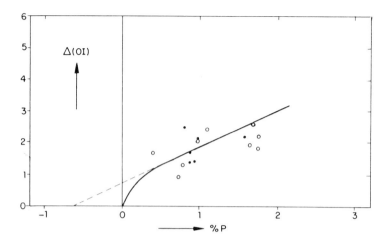

FIG. 12. Flame retardancy in nylon 6: (●) phosphazenes; (○) phosphine oxides.

polystyrene, polypropylene, and cellulose. The efficacy coefficient a may vary from 0.1 to 3.

From Table VI, it is evident that phosphorus is active in all fiber polymers, though less in polyamides than in the other types. Chlorine shows the best activity in polyester and is almost inactive in polyamides. Bromine, too, is the most active in polyester, is also active in polystyrene and polypropylene but, like chlorine, is almost inactive in polyamides. This means that in polyamides there is hardly any influence of flame retardants in the gaseous chain reaction; if their flammability is to be influenced, it has to be done by increasing their char-forming tendency.

Recent investigations have proved that the main influence of halogen-containing fire retardants is an accelerated polymer breakdown, manifesting itself in largely increased dripping. The increase in OI of thermoplastic polymers by flame retardants is mainly due to an enhanced dripping effect.

Conclusions

(1) There is a distinct functional difference between two classes of fire retardants, viz., those based on phosphorus and those based on halogen. The first class acts mainly in the solid phase and influences pyrolysis (char formation); the second class enhances polymer breakdown and may also influence the chain mechanism of combustion.

(2) A quantitative linear relationship describes the influence of flame retardants on OI above concentrations of the flame-retardant element of about 1%. This expression contains a constant which is dependent on the combination flame-retardant polymer and another constant which depends on the flame-retardant element only.

FLAME-RETARDANT MODIFICATIONS OF ESTABLISHED FIBERS

Introduction

As mentioned earlier, there are in principle two methods to distribute additives in polymers homogeneously, viz., "topical treatment" in the end stage or admixture as "spinning dope" to the polymer melt or solution.

A third method is the incorporation of the fire retardant into the skeleton of the polymer by copolymerization. In short, the three methods are (I) topical treatment, (II) spinning dope, and (III) copolymerization.

It is obvious that for natural fibers only method I can be used: to man-made cellulosics, methods I and II can be applied. For synthetic fibers all three methods may be used, although not always successfully.

Table VII gives a survey of the status of the art; only examples of semicommercial and commercial products are given.

The different fibers will now be discussed in some detail.

Natural Fibers

Natural fibers are beyond the scope of this paper, but for the sake of completeness a few words on them may be useful. As a matter of fact, only topical treatment is possible here. For *wool,* the so-called "mordant" treatment is commercially used in which the fiber is treated with anionic complexes of titanium or zirconium. For *cotton,* two techniques are in commercial use: treatments based on tetrabis(hydroxymethyl)phosphonium chloride (with amino resins or ammonia curing) and Ciba-Geigy's Pyrovatex. Table VIII gives a survey.

It is possible to make cotton fireproof but this impairs the quality of the fiber; notably the stiffness is increased and the moisture regain decreased.

TABLE VII
Methods for Flameproofing Conventional Fibers*

FIBER	NATURAL FIBERS		MAN-MADE FIBERS					
			CELLULOSICS		SYNTHETICS			
METHOD	WOOL	COTTON	RAYON	ACETATE	NYLON	POLY-ESTER	ACRYLIC	POLY-PROPYLENE
I TOPICAL TREATMENT ("FINISH")	(+)	++	+	+	(+)	+	(+)	(+)
II SPINNING DOPE	−	−	++	++	−	+	+	+
III COPOLY-MERIZATION	−	−	−	−	−	++	++	−

*Symbols: ++ important; + in use but less important; (+) of minor importance; − not used.

TABLE VIII

Flame Retardants for Cellulose Suitable in Finishing Operations

FLAME RETARDANT	USED IN COMBINATION WITH	TRADE MARK OF PROCESS	FIRM.
TETRAKIS (HYDROXYMETHYL) PHOSPHONIUM SALT $\begin{bmatrix} CH_2OH \\ \| \\ HOH_2C-P^{\oplus}-CH_2OH \\ \| \\ CH_2OH \end{bmatrix} A^{\ominus}$	$(A^{\ominus}=Cl^-)$ UREA AND METHYLOL MELAMINE	PERMA PROOF® ROXELL® FR−1®	TREES DALE LAB. HOOKER LYNRUS FINISHING Co
	$(A^{\ominus}=Cl^-)$ a UREA AND POLYVINYLBROMIDE b CYANAMIDE AND FIRE MASTER 200®	(IN DEVELOPMENT)	(SOUTHERN REGIONAL RESEARCH CENTER US DEP. OF AGRICULTURE)
	$A^{\ominus}=Cl^-$ $A^{\ominus}=$ PHOSPHATE/ACETATE $A^{\ominus}=$ OXALATE AMMONIA CURE	PYROSET TKC® PYROSET TKP® PYROSET TKS®	AMERICAN CYANAMID
	$(A^{\ominus}=Cl^-)$ UREA OR CYANAMIDE AND AMMONIA CURE	PROBAN®	ALBRIGHT & WILSON
	$A^{\ominus}=OH^-$ AMMONIA CURE	HOOKER FINISHING #5®	HOOKER
N-METHYLOL DIALKYL PHOSPHONO-PROPIONAMIDE $\begin{matrix} RO & O & O \\ \diagdown \| & & \| \\ P-(CH_2)_2 & C-NH-CH_2OH \\ \diagup \| & & \\ RO & & \end{matrix}$	UREA AND METHYLOL MELAMINE	PYROVATEX®	CIBA−GEIGY
	FORMATION IN SITU DURING FINISH FROM VINYLPHOSPHONATE AND N-METHYLOL ACRYLAMIDE (K-PERSULPHATE INITIATION)	FYROL 76®	STAUFFER

Cellulosic Man-Made Fibers

Rayon and acetate can be made flameproof by addition of spinning dopes. For rayon the most successful are the phosphazenes; for acetate the trisdibromopropylphosphate (TDBP) is used (see Fig. 13). The usual method is to add the fire retardant to the spinning solution; additions up to 20% by weight may be required to achieve adequate flame retardancy.

Polyamides

Nylons have self-extinguishing properties due to extensive shrinking and dripping during combustion. The real problem arises in blends with nonthermoplastic fibers: owing to the "scaffolding" effect, the blend burns vigorously (the nonthermoplastic, e.g., cellulosic, fiber will char and form a supporting structure which will hold the molten polymer).

No satisfactory means have been found to make polyamides flameproof. Spinning dopes degrade the molten polymer. Only with topical treatments has some success been achieved; treatments based on thiourea and amino resin have been introduced for this (see Fig. 13a). Their action seems to be that the melt temperature of the fiber is lowered, which facilitates dripping (!)

Polyester

Again, the real problem is in the blends. Topical treatments based on trisdibromopropylphosphate are commercialized (Eastman) (see Fig. 13b).

Spinning dopes have not been very successful until now. Most bromine compounds have an unsatisfactory heat resistance and will not withstand the

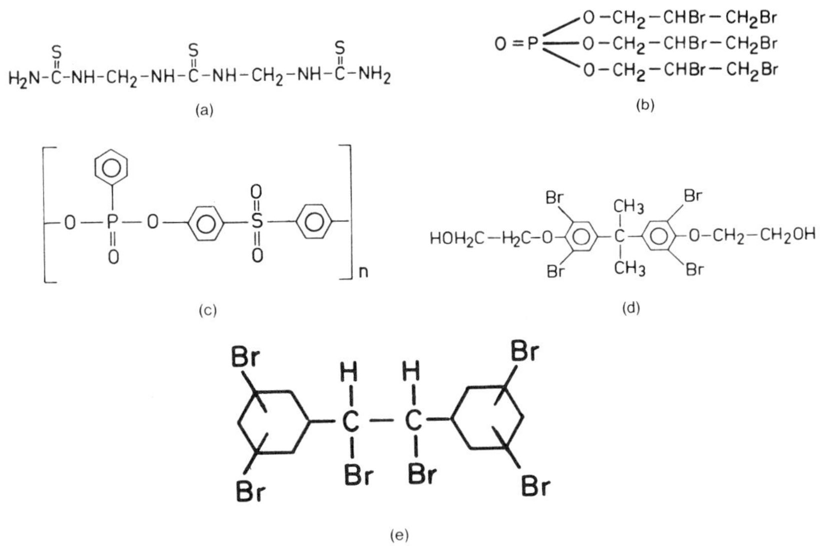

$H_2N-\overset{S}{\overset{\|}{C}}-NH-CH_2-NH-\overset{S}{\overset{\|}{C}}-NH-CH_2-NH-\overset{S}{\overset{\|}{C}}-NH_2$

(a)

$O=P\begin{cases}O-CH_2-CHBr-CH_2Br\\O-CH_2-CHBr-CH_2Br\\O-CH_2-CHBr-CH_2Br\end{cases}$

(b)

(c)

$HOH_2C-H_2C-O-\text{(ring, Br)}-\overset{CH_3}{\underset{CH_3}{C}}-\text{(ring, Br)}-O-CH_2-CH_2OH$

(d)

(e)

FIG. 13. Some important flame retardants for synthetic fibers: (a) thiourea-based; (b) tris (2.3-dibromopropyl)phosphate; (c) in Heim; (d) Bromdian-bisglycol ether; (e) bis(cyclohexenylethylene) hexabromide.

temperature of the spinning melt. The only commercial polyester fiber so far based on phosphorus additions is "Heim" launched by Toyobo (see Fig. 13c).

The most satisfactory method in the polyester field is copolymerization. Dacron D 489 and 900 of DuPont was the first example of an inherently flame-resistant fiber. The polymer is made by reacting tetrabromobisphenol A with 2 moles of ethylene glycol to yield bis(hydroxyethyl)tetrabromobisphenol A (see Fig. 13d) and polycondensing this compound with ethylene glycol and terephthalic acid. The OI of the copolymer is 0.26.

Acrylics

Acrylic fibers, which by definition contain more than 85% by weight of acrylonitrile, burn vigorously when ignited, in pure form and in blends.

As spinning dope TDBP can be used, but the most effective way to flameproof acrylics is by copolymerization. Vinyl bromide can be employed as a comonomer. Also, the modacrylics such as Dynel and Verel are copolymers for which comonomers like vinyl chloride and vinylidene chloride are used in ratios of 40/60 to 60/40.

Polypropylene

Due to its ease of melting and its high shrinkage, polypropylene is less flammable than one would expect. The only commercial flame-retardant polypro-

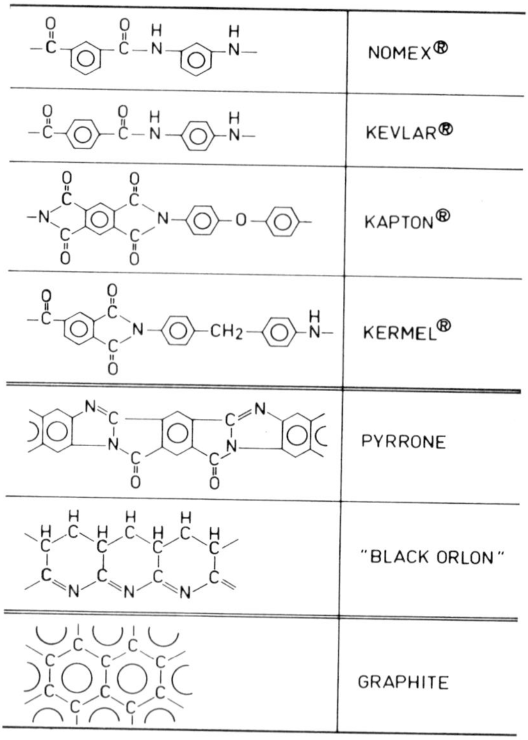

FIG. 14. Some important HT polymers.

pylene fibers are made on the basic of spinning dopes, e.g., by using biscyclo-hexenylethylene hexabromide (Phillips) as an additive (see Fig. 13e).

POLYMERS WITH INTRINSIC FLAME RESISTANCE

Structural Principles

In the foregoing, it has been shown that intrinsic flame resistance is a property that is closely connected with an aromatic structure. Aromatic structures enhance the char formation and at the same time prevent dripping. Another means to increase char formation is to apply crosslinking. A third method is the introduction of heavy metals into the polymer structure, which can act as catalysts for charring pyrolysis.

Developments have gone along these lines and the following classification can be made.

Class 1: linear aromatic polymers—including Nomex and Kevlar (Du-Pont).

Class 2: linear aromatic/heterocyclic polymers—Kermel (Rhone Poulenc) and Kaptan (DuPont) (the latter is not suitable for fibers).

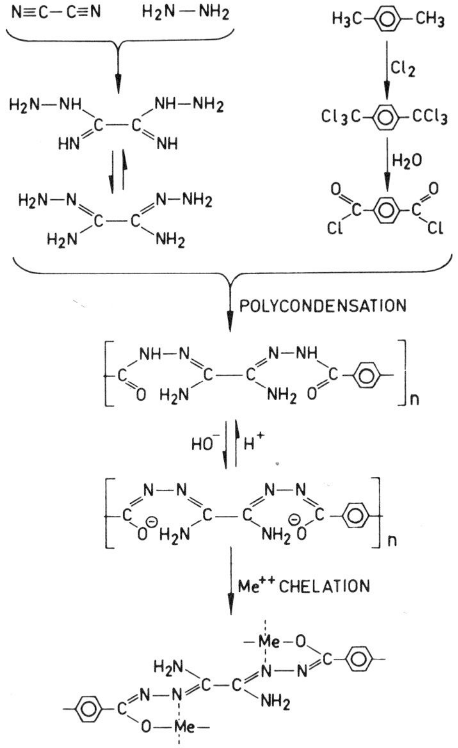

FIG. 15. Scheme showing the manufacturing process of Enkathem.

Class 3: condensed polycyclic hetero-aromatic polymers—these are the so-called "ladder polymers," culminating in the graphite fiber.

Class 4: crosslinked aromatic polymers—i.e., crosslinked after fiber forming.

Class 5: chelated aromatic polymers—in which metal atoms form part of the polymer structure.

Classes 1–3 form the heat (and flame-) resistant polymers. Some of the important representatives are shown in Figure 14.

A representative of class 4 is Kynol of Carborundum Co. This is a phenol formaldehyde novolak spun into threads and subsequently crosslinked with formaldehyde. This network polymer has a high flame resistance because of its strongly aromatic structure.

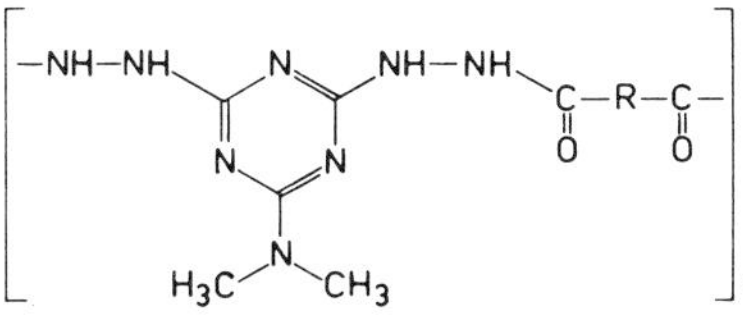

FIG. 16. Structure of a triazine derivative, an example of a chelate polymer that is almost colorless.

An example of class 5 is Enkatherm developed by the Akzo group [3, 13]. This is a poly(terephthaloyl oxalic acid-bisamidrazone) (PTO) which has been chelated (after fiber forming) with bivalent metal salts. The scheme of preparation is shown in Figure 15. In general, the chelates have a deep color which varies from yellow and orange to black. This is, of course, a drawback when application as a textile raw material is considered. However, exceptionally high OI values, varying from 0.35 to 0.50, can be attained in this way.

Recently, successful efforts have been made to prepare chelate polymers that are almost colorless. One of them (a triazine derivative) is shown in Figure 16. The outstanding flame resistance of this class of polymers is probably caused mainly by the atomic distribution of the metal over the macromolecule. Without chelation the polymer has an OI value of not more than 0.22. Upon heating, the atomically distributed metal has a tremendous influence on the decomposition. A considerable pyrolysis residue arises, which results in the high OI value.

CONCLUDING REMARKS

The development of flame-resistant fibers has been a largely pragmatic one; its further progress will be highly influenced and even determined by public regulations and legislations.

Flammability of a material is a structure property. It can be modified by the use of additives and by chemical aftertreatment. Some interesting successes have been achieved but more so in the cellulosic field than in the synthetic.

In the future, polymers with intrinsic flame resistance will have to be synthesized. The structural criteria that are necessary are now beginning to manifest themselves. The first representatives of this new polymer group are entering the market. Nobody can tell yet how far and how fast this development will go.

REFERENCES

[1] Frank-Kamenetzky, *Diffusion and Heat Transfer in Chemical Kinetics*, Plenum Press, New York, 1969.
[2] C. P. Fenimore, and F. J. Martin, *Mod. Plast., 43*, 141 (1966); *Combust. Flame, 10*, 135 (1966).
[3] D. W. Van Krevelen, *Angew. Makromol. Chem., 22*, 133 (1972).
[4] D. W. Van Krevelen, and P. J. Hoftyzer, unpublished results (1974).
[5] D. W. Van Krevelen, *Chimia, 28*, 504, 512 (1974).
[6] D. W. Van Krevelen, paper presented at the International IUPAC Symposium on Macromolecules, Madrid, 1974; *Preprints*, p. 833.
[7] D. W. Van Krevelen, *Polymer 16, 615* (1975).
[8] U. Einsele, *Z. Gesamte Textilind., 72*, 984 (1970).
[9] Ch. E. Hoke, *Soc. Plast. Eng. Technol. Pap., 19*, 548 (1973); *S.P.E.J., 29*, 36 (1973).
[10] J. J., Pitts, P. H. Scott, and D. G. Powell, *J. Cell. Plast. 1970*, 35.
[11] J. W. Lyons, *The Chemistry and Uses of Fire Retardants*, Wiley-Interscience, New York, 1970.
[12] E. P. Magré and C. D. W., Klos, Internal Report, Akzo R & E (1974).
[13] D., Frank, W. Dietrich, J. Behnke, A., Koböck, G. Scheick, and M. Wallrabenstein, *Angew. Makromol. Chem., 40/41*, 445 (1974).

THERMAL CONDUCTIVITY OF FABRICS AS RELATED TO SKIN BURN DAMAGE

JACK H. ROSS

Aeronautical Systems Division, Air Force Systems Command,
Wright-Patterson Air Force Base, Ohio

SYNOPSIS

The heat-flow process from a fire, through a clothing system to the underlying skin, was investigated. The goal was to optimize fiber and fabric to provide the maximum thermal protection. A number of laboratory heat sources, including a JP-4 jet fuel burner, were used in this study. Based on this investigation and the related flammability data generated in simultaneously conducted laboratory tests, the following conclusions can be drawn.

(1) Fiber thermal stability is the most important parameter in designing thermally protective fabrics. Only two fibers can meet this most critical parameter.

(2) The chemical stability at high temperatures is critical to the retention of structural stability in the fabric. Assuming equilibrium reactions between pyrolysis gases, the fire, and air environments, it was concluded that two fibers show no indication of combustion, while the presently used Nomex (for Air Force flight suits) was found to ignite around 550°C.

(3) Fabric weight, thickness, and optical properties are important parameters which can be adjusted by the fabric design and chemical additives to yield improved protective characteristics.

INTRODUCTION

The role of fire in our lives has been very inconsistent with respect to promoting increased sophistication in materials being incorporated into current planned transportation systems and residential buildings. Controlling fire, whether it be in burning of fuels to create various forms of power such as in aircraft engines, power plants, internal conbustion engines, or in the providing of heat for warmth and cooking, is most important. It is well known, even today, that fire when out of control, whether in a small area or a large building, can create untold havoc. Even worse is the explosive flame and heat growth that occurs when liquid fuels with high volatility ignite. This can lead to a catastrophic type of fire which consumes all in its path and cannot be controlled to any degree until the bulk of the fuel is consumed. (Three recent fires in fuel storage areas of oil refineries in the Philadelphia area demonstrate this fact.) In the case of aircraft, whether commercial or military, one of the major causes of deaths and injuries as a result of accidents is brought about by fire which frequently follows the accident. Past

Journal of Applied Polymer Science: Applied Polymer Symposium 31, 293–312 (1977)
© 1977 by John Wiley & Sons, Inc.

and present attempts to control fires consisting not only of the fuel, but the aircraft itself, have not been sufficiently successful to enable crew and passengers to escape unharmed (or with only minor burns). It is therefore necessary to provide protection, in the form of clothing for the crew, and to interior accessories, carpet, seats, wainscoating, etc., to prevent the penetration of heat and flames for sufficient time to allow escape, this normally being 6–30 sec.

A great deal of effort has been expended to study (1) the fibers used in clothing, (2) the fabric structure, (3) the properties of human skin as degraded by heat and flame, and (4) the heat-flame source itself. Most of the research in these areas have been carried out independently of the other. Only recently have these four areas been the subject of integrated research programs. The first effort [1] utilized both analytical and empirical methods to prove that heat is transferred through fabrics by convection, conduction, and radiation, and under certain circumstances by vaporization. Both laboratory and full-scale fires were used to generate data to verify the theory. The second case [2] was directed to the development of a theoretical and empirical mathematical relationship to define the fabric–skin system's response when exposed to a jet fuel (kerosene or its equivalent) fire. As in the work of Stoll and Chianta [1], this effort utilized both laboratory and full-scale fires to confirm a theoretical model. The third case, described in this paper, examines the practical application of these theoretical aspects in the optimization of fabrics for protective clothing, and uses the techniques described by Morse et al. [2] to determine some of the analytical data.

THEORETICAL BACKGROUND

The flame–fabric–skin system used is shown in Figure 1. Energy in the form of heat is transferred from the flame to or through the fabric and then to the underlying skin. The energy can be radiative, convective, or conductive, although the dominant component is in the form of radiant energy. Energy is received

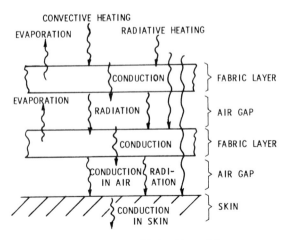

FIG. 1. Flame–fabric–skin model.

by the skin predominantly at the skin surface, and then the heat is conducted into the skin layers. A portion of the flame-emitted energy can be absorbed at depth within the skin dependent on the spectral characteristics of the energy emitted and the skin optical properties. The air, trapped between the fabric(s) and skin, is a thermal insulation layer and provides the bulk of the protection for the underlying skin. Energy transfer through the flame–fabric–skin system can be defined by one-dimensional heat-transfer analysis through the composite system. The end result is the net energy received at a given depth within the skin that will or will not cause a burn dependent on the heat rate and intensity.

The thermal hazard, or the energy source, is a JP-4 fuel fire evolving from an aircraft accident. The flame exhibits essentially black body characteristics with an emittance of 1.0. The flame temperature is set for nominal conditions at 982°C (1800°F) with a range of 871–1204°C (1600–2200°F) for the parametric analysis. Field studies, conducted with large JP-4 fuel fires and immersion of heat-flux sensors, have evidenced an average flux level of 2.2 cal/cm²-sec with intensities of up to 5.0 cal/cm²-sec, not uncommon.

In order to provide protection from a heat-flame source to an underlying skin surface, a fabric should exhibit specific characteristics, including the following.

Fabric Flammability

If a fabric burns while in the fire or continues to burn once removed from the fire, it presents a secondary hazard to the wearer attempting to escape from the primary hazard of the burning aircraft. Also, the higher the fabric ignition temperature, the better the protection offered the wearer.

Melting

When a thermoplastic fiber comes in contact with sufficient heat, it melts and becomes a molten flowing mass. If this molten fiber comes in contact with the skin, it sticks to the skin and transmits heat at a very rapid rate. If the molten fiber drips on, or contacts, the underclothing, it can stick to it and also cause it to ignite. A fabric that melts can also allow hole formation, thus allowing direct exposure of the skin to the flames.

Heat Transfer

Assuming that the fabric remains intact, it must then provide a barrier to heat transmission from the flames. A fabric that is nonporous and opaque to infrared radiant energy would be ideal for heat blockage, yet it would also retain too much body heat during normal use and thus be uncomfortable to wear. The hot fabric can also become a source of heat by emitting heat from the back side of the fabric.

Fiber Thermal Stability

If the fiber thermally degrades by charring, becoming friable, or by shrinking, it becomes partially ineffective as a thermal barrier. If it chars and becomes brittle, its heat-sink ability or thermal capacitance is increased along with its thermal conductivity, thus allowing the material to have a greater heat storage capacity and to increase peak temperature. Increased thermal capacitance can also be an asset to thermal protection by storing the heat and allowing the energy release to be spread out over a longer period of time, thus decreasing the potential of damage to the underlying skin. If the fiber (and fabric) should become friable, it readily falls apart due to the movement of the wearer and the protective barrier is eliminated. Shrinkage of the fabric is of considerable importance when considering thermal protection. The fabric heats up due to the flames and then shrinks to contact with the skin allowing a rapid flow of heat to the skin surface. The shrinking fabric voids the wearer of the protective insulating air gap between the fabric and the skin.

Fabric Optical Properties

Fabric reflectance, absorptance, and transmittance are also important to the overall heat transfer process. The more reflective fabric can greatly enhance the protective capacity of a fabric, while one that allows direct transmittance of the radiant heat reduces the fabric protection capacity. It would be more favorable for the fabric to absorb the energy rather than transmit it, yet the opposite is true when reflectance is compared to absorptance.

Multiple Fabric Layers

Using more than one fabric layer considerably increases the thermal protection offered by the clothing system. By introducing additional fabric layers, the thermal protection is increased by the added air gaps provided between the fabric layers. Two layers of lightweight fabric provide more protection than a single layer of fabric equal in weight to the two layers of fabric. Although this provides a solution to the thermal protection problem, it can also be too warm and reduce comfort in summer weight flight suits.

Blended Fabric

One of the ways of attempting to provide the desired properties is to construct the fabric from yarns of intimate fiber blends. Staple fibers of more than one fiber type are spun into yarn in order to utilize the desirable characteristics of the individual components. Hopefully this results in a blend of the favorable characteristics. This approach is examined partially here, although the blended fabrics were not extensively evaluated for mechanical properties. The concern here was to first determine the thermal protection characteristics of the fabrics. Several of the blended fabrics were eventually evaluated for mechanical characteristics after it was determined that they exhibited improved thermal properties.

Practical Fabric Considerations

Comfort

Many volumes can be found in the literature on the subject of comfort provided by wearing apparel, and one readily finds that many differing viewpoints exist as to how comfort is provided by a textile material. Factors such as fiber and fabric surface characteristics, finishes, construction, stiffness, etc., are all given credit towards providing comfort or the lack of comfort. Criteria for selecting a comfortable fabric include hand, moisture-absorbing characteristics, air permeability, and common sense. In the final analysis, what may be comfortable to one person may not suit another. Comfort is considered of great importance because it provides improved personnel performance and it reduces the possibility of personnel discarding the garment for a more comfortable choice which might not provide adequate thermal protection.

Durability

If a fabric is not durable, then it becomes nonfunctional. This becomes even more apparent under abrasive use conditions. Fibers used for military fabrics must resist abrasion, and must not rot due to mildew. The fabric must be non-tearing and it (the fabric) must be resistant to ultraviolet degradation.

Aesthetics

Fibers that provide good dyeability, that are pleasing to the eye, and provide a fabric that yields good military appearance in garments, are required. Even more important is that they allow for proper dyeing for camouflage and to resist surveillance techniques.

The model properties used for human skin are outlined by Morse et al. [2]. The data used for the thermal properties are a result of a considerable literature search. The skin density was held constant at 1.0 g/cm^3. The skin specific heat was 1.0 cal/g-°C for the outermost skin layer, the epidermis, and 0.5 cal/g-°C for the dermal or fatty tissue. Thermal conductivity values ranged from 5.5 cal/cm-sec-°C up to 14.0 cal/cm-sec-°C.

The burn predictions are based on the work of Henriques and Moritz as modified by Stoll and Green [3]. The empirical studies provide a time–temperature profile for a heat sink placed at some distance behind the fabric when exposed to a thermal energy source. Using appropriate calculations to provide data in terms of skin-surface temperature histories, the relationship derived by Stoll can be used to predict skin burns. The relationship is based on the classical Arrhenius equation that predicts the reaction rate as a function of temperature. Stoll, using the combined work of Henriques and Moritz and empirical studies, using porcine and human skin, derived the following relationship:

$$\frac{d\Omega}{d\theta} = 3.1 \times 10^{98} \, e - \left(\frac{75,000}{T(°K)}\right)$$

The damage rate constant was chosen such that a value of Ω equal to 1.0 is indicative of cell destruction in human skin. A second-degree burn is defined as destruction of the skin cells to the base of the epidermis. The thickness of the epidermal skin layer is set at 100 μ.

The skin optical properties influenced by a JP-4 fuel fire source and by the energy emitted by the backside of the fabric are discussed in detail elsewhere [2]. Better than 95% of the total emitted energy is considered to be absorbed at the skin surface. That part of the energy that is reflected is within the 0.2–2.0 μ wavelength range and essentially all of the energy beyond 2.0 μ is considered to be absorbed at the skin surface.

MATERIALS EVALUATED

When considering the need for polymers which can be formed into fibers to provide the degree of protection from the heat-flame source desired, the real problem was not how to obtain high-melting polymers but how to obtain thermally stable polymers that could be formed into fibers. Many polymers have been synthesized which have the desired thermal stability, but only certain ones are capable of being formed into fibers. The first commercial product capable of being formed into fibers was a wholly aromatic, all *meta*-oriented ordered copolymer [4], Nomex. Another polymer of this class was spun by Monsanto Research Corporation and had properties similar to Nomex, but was subsequently discarded for reasons other than the ability to provide thermal protection. The *para*-oriented form of this polymer has a higher decomposition temperature (550°C) than Nomex (425°C). The effect of heat on these fibers (all *meta*-oriented polymers and polymers containing *para*-oriented rings) is significantly different. In the all *meta*-oriented system, after heating, the polymer shrinks, splits apart and can ignite; while the *para*-oriented system is sufficiently flexible to be tied in a knot (the fiber). This latter polymer has evolved into a product labeled HT-4. Another class of polymers which are thermally stable and capable

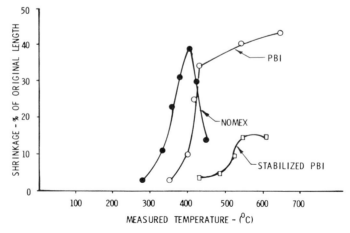

FIG. 2. Shrinkage test results in a nitrogen environment.

of being formed into fibers are the poly(m-phenylene dibenzimidazoles), one of which is called PBI. In this system, substitution of heterocyclic rings for a portion of the phenylene rings in poly(p-phenylene) has resulted in a very thermally stable structure. The addition of heterocyclic rings normally provides considerable resistance to burning when a part of aromatic polymers. PBI, despite some shrinkage during exposure to open flame, resists burning completely. To overcome the problem of shrinkage in these heterocyclic polymers, it has been found that the fiber, yarn, or fabric can be treated with an acid such as sulfuric acid. The optimum treatment evolved consists of heating the PBI for 60 min at 96°C (205°F) in an aqueous combination of 2% solution of sulfuric acid, 6% solution of benzyl alcohol, and 0.5% solution of Ultrawet 60L, followed by rinsing and air-drying. Not only was shrinkage reduced significantly (Fig. 2), but ultraviolet resistance was increased as was the affinity of the fiber for dispersed and cationic dyes.

The material (fabrics) evaluated in this program were chosen based on their advanced thermal stability characteristics when compared to some of the more conventional commercial fibers. Each has a unique set of fiber characteristics that determines the overall performance of that fiber type when exposed to an extreme thermal environment. The fabrics and some of their physical characteristics are listed in Table I.

Durette is manufactured by the Fire Safe Products, Inc., and is chemically modified, heat-treated aromatic polyamide (Nomex), to effect improved fabric thermal stability and fire resistance. Fypro, manufactured by Travis Mills Corp., is also a modified Nomex and exhibits similar thermal performance characteristics to those of Durette. Kynol, made by the Carborundum Co., is a phenolic fiber, described as being cross-linked in a novel manner to effect good thermal stability. Nomex is a fire-resistant aromatic polyamide and is now the only authorized fiber for Air Force flight garments. HT-4 is a modified aromatic polyamide based on a new group of fibers of the Aramid class being marketed by the du Pont Co. PBI is an Air Force developed fiber produced experimentally by the Celanese Research Co., and stabilized PBI is a chemically treated PBI to yield a fiber with reduced shrinkage upon thermal exposure.

The fabrics listed in Table I were used for the evaluation of the various fiber types. Fabric construction varied from sample to sample as did the fabric air permeability. One other noticeable different fabric characteristic was the relatively high bulk density of the HT-4 fabric. Analysis of the variance in these fabric properties is examined by how they affect the fabric thermal and optical characteristics. A table of data is provided showing the physical characteristics of the blended fabrics evaluated (Table II).

The optical property data was obtained using a Beckman DK-2A Recording Reflectometer modified with a Gier Dunkle integrating sphere having a xenon source (angle of incidence 15°, 0.5–2.5 μ wavelength range). These data do not cover the full wavelength range of interest for an 982°C black body source, but the data do indicate the relative difference of the fabric optical properties in the wavelength range that would show the widest range of variation for normal textile materials. The resultant data are shown in Table III.

TABLE I
Materials Evaluated: Fiber Type and Physical Characteristics

FIBER	WEIGHT OZ/YD	ENDS/INCH WARP/FILL	WEAVE	THICKNESS MILS	BULK DENSITY G/CC	AIR PERMEABILITY FT ^{3}/FT ^{2}/MIN
DURETTE	4.5	50/44	PLAIN	14.3	.42	220
FYPRO	4.1	132/83	2/1 TWILL	11.9	.46	82
KYNOL	5.4	49/50	PLAIN	20.7	.35	147
NOMEX	4.2	124/83	2/2 TWILL	12.7	.45	99
HT-4	4.5	56/37	PLAIN	10.5	.57	60
PBI	4.6	67/60	2/1 TWILL	14.2	.43	140
STABILIZED PBI	4.3	116/72	2/1 TWILL	13.9	.41	134

TABLE II
Fiber Blends Evaluated

FIBER BLEND %	WEIGHT OZ/SQ YD.	THICKNESS MILS.	DESIGN W-WOVEN K-KNITTED	YARNS PER INCH WARP/FILL	AIR PERMEABILITY CU FT./SQ FT./MIN.
50 KYNOL/50 VEREL	6.7	22.6	W-PLAIN	38X36	38
50 KYNOL/50 FORTREL	7.2	20.7	W-PLAIN	37X37	56
50 KYNOL/50 COTTON	6.6	22.5	W-PLAIN	38X36	12
50 KYNOL/50 NOMEX	6.7	21.1	W-PLAIN	38X37	33
50 KYNOL/50 PBI	4.6	14.5	W-2/1 TWILL	67X63	144
80 KYNOL/20 NOMEX	5.1	17.8	W-2/2 TWILL	74X68	126
80 KYNOL/20 NOMEX	4.5	14,6	W-PLAIN	68X56	99
80 KYNOL/20 NOMEX	5.2	17.4	W-2/2 TWILL	73X68	118
80 KYNOL/20 NOMEX	4.8	13.7	W-PLAIN	73X62	154
50 KYNOL/50 NOMEX	5.0	15.4	W-PLAIN	45X45	131

TABLE III
Fabric Optical Property Data

FIBER	TRANSMITTANCE %	REFLECTANCE %	ABSORPTANCE %
STABILIZED PBI	22.3	39.3	38.4
HT-4	10.9	46.5	42.6
FYPRO	15.8	41.5	42.7
KYNOL	15.6	55.6	28.8
DURETTE	27.4	50.3	22.3
NOMEX	18.0	43.0	39.0
PBI	15.6	42.9	41.5

Fabric mechanical property data are provided in Table IV. Of special note is the loss of mechanical properties when the Nomex and PBI were chemically and thermally treated to obtain greater thermal stability and reduced shrinkage. Specifically, Durette and Fypro (which are treated Nomex) show a severe loss in strength, if it is assumed that they should exhibit properties similar to the untreated Nomex fabric. The reduction in mechanical properties is even more noticeable after review of the abrasion data. The same rationale can be evidenced by examining the PBI fabric data and comparing it to the stabilized PBI fabric regarding strength after chemothermic treatment.

TABLE IV
Fabric Mechanical Properties Data

FIBER	BREAKING STRENGTH (LBS/IN) WARP FILL		ELONGATION (%) WARP FILL		TEAR STRENGTH (%) WARP FILL		ABRASION RESISTANCE (CYCLES TO DESTRUCTION) SHIEFER TEST
50 KYNOL/50 NOMEX	57	60	24	20	12	13	1130
50 KYNOL/20 NOMES	45	37	16	19	7	6	393
KYNOL	23	25	5	10	14	*	192
COTTON	154	67	16	12	5	5	834
DURETTE	85	70	34	30	10	9	419
FYPRO	66	44	21	19	5	4	534
NOMEX	110	66	34	29	18	14	1945
PBI	108	98	22	23	17	14	2102
STABILIZED PBI	71	44	13	7	9	9	2500-4300
HT-4	86	59	12	7	9	8	500-600

* Diagonal tear.

TABLE V
Fabric Moisture Regain at 72°F and 65% R.H.

FIBER	MOISTURE REGAIN %
COTTON	8.0-10.0
DURETTE	5.0
FYPRO	5.0
KYNOL	5.1
NOMEX	5.0
HT-4	5.5
PBI	12.0
STABILIZED PBI	13.2
50 KYNOL/50 NOMEX	5.4
80 KYNOL/20 NOMEX	5.9

The various fabrics were conditioned at 65% relative humidity and at 72°F and the fabric moisture regain was determined. The data indicate the shortcoming of the majority of man-made fibers in that they exhibit poor moisture regain values in comparison to the most commonly referred to cotton material also listed in Table V. The reason for considering this effect to be a shortcoming is the association of low moisture regain in fibers with the inability to absorb and pick up body moisture (this characteristic is usually called wicking) in order to provide comfort to the wearer. Two of the materials listed represent an unusual trend for a man-made fiber in that PBI and stabilized PBI both exhibit moisture regain in excess of cotton.

The materials were also characterized by the Vertical Flame Test (Table VI).

All of the materials exhibited good flame resistance with the exception of the
Kynol/Fortrel and Kynol/Cotton blends (Table VI). These two materials were
totally consumed due to the flammable nature of the cotton and Fortrel fibers.
The Kynol/Verel blend also tends to yield a longer char length in comparison
to the remaining fibers evaluated.

THERMAL EVALUATION TECHNIQUES

The laboratory evaluations were conducted using a Fire Athermancy Apparatus [5]. This equipment consists of a table-mounted shutter device which is
used to control the exposure of the fabric sample to the desired heat source.
Support equipment includes two heating sources, timers, and recording equipment. The first heat source consists of a bank of four quartz lamps combined
with a Meker burner fed by natural gas. By controlling the output of the lamps
and the flame-producing burner, one can regulate the amount of convective and
radiant heating.

The second heat source is a JP-4 burner developed to more closely simulate
the intended end-use environment. While the first source provides a controlled
thermal environment, it is not capable of providing the total heating effects that
would be seen in a JP-4 fuel fire. The latter source does provide the combined
thermochemical effects associated with a JP-4 fuel fire and can be operated with
reasonable control. By controlling the air–fuel ratio, one can achieve the desired
heating characteristics. The burner is gravity fed by raising or lowering the fuel
reservoir (Fig. 3).

TABLE VI
Vertical Flame Test Results of Fabrics

FIBER USED	FLAME TIME SEC	GLOW TIME SEC	CHAR LENGTH INCHES
DURETTE	0	5.0	1.6
FYPRO	0	3.0	1.3
KYNOL	0	1.7	0.1
NOMEX	0	9.8	4.6
HT-4	0	0.5	2.1
PBI	0	0.0	2.4
PBI (STABILIZED)	0	1.0	1.2
80 KYNOL/20 NOMEX	0	3.2	2.6
50 KYNOL/50 NOMEX	0	4.3	2.9
50 KYNOL/50 VEREL *	0	7.0	5.5
50 KYNOL/50 FORTREL**	67	0.0	TOTAL
50 KYNOL/50 COTTON	57	4.8	TOTAL
50 KYNOL/50 NOMEX·	0	5.0	3.1
50 KYNOL/50 NOMEX	0	1.4	1.9
80 KYNOL/20 NOMEX	0	1.6	2.9
80 KYNOL/20 NOMEX	0	2.0	3.3
80 KYNOL/20 NOMEX	0	2.4	3.0

* Modacrylic; trademark of Tennessee Eastman Co.
** Polyester; trademark of Celanese Fibers Co.

FIG. 3.   JP-4 burner at 2.6 cal/cm^2-sec output.

The fabric to be evaluated is mounted in the fabric housing as shown in Figure 4. An exploded view of the fabric housing is shown in Figure 5. The fabric is mounted with a set air gap between the backside of the fabric and the heat-sensing device. The fabric mounting clamp is that commonly used when mounting fabric for the Schiefer abrasion test, and allows the desired accuracy in setting the air gap between the fabric and the heat sensor shown extending from the fabric housing. The fabric housing is then mounted in the Fire Athermancy Device as shown in Figure 6. Figure 7 shows the entire Fire Athermancy Device with timers, burner, and recorder in place. The heat-transfer measurements for this study were accomplished using an Aerotherm sensor which provides a time–temperature profile that can then be reduced to resultant skin-burn data by computer techniques [2].

During a normal test sequence, the fabric is exposed for the first 3 sec of the test and data are recorded for at least 10 sec. This results in 7 sec of postexposure data in order to examine the effects of latent heat. The total exposed surface area of the fabric sample is initially set at 4.5 in.^{2} (assuming a 0.5-cm air gap). In the

FIG. 4. Fabric holder with sample mounted.

FIG. 5. Exploded view of fabric sample holder showing heat sensor.

above configuration, the fabric is allowed to shrink to a fabric area of 3.8 in.^{2} or 85% of its original size. When the shutter opens, it provides a 4.0-in. diameter opening. The JP-4 burner is 2.0 in. in diameter. The heat-flux profile across the fabric surface is at its maximum of 2.6 cal/cm^2-sec. At any one location in the heating profile, the average heat flux remains constant within 0.1 cal/cm^2-sec. The vertical distance between the top of the graphite barrel and the heat sensor is set at 1.5 in. The quartz lamp/Meker burner source is 8.0 in. long by 3.0 in. wide in the plane of the lamp array. The top of the lamp housing is set at a vertical distance of 2.0 in. from the sensor surface. Additional information on the testing procedures, including additional heat sources and test procedures for a thermal shrinkage test, is described by Ernst [5].

RESULTS AND DISCUSSION

Table VII lists the data comparing fiber types as they affect heat transmission when exposed to the combined quartz lamp and Meker burner source for 3 sec

FIG. 6. Sample holder in place on Fire Athermancy Device (shutter open).

at varied source intensities. For the 1.9 cal/cm²-sec exposures, the quartz lamps were turned off. The stabilized PBI provided the best heat blocking of the fabrics evaluated as indicated by the amount of heat throughput at the end of a 10-sec recording cycle. The source intensity was increased for each fiber type until a burn (indicated by a value of Ω = to 1.0 at 100 μ into the skin) occurred. HT-4 also performed in a superior manner as compared to the majority of the fabrics tested. Stabilized PBI and HT-4 exhibited superior thermal stability characteristics and burn protection as evidenced by the data provided. It is interesting to note that the optical property data for the stabilized PBI and HT-4 fabrics would warrant that the HT-4 fabric should have performed in a superior manner because of the higher transmittance value associated with the Stabilized PBI fabric (Table III).

The unstabilized PBI exhibited good protection at the lower exposure levels. However, at 2.6 cal/cm²-sec, the fabric shrank to the sensor surface, causing a rapid rise in the sensor surface temperature, resulting in a total heat throughput

FIG. 7. Complete test device with sample holder and JP-4 burner in place with recorder.

TABLE VII
Effect of Fiber Type on Heat Transmission through Fabric

FIBER	WEIGHT OZ/YD^2	THICKNESS MILS	$t_{1.9}$ SEC	Q_{10} CAL/CM^2	$t_{2.6}$ SEC	q_{10} CAL/CM^2	$t_{3.4}$ SEC	Q_{10} CAL/CM^2
STABILIZED PBI	4.3	13.9	*	1.3	*	2.1	3.4	2.7
HT-4	4.6	10.5	*	1.5	*	2.3	3.3	2.9
KYNOL	5.4	20.7	*	2.0	4.2	2.9	---	---
PBI	4.6	14.2	*	2.0	2.8	5.1	---	---
DURETTE	4.5	14.3	*	2.1	3.5	2.6	---	---
FYPRO	4.1	11.9	3.1	3.5	---	---	---	---
NOMEX	4.2	12.7	2.2	4.9	---	---	---	---

Note: t_x = time to $\Omega = 1.0$ at $100\ \mu$ at a source intensity of X cal/cm^2-sec; Q_{10} = total heat received by the sensor in 10 sec; heat source, combined Meker burner and quartz lamp.

* Ω maintained below a value of 1.0 over the 10-sec period. Time of all exposures, 3 sec.

of 5.1 cal/cm^2. The shrinking fabric voided the protective air gap between the fabric and the sensor. The Nomex fabric also shrank to the sensor surface, only it shrank during the less intense exposure at 1.9 cal/cm^2-sec.

Kynol forms a carbonaceous char on the fabric surface upon exposure to a high-intensity thermal source. The exposed surface becomes blackened and the underside of the fabric remains relatively unchanged with only slight discoloration. One of the most desirable characteristics of the Kynol fabric is its outstanding dimensional stability, even at 2.6 cal/cm^2-sec (Table VII). When exposed to the heat, the fabric shows no indication of shrinking. It was expected that the Kynol fabric would have provided better heat blockage in comparison to some of the other materials because of the relatively higher weight and greater thickness.

The Durette fabric provided complete protection at 1.9 cal/cm^2-sec exposure level. However, the fabric did exhibit a tendency to shrink when the source intensity was raised to 2.6 cal/cm^2-sec. The fabric shrank to just short of coming in contact with the sensor surface. Both Fypro and Nomex performed poorly in comparison to the other materials evaluated, although they allowed the heat passage by differing mechanisms. The Nomex fabric shrank to contact with the sensor and then split apart, allowing direct exposure of the underlying sensor. The rapid heat loss from the hot fabric as well as the direct exposure of the sensor resulted in the relatively high heat throughput of 4.9 cal/cm^2. The Fypro material remained intact and exhibited little shrinkage. The optical characteristics of the Fypro fabric indicate that it should perform equally with the majority of the fabrics evaluated. The only noticeable deleterious effect on the Fypro material was that it became severely weakened and partially friable upon exposure to the heat. The color (black) might be a partial explanation for the data.

At this point in the overall effort, it was decided to examine these materials in the intended end-use environment, i.e., in a JP-4 fuel fire, to determine whether or not correlation exists between the laboratory results and the end-use performance required.

The intense thermal environment created by a large-scale fire, the movement of the man through the fire, and the resulting thermochemical influences on the various fiber types have demonstrated that laboratory evaluations were not capable of predicting the total thermal performance characteristics of the materials studied. In several cases, the materials evaluated in the laboratory were found to be completely nonflammable, while exposure to JP-4 fuel fire in the fire pit caused the materials to afterflame extensively once removed from the fire source. Stabilized PBI and HT-4 did provide the best protection in both the laboratory evaluation and in the field studies. The untreated PBI was the next best, while the remaining fabrics deteriorated extensively and allowed extensive

TABLE VIII
Heat Transmission Data by Fiber Type

FIBER	WEIGHT OZ/YD^2	THICKNESS MILS	$t_{2.1}$ SEC	Q_{10} CAL/CM^2
STABILIZED PBI	4.3	13.9	*	1.5
HT-4	4.6	10.5	*	1.8
PBI	4.6	14.2	3.6	2.8
NOMEX	4.2	12.7	2.0	3.3

Note: t_x and Q_{10} are the same as indicated in the note to Table VII; heat source, JP-4 fuel burner.

* See footnote to Table VII.

damage to occur under flight suits made from Kynol, Fypro, Durette, and Nomex fabrics. It is of value to note that the fabric weight required to prevent a burn at the 2.6 cal/cm^2-sec exposure level was 4.3 oz/yd^2 for the stabilized PBI and 4.6 oz/yd^2 for the HT-4 fabrics. Emphasis was then directed towards developing a test technique that would more closely approximate the thermochemical effects seen in the field evaluations. The above resulted in the development of the JP-4 burner.

Four of the candidate fabrics were examined using the JP-4 burner. The four materials chosen for evaluation were those materials, i.e., stabilized PBI, HT-4, PBI, and the standard material used in Air Force flight suits (Nomex), that performed best in large-scale fire evaluations. The resulting data are listed in Table VIII.

Intimate fiber blending (i.e., blending of the staple fibers prior to spinning the yarn) is an attempt to derive the more desirable characteristics of the constituent members. These characteristics include economics, durability, nonflammability, comfort, etc. For the blended fabrics listed in Table II, Kynol and Nomex were chosen as those materials that are available and economically reasonable in cost which also represent fibers that exhibit differing yet desirable thermal characteristics. Since Kynol is nonflammable and nonshrinking when exposed to a flame, it was blended with some of the more common fibers, such as Verel, Fortrel, and cotton, to add toughness and comfort characteristics. Kynol was also blended with Nomex in varying weights to utilize the fire-resistant qualities of both fibers, yet combining the nonshrinking characteristics of Kynol with the Nomex fiber which is more susceptible to thermal shrinkage. The Nomex fiber would provide toughness and durability to the less durable Kynol fiber. An attempt was also made to blend Nomex and PBI fibers. Nomex may be less expensive than PBI and therefore would assist in reducing the cost of a fabric using the more expensive PBI fiber; at the same time, the nonflammable PBI fiber would improve the thermal stability of the blended fabric. It would be desirable that the best qualities of the two constituents would be most prevalent in the final fabric, although more often than not the weaker characteristics of the composite system have the greatest affect on the resultant properties.

The blending of Kynol with Fortrel and cotton resulted in flammable combinations as evidenced in the data provided in Table VI. The Kynol–Verel blend was also somewhat less fire-resistant than the remaining fibers evaluated, although it was self-extinguishing. Fortrel is a thermoplastic fiber which degraded the nonflammability characteristics of the Kynol fiber. Verel is a modacrylic which by itself will not support combustion, although it will burn in the presence of a flame. The latter combination of materials clearly indicates how the weaker characteristics of one of the constituent fibers can cause the desirable property of nonflammability to be degraded. The char length of 100% Kynol was 0.1 in., while the blended fabric of Kynol–Verel exhibited a char length of 5.5 in. in a 50/50 fiber blend (Table VI). It is interesting to note (Table IX) that the heat transmission data for the more flammable Kynol–cotton blend shows it to provide the best thermal protection of these three blend combinations. This clearly demonstrates the importance of having a nonshrinking fiber (Kynol). By

TABLE IX

Effect of Fiber Blending on Heat Transmission through Fabric

FIBERS (%)	WEIGHT OZ/YD²	THICKNESS MILs	$t_{1.9}$ SEC	Q_{10} CAL/CM²	$t_{2.6}$ SEC	Q_{10} CAL/CM²	$t_{4.4}$ SEC	Q_{10} CAL/CM
KYNOL 50/VEREL 50	6.7	22.6	7.8	2.6	3.4	3.8	2.4	4.6
KYNOL 50/FORTREL 50	7.2	20.7	*	2.2	2.8	3.1	1.6	4.1
KYNOL 50/COTTON 50	6.6	22.5	*	1.3	*	2.0	4.1	2.7
KYNOL 50/NOMEX 50	6.7	21.1	*	1.3	*	2.3	3.1	3.6
NOMEX 50/PBI 50	4.6	14.5	2.7	4.6	2.5	4.8	1.7	6.4
KYNOL 80/NOMEX 20	5.1	17.8	*	2.3	2.7	3.2	1.8	6.2
KYNOL 80/NOMEX 20	4.5	14.8	6.0	2.5	2.1	3.7	2.2	4.6
KYNOL 80/NOMEX 20	5.2	17.4	*	2.1	2.6	3.7	1.9	4.7
KYNOL 80/NOMEX 20	4.8	13.7	7.2	2.5	2.9	4.0	2.2	5.6
KYNOL 50/NOMEX 50	5.0	15.4	4.4	2.8	2.8	4.2	---	---

Note: t_x, Q_{10}, and the heat source used are the same as indicated in the note to Table VII.

* See footnote to Table VII.

maintaining the insulating gap between the fabric and the skin, the fabric provides good protection from direct heat passage. Even though the flammable nature of the above fabric makes it undesirable for the application being studied, it does emphasize the importance of fiber thermal stability.

The blends of Kynol and Nomex represent an interesting combination of materials. The addition of Nomex to the Kynol demonstrated a definite improvement in the durability of the blended fabric as compared to a 100% Kynol fabric. Increasing the Nomex fiber loading served to further improve the abrasion resistance of the blend (Table IV) as well as the strength properties. The economics of the blended material are predicted to be an improvement based on

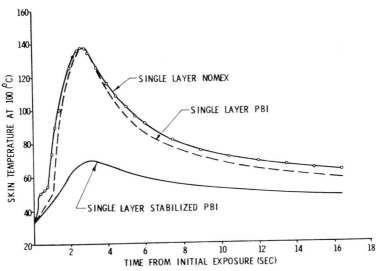

FIG. 8. Skin temperature response for single fabric layers (nominal exposure).

the low (potential) cost of the raw materials used in preparing the Kynol fiber as compared to the cost of Nomex fiber (and its starting materials). The vertical flame test data (Table VI) indicate that the addition of the Nomex to the Kynol does not greatly retard the superior nonflammability properties of the Kynol. However, the heat transfer of the Nomex/Kynol blends is generally greater than 100% Kynol.

The data in Figures 8 and 9 are plotted to show the temperature at 100 μ depth

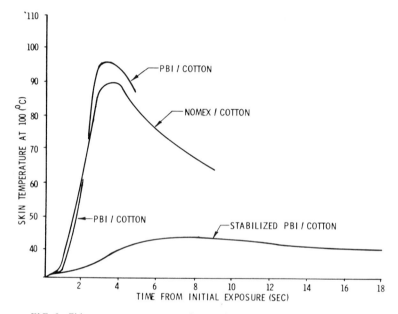

FIG. 9. Skin temperature response for two fabric layers (nominal exposure).

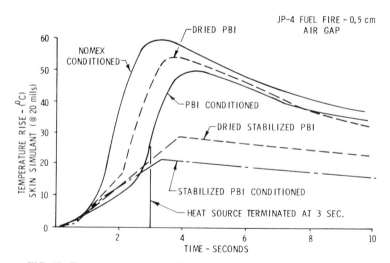

FIG. 10. Temperature response of skin simulant covered by various fabrics.

into the skin. Three specific fibers have been treated in this fashion. These are Nomex, PBI, and stabilized PBI. The data in Figure 8 were obtained using a model fire, a skin model, and a computer program to calculate the temperature rise in the skin at 100 μ depth. The skin surface is assumed to be totally absorbing, circulatory effects are assumed to be constant during the heating period, and an adiabatic back wall boundary condition used behind the 2-mm thick final node of fatty tissue. The heating time covered the first 3 sec of the cycle shown. The peak attained is due to the shrinkage of the fabric against the sensor eliminating the air gap which is capable of preventing a rapid temperature rise as evidenced by the stabilized PBI fabric. When we consider that underwear is normally used under the protective garment, the skin temperature using a knit cotton fabric as underwear (Fig. 9) is significantly lower. In these calculated curves, the structural failure of the fabric is not considered. We know from full-scale fire tests that non-PBI fabrics split apart and promote the spread of the fire to the underlayer of fabric. Once the fabric has ignited, the flame may propagate to portions of the clothing which originally remained intact. On the other hand, PBI and stabilized PBI fabrics do not exhibit this tendency to fail structurally (nor does HT-4, but it becomes friable).

In a test using the JP-4 burner already described, the data obtained in Figure 10 confirm the temperature rise in skin simulant covered by Nomex, PBI, and stabilized PBI. In addition, fabrics that had been predried as well as fabrics that had been conditioned at standard textile testing conditions were compared. If the HT-4 (the *para*-oriented aromatic polyamide) were to be added to this figure, it would compare to the dried stabilized PBI.

CONCLUSIONS

The following conclusions are based on the study described and on 10 years of related experience in the development and evaluation of fabrics for personnel thermal protection from a liquid fuel fire environment:

(A) The JP-4 burner modification to the Fire Athermancy Device provides a closer cross section of the fire properties that occur as a result of aircraft accidents and resulting fires.

(B) Fiber thermal stability is the most important parameter for designing a thermally protective fabric. The fabric must not melt, burn, or shink to provide maximum protection. Two fibers presently exist that can meet this all important objective. These are stabilized PBI and HT-4.

(C) In fabric design, three parameters exhibit significant potential for ensuring adequate thermal protection when utilizing the above fibers. (1) Fabric weight must be sufficient to provide a thermally protective barrier; for example, woven fabrics made from either stabilized PBI or HT-4 should be a minimum of 4.5 oz/yd^2 to provide thermal protection from a 3-sec exposure at 2.6 cal/cm^2-sec. (2) Fabric optical properties are an area where considerable improvement might be accomplished to achieve improved thermal protection. (3) Fabrics must be nonshrinking in a fire environment.

REFERENCES

[1] A. M. Stoll and M. A. Chianta, *Trans. New York Acad. Sci.*, 649 (1971).

[2] H. L. Morse, et al., *Analysis of the Thermal Response of Protective Fabrics*, AFML-TR-73-17, January, 1973.

[3] A. M. Stoll and L. C. Greene, *J. Appl. Physiol.*, *14*(3), 1959.

[4] J. Preston, paper presented at the Third International Chemical Fiber Symposium, Dresden, East Germany, December, 1970.

[5] D. Ernst, *Laboratory Test Techniques for Evaluating the Thermal Protection of Materials When Exposed to Various Heat Sources*, AFML-TR-74-118, August, 1974.

MEDICAL APPLICATIONS OF POLYMERIC FIBERS

ALLAN S. HOFFMAN

Center for Bioengineering and Department of Chemical Engineering,
University of Washington, Seattle, Washington 98195

SYNOPSIS

A survey is presented of the wide variety of medical applications of synthetic and natural fibrous materials, both "nonbiodegradable" (or slowly biodegrading) fibers and biodegradable fibers. The most important classes of medical polymers which are used in fiber form include polyesters, poly-(fluorocarbons), polyamides, polyolefins, carbon, polypeptides, and polysaccharides. The forms they are used in include continuous filaments and yarns, knitted and woven fabrics, nonwoven felts, flocking fibrils, and chopped fibers for reinforcement. They are sometimes used in unsupported forms and sometimes laminated on the surface of or within other polymeric matrices. Their uses as surgical sutures, in tissue substitutes, and parts of instruments, devices, or artificial organs will be discussed.

INTRODUCTION

There exist a number of significant applications of polymeric fibers in medicine, starting from their early use as sutures to more recent uses as tissue ingrowth and anchoring matrices [1–3]. Many of these applications are special to textiles, which can be readily fabricated in a wide range of microporous structures. Indeed, fibers or textiles represent a particularly broad class of compositional, morphological and "geometric" possibilities. Probably only a relatively small fraction of the morphological and geometric variations have been assayed for biomedical uses. For example, there are different geometric shapes possible for extruded fiber cross sections, different compositional morphologies possible for co-extruded biocomponent fibers, and many different geometries in which such fibers or their combinations may be woven or knitted together or matted into felts. Most of these variables would have significant effects on biomolecule–cell interactions such as inflammation, tissue ingrowth, encapsulation, etc.

Despite the fact that many of these variations have not yet been investigated for biomedical applications, there is an extensive literature on uses of polymeric fibers in medicine. This review will cover such uses in four major areas: (1) general surgical applications, (2) cardiovascular system applications, (3) musculoskeletal system applications, and (4) cutaneous and percutaneous applications. The important considerations of sterilization, biodegradation, and biocompatibility will first be briefly discussed.

STERILIZATION, BIODEGRADATION, AND "BIOCOMPATIBILITY" CONSIDERATIONS

All materials coming in contact with body fluids should first be sterilized. Bruck [4] has emphasized the need to establish individualized sterilization protocols for each material–application combination. Lee and Neville [1] summarize the various options available for sterilizing polymeric biomaterials.

When such materials are implanted in a living species or contacted with body fluids, acute (as well as chronic) inflammatory responses can occur. Depending on the stability (or instability) of the chemical bonds in the polymer material, such "foreign" materials can incite vascular responses of hyperemia and edema, leading to abscess or cellulitis development and cellular responses of ingestion, sequestration, and fibrous or mineral depositions, leading to granuloma development, tissue ingrowth, and local or overall fibrous encapsulations. Within the cardiovascular system, thrombotic or embolytic events are also encountered. Some of these responses are benign or controllable and some represent a problematical rejection of the foreign material. Most responses are very sensitive to the geometry of the textile implant (see below).

It is particularly important to assay the toxicity of implanted biomaterials when they contain extractable molecules from fiber additives [5, 6] or from degradation of the polymer chain [7]. Some of the commonly encountered polymeric fibers are essentially nonbiodegradable (or very slowly biodegrading)

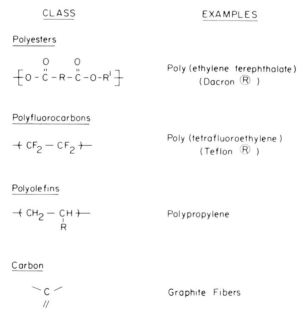

FIG. 1. Nonbiodegradable (or slowly absorbing) fiber-forming polymers.

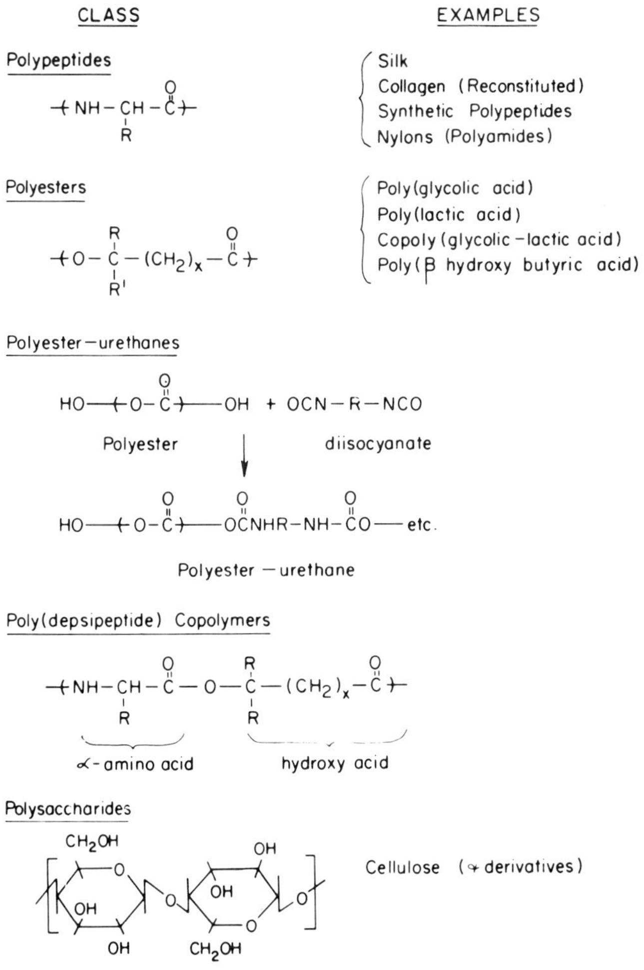

CLASS   EXAMPLES

Polypeptides

$$+NH-\underset{\underset{R}{|}}{CH}-\overset{\overset{O}{\|}}{C}+$$

⎧ Silk
⎨ Collagen (Reconstituted)
⎩ Synthetic Polypeptides
  Nylons (Polyamides)

Polyesters

$$+O-\underset{\underset{R'}{|}}{\overset{\overset{R}{|}}{C}}-(CH_2)_x-\overset{\overset{O}{\|}}{C}+$$

⎧ Poly(glycolic acid)
⎨ Poly(lactic acid)
⎩ Copoly(glycolic-lactic acid)
  Poly(β hydroxy butyric acid)

Polyester−urethanes

$$HO-(O-\overset{\overset{O}{\|}}{C})-OH \; + \; OCN-R-NCO$$

Polyester   diisocyanate

↓

$$HO-(O-\overset{\overset{O}{\|}}{C})-O\overset{\overset{O}{\|}}{C}NHR-NH-\overset{\overset{O}{\|}}{C}O-etc.$$

Polyester −urethane

Poly(depsipeptide) Copolymers

$$+NH-\underset{\underset{R}{|}}{CH}-\overset{\overset{O}{\|}}{C}-O-\underset{\underset{R}{|}}{\overset{\overset{R}{|}}{C}}-(CH_2)_x-\overset{\overset{O}{\|}}{C}+$$

α-amino acid   hydroxy acid

Polysaccharides

Cellulose (& derivatives)

FIG. 2. Bioabsorbable fiber-forming polymers.

and some are specifically used in *in vivo* situations—as for sutures—where biodegradability is desired. Figures 1 and 2 summarize these two categories of fiber-forming polymers.

Leininger [8] has surveyed the biocompatibility of polymers used as surgical implants, many of which are in fiber form. One must distinguish two major regions of the body where such considerations are relevant: (1) the cardiovascular system, i.e., blood, and (2) all of the extravascular regions, e.g., intramuscular, subcutaneous, intraperitoneal, etc. The problems of blood compatibility are particularly relevant to fiber applications in the cardiovascular system, such as for vascular prostheses, prosthetic heart valves, and components of heart-assist devices (see section on Cardiovascular System Applications). Extravascular biocompatibility is important to all of the many other polymeric fiber applications (see sections on General Surgical Applications, Musculoskeletal System Applications, and Percutaneous and Cutaneous Applications).

There are a wide variety of *in vitro* and *in vivo* methods which have been proposed for evaluating the biocompatibility of foreign materials, and it is not possible to review them critically in this article. Several reviews covering this topic have been published [1, 2, 6, 8–11].

GENERAL SURGICAL APPLICATIONS

Sutures and Ligatures

Sutures are threads used to close wounds while ligatures are threads used to tie off bleeding vessels. Similar materials are used for both applications, and the term, "suture" will be used here to denote either of the two.

There are two major classes of sutures: adsorbable (or biodegradable), used mainly to close internal wounds; and nonabsorbable (or nonbiodegradable or very slowly biodegradable), used mainly for exposed or cutaneous wound closure. Some typical polymer compositions are shown in Figures 1 and 2.

Among the absorbable sutures, one of the longest in use is the collagenous "catgut" type, derived directly from sheep intestinal mucosa, which can also be crosslinked by chromium ions to yield "chromic catgut," a more slowly absorbing suture. (Loss in strength of catgut is seen in about 3–7 days while chromic treatment can extend this from 10–40 days depending on the extent of treatment [1].) Such sutures can create inflammatory reactions which delay wound healing [1]. This may not be a serious problem since they are eventually totally absorbed. However, because of the problems of supply, preparation, quality control, packaging, as well perhaps as the inflammatory response, attempts have been made to fabricate under controlled conditions a chemically reconstituted collagen suture derived from purified beef tendons. These have met with limited success due to problems of quality control, and at present there is no such suture on the market.

Other, more slowly absorbing sutures from naturally occurring polymers include silk and cotton (or linen). They may gradually lose strength due to absorption over a period of months, rather than days, but they do eventually become totally absorbed [1].

There are inherent disadvantages in using these naturally occurring poly-

TABLE 1
Tensile Strength of Size 3/0 Sutures [12]

	Diameter (inches)	Straight Pull (psi)	Knot Pull (psi)
Catgut[a]	.0125	54,600	30,900
PGA	.0100	77,800	49,700
Silk	.0104	62,400	40,000
Polyester[b]	.0097	115,000	50,100

[a] Medium chromic.

[b] Dacron polyester fiber; trademark of DuPont.

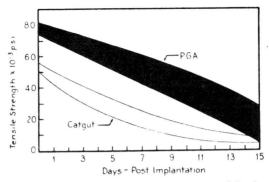

FIG. 3. Tensile strength of PGA and medium chromic catgut sutures following subcutaneous implantation in rabbits [12].

peptides or polysaccharides, especially those due to source variability in composition and properties. Thus, there have been attempts made to synthesize under controlled conditions a reproducible, useful, synthetic bioabsorbable polymeric suture [12]. Four primary requirements have been delineated for the ideal bioabsorbable polymeric suture [12]: (1) absorbability; (2) negligible toxicity; (3) high strength, knot-tying and knot-holding qualities; and (4) readily sterilized.

Among the many different synthetic biodegradable polymers possible [7], only two classes of polymers have been seriously investigated for this application: synthetic polypeptides [13] and polyesters derived from aliphatic hydroxy carboxylic acids, especially glycolic [12] and lactic acids [14]. One major disadvantage of synthetic polypeptides is their tendency to incite undesirable immunogenic reactions [15]. It was for this reason that Frazza and Schmitt [12] centered their attention on polyesters derived from hydroxy acids as glycolic acid. The products of degradation of such polymers are mainly the hydroxy acids themselves, many of which are already endogenous metabolites. Two synthetic, absorbable polyester sutures now are marketed, based on poly(glycolic acid) (PGA) (Dexon, Davis and Geck Co.) and a copolymer of glycolic (mainly) and lactic acids (Polyglactin, Johnson and Johnson Co.). Table I compares the strength properties of PGA with some other common sutures. Figure 3 shows the rate of loss in strength for PGA and chromic catgut after subcutaneous implantation in rabbits. One advantage cited for PGA over catgut is the retention of a high percentage of its initial tensile strength during the first 7–11 days after implantation; this is the most critical period of wound repair [12].

Pure poly([L+]lactic acid) (PLA) degrades to L+ lactic acid, but at much too slow a rate to be a useful absorbable suture, despite the easier processing conditions over those for PGA. Thus, it has either been copolymerized with glycolic acid, or polymerized in the dl form to yield a more rapidly absorbing polymer. Kulkarmi [14] describes poly(dl-lactic acid).

Topographical changes in absorbing sutures (catgut, PLA, PGA, and two proprietary polyesters) have been studied in the SEM following various periods of subcutaneous implantation in the rat [16]. Specific, localized structural defects were noted to form in most of the sutures, although it was concluded that such defects did not necessarily correspond in time to the loss in strength of the suture.

Among the nonabsorbable (or slowly absorbing) polymeric sutures which have been studied are Dacron (DuPont Co., Inc.), nylon, Orlon (DuPont Co., Inc.), polyfluorocarbons as Teflon TFE (DuPont Co., Inc.) and Teflon FEP, and polyolefins as polyethylene and polypropylene. Teflon-coated Dacron (Polydek, Deknatel Corp.) and nylon (Tevdek, Deknatel Corp.) [1] and silicone-treated silk sutures [17] have also been studied.

Such sutures would presumably be used mostly for closing cutaneous or oral incisions, so that after sufficient wound healing they might easily be removed. Homsy et al. [17] have studied the instantaneous stress and total work required to withdraw various monofilamentary sutures after subcuticular stitches had been left for various periods of time in porcine subjects. Figure 4 shows results after 3 weeks implantation. Taking into account additional factors as suture strength, suppleness, and knot durability, they concluded that nylon and Teflon FEP are the sutures of choice for cutaneous closure.

Tissue compatibility (*in vivo*) and cell culture (*in vitro*) studies have compared both absorbable and nonabsorbable suture materials. Postlethwait et al. [18] showed that nylon, Dacron, and Orlon exhibited a lower tissue inflammatory response than did silk or cotton. However, studies by the same author using the technique of Sewell et al. for quantitation of tissue response to sutures [19] showed low inflammatory response for all sutures studied (PGA, catgut, silk, and Dacron) [20]. Leininger [8] has summarized the *in vivo* and *in vitro* biocompatibility studies on suture materials.

Problems of infection of both absorbable and nonabsorbable sutures used in contaminated tissues were considered by Edlich et al. [21]. They concluded that

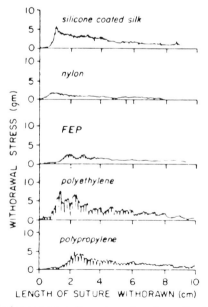

FIG. 4. Stresses required to remove various sutures 3 weeks postimplantation in rabbit [17].

the chemical composition of the suture appears to be the most important determinant of early infection, with nylon and polypropylene exhibiting less incidence of infection than other nonabsorbable sutures. Among the absorbable sutures, PGA appeared to elicit the least inflammatory response in contaminated tissues. Multifilament sutures of cotton and silk potentiated more infection than did any nonabsorbable suture, in agreement with the earlier results of Postlethwait et al. [18]. Coating the multifilament suture with silicone, wax, or Teflon TFE did not alter the incidence of early infection. Indeed, they conclude that all sutures potentiate some degree of infection in contaminated subdermal tissue, and probably no suture should be used in such circumstances.

However, one possible solution to such problems of infection has been suggested by Tollar et al. [22], who coated a wide variety of sutures with a hydrophilic, crosslinked poly(2-hydroxyethyl methacrylate) gel (Hydron "hydrogel," National Patent Development) into which they imbibed antibiotics. They noted that neither the knotting ability nor any other mechanical properties of the suture were influenced by the gel coating.

Reconstructive and Repairative Surgery on Soft Tissues

A large number of Medical Grade Silastic (Dow Corning Co., Inc.) implants for reconstructive surgery are currently marketed [23, 24]. In many cases, the Silastic implant may be reinforced by an internal, woven Dacron mesh and/or backed by a Dacron felt which is cemented to the implant with Medical Grade Silastic Adhesive (Dow Corning Co., Inc.). The latter is used to permit anchoring of the implant by the surrounding tissues as fibroblasts penetrate the felt backing and proliferate fibrous tissue within and outside of the felt. In the absence of the felt, it is likely that the smooth surface of the Silastic would elicit a smooth fibrous encapsulation around the whole implant, which would not prevent its movement within the body (which is often undesirable).

Another approach to anchoring such implants by fibrous tissue proliferation within the implant has been suggested by Toranto et al. [25]. They filled a Silastic capsule or a clear dental acrylic with less than 5% by weight of small particles of absorbable sutures as PGA, catgut, and chromic catgut. As the suture particles absorbed, the pores left behind permitted fibrous ingrowth, with a minimum of fibrous encapsulation.

Dacron velour-lined Silastic [26] and nylon 66 velour-lined polyurethane [27] have been suggested as bladder prostheses. Merendino et al. [28] has used a Teflon mesh for repair of diaphragmatic hernias. Schmitt and Polistina [29] have designed a variety of absorbable gauze, felt, or velour surgical dressings of PGA textiles.

Clark et al. [30] have studied two tissue reactions, the inflammatory response and the extent of encapsulation in subcutaneous rat implantations for four porous polymeric systems: a polyurethane (PU) foam, nonwoven Teflon TFE, and two woven polyester textiles. They observed such severe tissue reactions to the PU and Teflon TFE that no correlation could be made between the characteristics measuring surface porosity and the tissue response. However, they were able

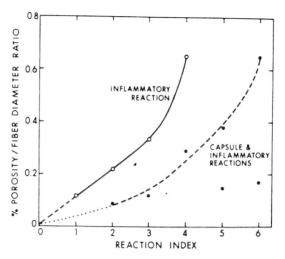

FIG. 5. Reactive index of subcutaneous tissue of the rat to polyester fabrics as a function of the ratio of the percent open area of the porous surface to the fiber diameter in microns [30,71].

to observe an effect of the ratio of percent open area or porosity to fiber diameter for the polyester textiles, with the least capsule and inflammatory reaction at a percent porosity/fiber diameter ratio ≤ 0.1 (Fig. 5). Based on these observations, they designed a specific fabric with a 2.3% open area, a pore size of 32 μm, and an elliptical, 180 $\times$ 20 μm yarn, which exhibited a tissue response equal to the best seen in their earlier studies. Such investigations emphasize the importance of the fiber–textile system composition, morphology, and geometry to its ultimate tissue response and biocompatibility. It should be noted, however, that much remains to be learned about the influence on biological response of such important factors as woven, knitted or nonwoven textile structures, filament diameter, and cross section yarn construction, thickness and depth of the pile, sheared versus looped velour structures, length to diameter and density of flocked fibrils, etc.

CARDIOVASCULAR SYSTEM APPLICATIONS

Vascular Prostheses

One of the most widely used and successful applications of textiles in medicine is as vascular prostheses for the larger, high flow-rate arteries [1–3]. Most are woven, knitted, or microporous tubular structures which are supple and resemble the softness and flexibility of natural blood vessels. Since they are porous structures, such prostheses must be "plugged" in order to prevent blood loss after implantation. This is normally accomplished by either pre-clotting in blood before implantation, or sometimes by allowing the clot to form *in situ*. The two major synthetic polymer materials used as vascular grafts are Dacron and Teflon TFE. Both are highly thrombogenic, and there is no problem in getting a fibrin deposit on and within the fibrous structures. Often the woven or knitted fabrics are made as crimped tubes, in order to avoid sticking of adjacent or opposite areas of the tubes to each other during fibrin deposition [31]. Irregular crimping may be better in this action than regular crimping, although the latter may be more aesthetically appealing to the surgeon.

Sometimes excessive or irregular initial or secondary fibrin deposition can lead to local formations of thrombus, an undesirable event. What is desired is the eventual lysis of the fibrin and replacement by a monolayer of normal endothelial cells on the inside (intimal surface) along with an anchoring fibrous ingrowth of normal connective tissue on the outside (adventitial surface). It is toward these goals that a number of surgical–industrial engineering teams have been striving, and in parallel with these efforts there has ensued a lively debate, starting in the operating room and extending to the professional literature and the marketplace.

Conventional, smooth-textured woven and knitted Dacron grafts have been commercially available since the mid-1950s. The good tensile strength, resistance to deterioration, ease of handling, and retention of properties after implantation of Dacron have led to reasonable satisfactory results after thousands of procedures.

The DeBakey Ultra-Lightweight Dacron knitted graft (USCI, Inc.) was introduced in 1957 and has since been utilized in upwards of 50,000 implants [32]. Cooley Graft (Meadox Medicals, Inc.) [33] is available in both knitted and woven fabrics in an attempt to tailor the construction to the use. The knitted graft is made with a fine yarn in order to reduce pore size while increasing porosity. The small pore size enhances the preclotting procedure, while the high porosity is designed to provide anchorage for the neointima and encourage tissue ingrowth. The woven graft is designed for use with a fully heparinized patient where prevention of implant bleeding is of primary concern.

Wesolowski et al. [34] concluded in 1961 that porosity was the primary determinant of success or failure of the vascular graft. Later studies [35] of the various geometric and compositional factors involved in long-term failures in human vascular prosthetic grafts led to the design of a finely knitted Dacron polyester prosthesis, which they named the Wesolowski Weavenit (Meadox

TABLE II
Comparison of Some Dacron Arterial Prostheses [39]

TYPE OF PROSTHESIS	INTERSTICES PER SQUARE CENTIMETER	FABRIC THICKNESS, μ	MEAN WATER POROSITY, ML/SQ CM/MIN AT 120 MM HG
POROUS (SMOOTH-WALLED)			
DE BAKEY KNITTED DACRON 6010	1,480	380	3,000
WESOLOWSKI WEAVENIT	1,800	284	4,800
MICROKNIT	3,360	209	8,600
POROUS (FILAMENTOUS-WALLED)			
EXTERNAL VELOUR	600	700	3,000–5,000

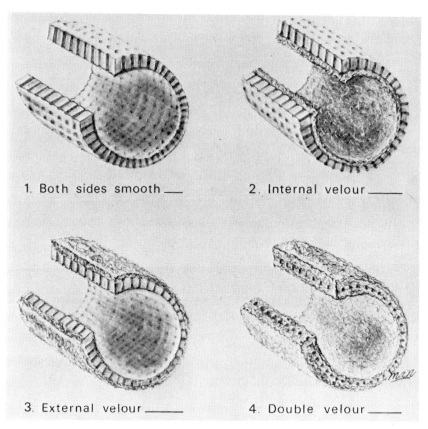

1. Both sides smooth ___

2. Internal velour _____

3. External velour _____

4. Double velour _____

FIG. 6. Diagrammatic views of the surfaces presented to the blood and perigraft tissues by conventional knitted prostheses with "smooth" walls and by knitted prostheses having velour surfaces inside, outside, or both [50].

Medicals, Inc.) [36]. It has been designed according to the same philosophy as was the Cooley Graft. Unlike previous prostheses knitted at 20 needles/in., the Weavenit fabric is made at 40 needles/in., thereby eliminating the need for relatively bulky yarns to fill the interstices. The result is a fine-knit fabric with the feel of a woven material. Good handling qualities would be expected to help minimize failures due to technical errors. The smaller mass is expected to minimize tissue reaction and capsule thickness. The graft is thin-walled, non-fraying, conformable at anastomoses and has high porosity ($\sim$4000 cc H_2O/cm^2/min at 120 mm Hg vs. $\sim$3000 for Cooley Graft). The healing of the implanted prosthesis exhibits a very orderly pattern, no doubt attributable in large part to its design. Microknit (Golaski Laboratories, Inc.) [36] Dacron graft is an extension of the Weavenit design. This very light, fine, knitted graft has a thickness. averaging 200 μ, compared to 230–290 μ for Weavenit and 320–380 μ for the DeBakey graft, while its porosity is very high, on the order of 8600 cc H_2O/cm^2/min.

All of the above four Dacron prostheses are smooth-walled, and although thread size, interstices size, texturization or knit pattern may vary (Table II and

FIG. 7. External and internal surfaces of porous grafts (20×): (A,B) DeBakey knitted Dacron; (C,D) Wesolowski Weavenit; (E,F) Microknit; (G,H) external Dacron velour 39.

Figs. 6–8), it has been stated by Sauvage et al. [37] that "all fail to achieve the final, vital step of full-wall healing—fibrous tissue healing and endothelialization of the flow surface of the graft".

Based on this conclusion, Sauvage et al. [37, 38] developed a texturized knitted Dacron fabric with an external velour surface (USCI Sauvage External Velour Prosthesis) [32] (Figs. 6–8). The velour fills the interstices of the underlying fabric, reducing implant bleeding while maintaining a porosity of from 2000 to 3500 cc/cm^2/min [32, 37]. The external velour surface and high porosity

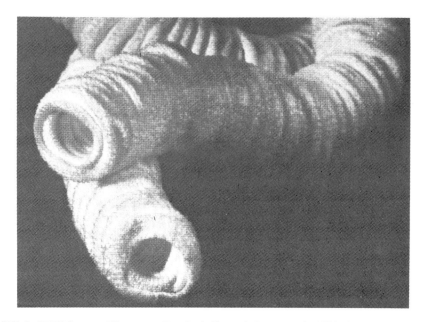

FIG. 8. USCI Sauvage Filamentous Prosthesis (formerly known as the USCI Sauvage External Dacron Velour Prosthesis) [32].

facilitate secure attachment of the outer fibrous capsule, albeit at the cost of sizable wall thickness (700 μ). Newer design specifications include a total thickness of only 400 μ and a water porosity of 1500 cc/cm^2/min (USCI Sauvage Filamentous Velour Prosthesis). In such double velour prosthetic materials, it is proposed that cellular migration along the velour "trellis" and through the interstices eventually will lead to replacement of the lumenal thrombus accumulation with fibrous tissue, and an intact endothelialized flow surface. Promising indications of such full-wall healing in clinical applications have been reported [37–39]. Double velour Dacron grafts are now also offered by Meadox Medicals, under the trademark Cooley Double Velour Graft and Mircrovel Double Velour Graft [36]. There is evident need for further animal studies to identify the optimum single or double velour structure [40].

Other considerations which are important to the design of a vascular graft involve suturing the graft to the natural vessel wall, where thrombus formation can initiate, and the flow and thrombosis characteristics within bifurcation grafts [41].

Teflon TFE has been used in the past as a woven or knitted prosthetic (e.g., Edwards Teflon Prosthesis, USCI), but has not been used as much recently relative to Dacron. Woven Teflon TFE grafts have very low porosity, which prevents resorption of degenerated fibrous tissue, leading to calcification, distal embolization, and/or thrombotic occlusion [34]. Boyd and Midell [42] cite the wide disparity between woven Teflon TFE and knitted Dacron failure rates as sufficient grounds for restricting Teflon TFE to use as a temporary shunt, its use as a permanent graft being "dangerous."

Recently, however, a new type of "expanded" Teflon TFE has become

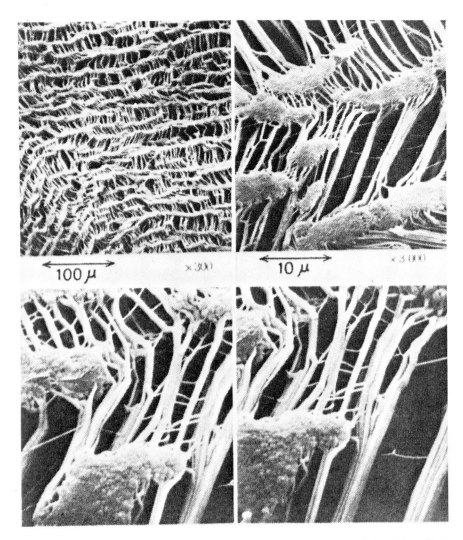

FIG. 9. Electron microscopic pictures of expanded porous polytetrafluoroethylene (Gore-Tex) [44].

available, called Gore-Tex (W. L. Gore & Assoc., Inc.). Gore-Tex is a highly porous Teflon TFE material, consisting of a network of very small nodules interconnected by thin fibrils (Fig. 9). The ratio of pore to matrix volume can be adjusted essentially continuously from 0 to 96%. As a vascular prosthetic material, it is essentially nondegradable, highly thrombogenic, easily cut, resistant to fraying, uniformly expansive, and it can be made with sufficient body to maintain its cross-sectional shape. Sometimes problems may be encountered with its low tear strength and tendency to stick to itself when being preclotted.

This material is being investigated as a venous or small caliber arterial prosthesis with generally encouraging results [43–49]. For example, Soyer et al. [43]

Graft Type	Stiffness	Fraying	Graft Bleeding
Teflon weave	++++	++++	+
Coarse Dacron knit	+++	+	+++
Teflon knit	++++	0	++++
Fine Dacron knit	+++	0	+++
Autogenous vein	0	0	0

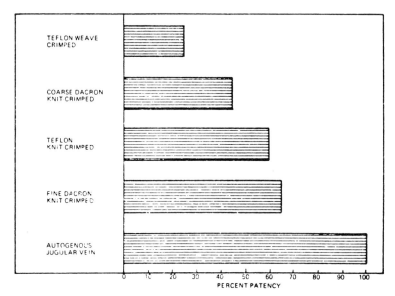

FIG. 10. Comparison of undesirable physical characteristics of five small artery prostheses [52].

used Gore-Tex as a venous prosthesis in piglets. They observed no fibrous ingrowth through the graft walls, and endothelialization only near the anastomoses. However, implants in two human cancer patients demonstrated patency after 5 months, indicating potential as a temporary venous replacement.

Matsumoto et al. [44] compared the use of smooth, noncrimped grafts of expanded Teflon TFE (85% porosity) with ultralightweight woven Teflon TFE in small canine arteries. All grafts of expanded Teflon TFE remained patent up to 4.5–11 months, whereas the woven grafts all occluded within 101 days. The patent grafts showed evidence of fibroblastic penetration through the graft walls, and a thin, fibrous, well-attached neointima with apparent endothelialization. This contrast between the porous and woven Teflon TFE materials again demonstrates that porosity and not material is probably the primary determinant of ultimate success or failure of synthetic grafts.

A study of the relationship between tissue ingrowth and internodal distance in expanded Teflon TFE was conducted by Volder et al. [45], who implanted 32 grafts in sheep carotid arteries. All grafts remained patent during the study (up to 6 months), but a marked difference was noted in the histologic sections taken from the two groups under study. The long-fibril-length (30–100 μ)

samples showed a regular, well-developed neointima covered by a smooth endothelium-like layer, and complete penetration to the lumen by fibroblasts and capillaries. The other group of short-fibril-length (5–20 μ) material showed irregular coverage, incomplete ingrowth, and no cellular neointima. This correlates with observations by Sauvage et al. [50]. Thus it appears that Teflon TFE, sometimes avoided as a vascular prosthetic material, may find future use in a different structural form.

A number of studies have been published which compare knit or woven Dacron and Teflon TFE prostheses (e.g., refs. [51–53] and Fig. 10). It should be mentioned, however, that the factors of fiber and yarn diameter, cross-sectional shape, and knit or weave texture and porosity are difficult to maintain exactly the same in these two different polymeric fibers. Thus, true *material* comparisons based on their obvious differences in wettability and other (less understood) differences in reactivity toward blood components may be overshadowed by differences in the geometric or mechanical factors.

Philips et al. have compared Dacron and Teflon TFE prostheses with the autogenous jugular vein for use as femoral artery grafts. Although the knit and crimped Dacron performed slightly better than the knit and crimpled Teflon TFE, the autogenous vein was clearly the best of those studies (Fig. 10). Wesolowski et al. [51] and Haimovici et al. [54] have also emphasized the better performance of a natural tissue graft.

In order to encourage endothelialization on the lumenal side of a vascular prosthesis (forming a "pseudointima"), a number of coated materials have been investigated. Hall et al. [55] originally proposed the use of velour-lined materials. In the DeBakey Vascular-D graft (USCI) [32], the Dacron velour lining the inner walls of this graft provides a textured, highly filamentous surface to which the inner capsule can adhere, which is given credit for the impressive long-term results (70% patency after 5 years) in bypasses of small diameter arteries.

A compound graft material of reticulated polyurethane foam (Scottfoam by Scott Paper Co.) glued to knitted Dacron fabric has been investigated by Berkowitz et al. [56]. Experimental results in dogs indicated rapid coverage with a smooth, thin, endothelium-like surface, with stability and speed of coverage superior to control grafts of Dacron velour. The open three-dimensional structure of the polyurethane lattice provided excellent anchorage for the pseudointima and the speed of surface coverage may account for the degree of thromboresistance exhibited.

Sharp and co-workers [57, 58] have reported on the use of an electronegative carbon-filled (28%) polyurethane, backed with Dacron or nylon velour (Electrolour, Goodyear Tire & Rubber Co.) as a graft material. Flocked fibers were also studied as linings since the nonporous polyurethane backing prevents nutrition via external ingrowth of capillaries, the thin neointima depends on direct nutrition from blood in the lumen. Experimental results in dogs showed excellent patency rates, thin well-attached neointimas, and evidence of complete endothelialization in 3 weeks. Whether similar results could be expected in humans, even at a much slower rate, is not known. The use of negatively charged surfaces to inhibit excessive thrombus deposition may also find form in hydrogel surface coatings on fiber surfaces in vascular prosthetics. Such coatings could still be permeable to biomolecules and cells, and may encourage more rapid cellular ingrowth [50]. Hagesawa et al. [59] have radiation grafted methyl methacrylate to Teflon prostheses with promising results. Chemical modification of fiber surfaces could significantly affect the cellular penetration and subsequent en-dothelialization or fibrous ingrowth in synthetic fabrics, and thus should be a fruitful avenue for future research.

Ideally, one would like fresh vascular tissue and pseudointima to develop in parallel with a gradual disappearance of the fibrous prosthetic "scaffold." With this in mind, Schmitt and Polistina [60] have proposed a textile of degradable PGA and nondegradable fibers, and Ruderman et al. [61] have published on the combination of PLA and Dacron as a vascular graft. Chvapil and Krajicek [62] have proposed a porous collagen vascular graft. Reichle et al. [63] have noted the desirability of forming a collagenous intimal coating on Dacron prostheses.

Others have attempted to seed various cells on the inner surface of a synthetic vascular graft to yield true endothelial or pseudo-endothelial linings. Kahn and Burkel [64] used (1) a nonwoven polypropylene fabric coated with poly(mo-nochloro-p-xylylene) (Parylene C of Union Carbide Co.), (2) a nylon microfi-lament coating extrusion coated onto a polyurethane backing sheet, and (3) gold filaments in a nonwoven felt. The polypropylene and gold fabrics were given the best ratings for WI-38 diploid cell ingrowth. Mansfield et al. [65] have an ex-tensive program on the cell seeding of cultured, autologous endothelial cells on different microfibrillar surfaces, such as Parylene C coated polypropylene fibers and various polyester fibers. They have concluded that endothelial cell coatings alone do not have the capability of thinning and organizing thrombotic deposits, but they do prevent such deposits from forming. Another way to prevent thrombotic deposition could be to heparinize the synthetic graft [66].

Prosthetic Heart Valves

Dacron and, to a much lesser extent, Teflon TFE, nylon, polypropylene, and rayon have been used in prosthetic heart valves as "sewing rings" (to suture the valve to the tissue) and/or as scaffolds for long-term fibrous tissue ingrowth (e.g., refs. [1, 67, 68] and Table III).

The caged or double-caged ball valves have metallic struts which have been covered with Dacron [69] or polypropylene [70] fabrics for the purpose of de-veloping an ingrown tissue layer on the struts and minimizing the formation of emboli. However, the action of the ball may abrade off the fabric covers, an undesirable event.

The idea of encouraging tissue ingrowth into operating parts of the prosthetic heart valve may still be worth pursuing, and to this end Clarke et al. [71] have designed a fabric of highly purified polyester filaments. The filaments used are $10\ \mu$ in diameter and are woven into a $22-38\ \mu$ thick fabric with $30\ \mu$ size pores, using 15 filaments per band of yarn (made by Fabric Research Labs.). Tests are under way using this material in prosthetic heart valves.

TABLE III
Materials that Have Been Used in the Fabrication of Prosthetic Heart Valves [68]

Metals (Fixation Rings, Cages, Poppets)	Nonmetals (Mostly for Poppets)	Fabrics (Coverings for Fixation Rings, Struts)
Aluminum alloys	GBH: graphite-benzalkonium-	Dacron[a]
Magnesium	heparin	Nylon
Magnesium alloys	Nylon	Polypropylene
Nickel-cobalt alloys	Polypropylene	Teflon[a]
Stainless Steel	Polyethylene	Rayon
Inconel	Polycarbonate (Lexan)	
Hasteloy	Polymethyl methacrylate	
SS 302, 304, 309	(Lucite)	
Stellite 21:[a]	Polytetrafluoroethylene[a]	
cobalt-chromium-	(Teflon, Halon)	
molybdenum alloy	Polyformaldehyde[a] (Delrin)	
Titanium[a]	Pyrolytic carbon[a]	
	Silicone rubber[a] (Silastic)	

[a] Most popular at present.

There is also a very successful application of heterograft pig heart valves as human prostheses [72, 73]. The fresh valves may be enzyme and formaldehyde or glutaraldehyde treated to reduce immunogenic reactions, mounted on a polypropylene stent plus a silicone rubber foam rubber cushion, and covered around the sides and edges with a high-porosity Dacron cloth for suture and tissue anchoring [72]. Recent trends are toward the use of glutaraldehyde fixation of the heterograft.

Heart Assist Devices

There are a number of different approaches to the design of heart assist devices which utilize polymeric fibers in their construction. A variety of clinical indications for assisted circulation have been enumerated, based on whether the failure is respiratory, right or left heart, or circulatory [74]. Many of the principles and designs of assist devices have been summarized [1, 75–77].

Most of these devices involve flexible polymer diaphragm blood-pumping chambers, usually in a rigid housing. The use of velour linings or flocked surfaces has been suggested for the interfaces of such membranes with the circulating blood [1, 56, 75–77]. Most often the fiber used is Dacron, as an exposed flocked fibril intima on a polyurethane diaphragm [78], as a flocked fibril intima on a Silastic diaphragm, overcoated with an adhesive layer of Silastic adhesive [79], or as a velour-lined intima [75]. Kolff and Lawson [77] have discovered Dacron fibrils in the brain of one of their calves, and therefore they have adhered their fibrils with a layer of Silastic adhesive. They remain "very suspicious of fibrils" [77]. Others have used polyurethane adhesives for the same purpose. Kolff and Lawson also claim that both Dacron velour or flocked Dacron fibrils continue to build up fibrin on the pumping diaphragm surface, where there is motion, and after 20–30 days this fibrin layer begins to calcify, stiffening the diaphragm and leading to pump failure. Therefore, Kolff et al. have switched to smooth, fiber-free polyurethane bladders.

Dacron mesh has been used as reinforcement in construction of device components and housing, and Dacron fabrics have been used as suture and tissue anchoring matrices [1, 75–77]. Miller et al. [80] describe another fiber system which has been developed for a lining on blood contacting surfaces in artificial organs. It is a 25 μ thick, nonwoven polypropylene fiber web-coated and bound together with Parylene C [80]. The polypropylene fibers are only 1 μ in diameter; the web is made by co-extrusion as a thin oriented film of the polypropylene with an incompatible ethylene copolymer salt which is later extracted; the "remains" of the film are then drawn transverse to the extrusion machine direction and the nonwoven fabric is produced. Newayser et al. [81] describe other nonwoven fabrics prepared from submicron-sized filaments; these fabrics have been shown to support the growth and propagation of fibroblast cell linings for use in vascular assist devices. The fibrous mats were composed of poly(tetramethylene ter-ephthalate) (6 PRDE of Eastman) or Parylene C-coated nylon 6-6. Carbon was also deposited on the nylon fibrils. *In vitro* cell culture techniques showed promising results.

MUSCULOSKELETAL SYSTEM APPLICATIONS

Orthopedic and Dental Applications

Homsy [82] has proposed a useful *in vitro* biocompatibility screening protocol for biomaterials. The most biocompatible fibers by this two-step protocol, involving extraction and tissue culture response tests, were found to be graphite and Teflon TFE. Based on these studies, he has developed a material to serve as a stabilizing interface between skeletal implant prostheses and adjacent hard and soft tissue. It is basically a nonwoven graphite–Teflon TFE fibrous mat which acts as a matrix for tissue ingrowth for stabilization of orthopedic or dental implants, or for reconstructive maxillofacial and oral surgery [83–85]. This material (Proplast, Vitek/Dow-Corning) may contain 4–20% (vol.) of graphite fibers, 4–10% (vol.) of Teflon TFE fibers, 2–10% (vol.) of Teflon TFE particles, and 90–60% (vol.) of a soluble salt, such as NaCl (which is subsequently washed out to provide the microporosity) [85]. The material appears to have a promising future as a tissue ingrowth matrix.

Graphite fibers have also been used as reinforcing fillers in poly(methyl methacrylate) resin matrices for implantation as internal fixation plates; this material may match the modulus of bone better than the metal plates currently used [86]. High-density polyethylene has also been reinforced with carbon fibers in order to match more closely the modulus of bone as well as to strengthen the polyethylene for orthopedic uses [87]. Quartz and graphite fibers have been used to reinforce epoxy matrices for similar uses [88].

Certain joints in the hand may be deteriorated by progressive rheumatoid arthritis. Silicon rubber/Dacron fabric arthroplastics have been used as joint prostheses to alleviate pain and increase hand function [89]. Intervertebral disks, which may herniate due to improper stressing, may be replaced by a silicone rubber/Dacron fabric prosthesis [90].

Artificial Tendon

Replacement of injured or defective tendons is very important, especially if the patient is subsequently able to become a more productive and useful individual. A number of synthetic polymer systems have been suggested for replacement of the tendons, especially the flexor tendon in the hand [1]: Teflon TFE, Dacron, Mersilene (polyester) nylon, silk, and metal fibers have been used as components of such prostheses [1, 55, 91–95]. The major problems have been to anchor the tendon in the muscle at one end and bone at the other, and to facilitate "gliding" by avoiding scar formation and adhesions. Proplast may be very useful in the future for anchoring tendon prostheses. The most successful artificial flexor tendon to date is the Hunter tendon, which is composed of a Silastic-coated Dacron fabric with exposed Dacron fabric at the end for fixation [91, 92]. The strength of the bone ingrowth into the Dacron fabric was found to be primarily dependent on the internal fabric porosity and not the surface morphology [95].

PERCUTANEOUS AND CUTANEOUS APPLICATIONS

Percutaneous Systems

Percutaneous shunt systems have been developed in order to gain access to the circulation for routine dialysis of the blood of kidney patients [1, 96]. Some contain a short Dacron felt or velour cuff adhered to the Silastic shunt tube as it exits from a blood vessel; the cuff terminates before the tube exits from the skin. This is used to encourage subdermal connective tissue ingrowth, which anchors the shunt and opposes motion of the shunt tip within the blood vessel. This minimizes the chances of infection and thus lengthens the useful life of that particular shunt. Others have used subcutaneous Dacron Weavenit skirts [97] for similar purposes. Payne and Berne have summarized the state-of-the-art for blood access systems [98].

Percutaneous carbon buttons (for electrical stimulation to alleviate pain) may also be anchored under the skin with Dacron fabric [1].

Artificial Skin

Artificial skin is for use as a temporary burn dressing. It should reduce pain; prevent bacterial invasion; prevent excessive fluid, electrolyte, and protein loss; permit heat to transfer out of the burn wound area; and be nontoxic, nonimmunogenic, etc. [1, 99]. A variety of velour-lined materials have been proposed for this application [55]. The most successful material developed by Hall et al.

[99] appears to be a nylon velour on a synthetic polypeptide backing. Dressler et al. [27] have utilized a nylon velour on a polyurethane backing as an artificial skin.

REFERENCES

[1] H. Lee and K. Neville, *Handbook of Biomedical Plastics,* Pasadena Technology Press, Pasadena, Calif., 1971.
[2] D. F. Williams and R. Roaf, *Implants in Surgery,* Saunders Co., Ltd., Toronto, 1973.
[3] R. L. Kronenthal, Z. Oser, and E. Martin (Eds.), *Polymers in Medicine and Surgery* (*Polymer Science and Technology, Vol. 8*), Plenum Press, New York, 1975.
[4] S. D. Bruck, *J. Biomed. Mater. Res., 5,* 139 (1971).
[5] J. Skelton, *Biomater. Med. Dev. Artif. Organs, 2,* 345 (1974).
[6] L. J. DeMerre and C. W. Bruch, *Biomater. Med. Dev. Artif. Organs, 2,* 397 (1974).
[7] R. L. Kronenthal, in *Polymers in Medicine and Surgery,* R. L. Kronethal, Z. Osen, and E. Martin, Eds., Plenum Press, New York, 1975, p. 119.
[8] R. I. Leininger *CRC Crit. Rev. Bioeng., 1,* 333 (1972).
[9] S. D. Bruck et al., *Biomater. Med. Dev. Artif. Organs, 1,* 191 (1973).
[10] C. A. Homsy, *J. Biomed. Mater. Res., 4,* 341 (1970).
[11] J. Autian in *Polymers in Medicine and Surgery,* R. L. Kronenthal, Z. Oser, and E. Martin, Eds., Plenum Press, New York, 1975.
[12] E. J. Frazza and E. E. Schmitt, *J. Biomed. Mater. Res. Symp. No. 1,* 43 (1971).
[13] J. M. Anderson and D. F. Gibbons, *Biomater. Med. Dev. Artif. Organs, 2,* 235 (1974).
[14] R. D. Kulkarmi, *J. Biomed. Mater. Res., 5,* 169 (1971).
[15] M. Sela, *Science, 166,* 1365 (1969).
[16] R. J. Ruderman et al., *J. Biomed. Mater. Res., 7,* 215 (1973).
[17] C. A. Homsy, J. Fissette, W. S. Watkins, H. O. Williams, and B. S. Freeman, *J. Biomed. Mater. Res., 3,* 383 (1969).
[18] R. W. Postlethwait et al., *Surg. Gynecol. Obstet., 108,* 555 (1959).
[19] W. R. Sewell et al., *Surg. Gynecol. Obstet., 100,* 483 (1955).
[20] R. W. Postlethwait, *Arch. Surg., 101,* 489 (1970).
[21] R. F. Edlich, P. H. Panek, G. T. Rodcheaver, L. D. Kurtz, and M. T. Edgerton, *J. Biomed. Mater. Res. Symp. No. 5* (Part 1), 115 (1974).
[22] M. Tollar, M. Štol, and K. Kliment, *J. Biomed. Mater. Res., 3,* 305 (1969).
[23] Dow Corning Co., Midland, Mich. 48640.
[24] Surgitek, Inc., Racine, Wis. 53404.
[25] I. R. Toranto, K. E. Salyer, and W. B. Nickell, *J. Biomed. Mater. Res. Symp. No. 5* (Part 1), 127 (1974).
[26] T. H. Stanley and J. K. Lattimer, *J. Biomed. Mater. Res, 6,* 533 (1972).
[27] D. P. Dressler et al., *J. Biomed. Mater. Res. Symp. No. 1,* 169 (1971).
[28] K. A. Merendino et al., *Amer. J. Surg., 110,* 416 (1965).
[29] E. E. Schmitt and R. A. Polistina (to American Cyanamid Co.), U.S. Pat. 3,875,937 (April 8, 1975).
[30] R. E. Clark, J. C. Boyd, and J. F. Moran, *J. Surg. Res., 16,* 510 (1974).
[31] C. Davis, personal communication (March 10, 1975).
[32] *Prostheses for Surgery,* Catalog of USCI, 1974.
[33] *Vascular Prostheses* and *Meadox Cardiovascular Fabrics,* brochures of Meadox Medicals, Inc., 1975.
[34] S. A. Wesolowski et al., *Surgery, 50,* 91 (1961).
[35] S. A. Wesolowski et al., *J. Cardiovasc. Surg., 5,* 544 (1964).
[36] L. R. Sauvage et al., *Surgery, 65,* 78 (1969).
[37] L. R. Sauvage et al., *Arterial Prosthesis Healing,* report issued by the Reconstructive Cardiovascular Research Center, Providence Medical Center, Seattle, Wash., 1974.
[38] L. R. Sauvage et al., *Surgery, 70,* 940 (1971).
[39] L. R. Sauvage et al., *Arch. Surg., 109,* 698 (1974).

[40] S. M. Lindenauer et al., *Trans. Amer. Soc. Artif. Int. Organs, 20*, 314 (1974).
[41] B. F. Buxton et al., *Amer. J. Surg., 125*, 288 (1973).
[42] D. P. Boyd and A. I. Midell, *Surg. Clin. North Amer., 53*, 351 (1973).
[43] T. Soyer et al., *Surgery, 72*, 864 (1972).
[44] H. Matsumoto et al., *Surgery, 74*, 519 (1973).
[45] J. G. R. Volder et al., *Trans. Amer. Soc. Artif. Int. Organs, 19*, 38 (1973).
[46] C. D. Campbell et al., *Trans. Amer. Soc. Artif. Int. Organs, 20*, 86 (1974).
[47] Y. Fujiwara et al., *J. Thorac. Cardiovasc. Surg., 67*, 774 (1974).
[48] D. E. Smith et al., *J. Thorac. Cardiovasc. Surg., 69*, 589 (1975).
[49] L. Norton and B. Eiseman, *Surgery, 77*, 280 (1975).
[50] L. R. Sauvage et al., *Surg. Clin. North Amer., 54*, 213 (1974).
[51] F. H. Honigman et al., *J. Thorac. Cardiovasc. Surgery*, 61, 255 (1968).
[52] C. E. Phillips et al., *Arch. Surg., 82*, 38 (1961).
[53] S. A. Wesolowski et al., *Ann. N.Y. Acad. Sci., 146*, 325 (1968).
[54] H. Haimovici et al., *Surg. Gynecol. Obstet., 131*, 1173 (1970).
[55] C. W. Hall, D. Liotta, J. J. Ghidoni, M. E. DeBakey, and D. P. Dressler, *J. Biomed. Mater. Res., 1*, 179 (1967).
[56] H. D. Berkowitz et al., *Trans. Amer. Soc. Artif. Int. Organs, 18*, 25 (1972).
[57] W. V. Sharp et al., *J. Biomed. Mater. Res. Symp. No. 1*, 75 (1971).
[58] W. V. Sharp et al., *Trans. Amer. Soc. Artif. Int. Organs, 18*, 232 (1972).
[59] T. Hagesawa et al., *Surgery, 74*, 696 (1973).
[60] E. E. Schmitt and R. A. Polistina (to American Cyanamide Co.), U.S. Pat. 3,463,158 (Aug. 26, 1969).
[61] R. J. Ruderman et al., *Trans. Amer. Soc. Artif. Int. Organs, 18*, 30 (1972).
[62] M. Chvapil and M. Krajicek, *J. Surg. Res., 3*, 358 (1963).
[63] F. A. Reichle et al., *Surgery, 74*, 945 (1973).
[64] R. H. Kahn and W. E. Burkel, *In Vitro, 8*, 451 (1973).
[65] P. B. Mansfield et al., *Trans. Amer. Soc. Artif. Int. Organs, 21*, (1975).
[66] P. B. Mansfield et al., *Trans. Amer. Soc. Artif. Int. Organs, 21*, (1975); L. S. Hersh, V. L. Gott, and F. Najjar, *J. Biomed. Mater. Res. Symp. No. 3*, 85 (1972).
[67] K. A. Merendino, Ed., *Prosthetic Valves for Cardiac Surgery*, C. C Thomas, Springfield, Ill., 1961.
[68] E. J. Roschke, *Biomater. Med. Dev. Artif. Organs, 1*, 249 (1973).
[69] B. Bull et al., *Surgery, 65*, 640 (1969).
[70] J. W. Boretos, D. E. Detmer, and N. S. Braunwald, *J. Biomed. Mater. Res., 6*, 185 (1972).
[71] R. E. Clark et al., *Biomater. Med. Dev. Artif. Organs, 2*, 379 (1974).
[72] *Flexible Stented Xenograft Heart Valves*, brochure of Hancock Laboratories, Inc., 1974.
[73] A. Huc, C. Planche, M. Weiss, Ph. Mannschott, F. Chapin, and D. Chabrand, *J. Biomed. Mater. Res., 9*, 79 (1975).
[74] P. Galetti, *Disc. Trans. Amer. Soc. Artif. Int. Organs, 8*, 89 (1962).
[75] J. H. Kennedy et al., *Biomater. Med. Dev. Artif. Organs, 1*, 3 (1973).
[76] J. L. Peters et al., *Biomater. Med. Dev. Artif. Organs, 3*, 1 (1975).
[77] W. J. Kolff and J. Lawson, *Trans. Amer. Soc. Artif. Int. Organs, 21*, 620 (1975).
[78] C. G. LaFarge et al., *Trans. Amer. Soc. Artif. Int. Organs, 18*, 186 (1972).
[79] H. Oster et al., *Surgery, 77*, 113 (1975).
[80] W. A. Miller et al., *Text. Res. J., 43*, 728, (1973).
[81] E. S. Newayser et al., *Trans. Amer. Soc. Artif. Int. Organs, 19*, 168 (1973).
[82] C. A. Homsy, *J. Biomed. Mater. Res., 4*, 341 (1970).
[83] C. A. Homsy et al., *J. Amer. Dent. Assoc.*, 86, 817 (1973).
[84] C. A. Homsy, *Orthop. Clin. North Amer., 4*, 295 (1973).
[85] Vitek, Inc., Brit. Pat. 1,390,445 (April 9, 1975).
[86] S. L. Y. Woo, W. H. Akeson, B. Levenetz, R. D. Cootts, J. V. Matthews, and D. Amiel, *J. Biomed. Mater. Res., 8*, 321 (1974).
[87] E. Sclippa and K. Piekarski, *J. Biomed. Mater. Res., 7*, 59 (1973).
[88] S. Musikant, *J. Biomed. Mater. Res. Symp. No. 1*, 225 (1971).
[89] J. L. Goldner and J. R. Urbaniak, *J. Biomed. Mater. Res. Symp. No. 4*, 137 (1973).

[90] J. R. Urbaniak et al., *J. Biomed. Mater. Res. Symp. No. 4*, 165 (1973).

[91] J. M. Hunter, C. Steindel, R. Salisbury, and D. Hughes, *J. Biomed. Mater. Res. Symp. No. 5* (Part 1), 155 (1974).

[92] J. M. Hunter, D. Subin, F. Minkow, and J. Konikoff, *J. Biomed. Mater. Res. Symp. No. 5 (Part 1)*, 163 (1974).

[93] R. E. Salisbury, D. McKeel, B. A. Pruitt, Jr., A. D. Mason, Jr., N. Palermo, and C. W. R. Wade, *J. Biomed. Mater. Res. Symp. No. 5* (Part 1), 175 (1974).

[94] R. N. King, H. K. Dunn, and K. E. Bolstad, *J. Biomed. Mater. Res. Symp. No. 6*, 157 (1975).

[95] H. K. Dunn et al., *J. Biomed. Mater. Res. Symp. No. 4*, 109 (1973).

[96] *Arteriovenous Cannulation Systems for Hemodialysis*, brochure of Extracorporeal Medical Specialties, Inc., 1972.

[97] H. P. McDonald et al., *Trans. Amer. Soc. Artif. Int. Organs.*, *14*, 176 (1968).

[98] J. E. Payne and T. V. Berne, *Biomater. Med. Dev. Artif. Organs*, *1*, 241 (1973).

[99] C. W. Hall et al., *Trans. Amer. Soc. Artif. Int. Organs*, *16*, 12 (1970).

NOVEL USES OF FIBERS AS TENDONS AND BONES

R. N. KING, G. B. McKENNA, and W. O. STATTON

*Department of Materials Science and Engineering,
University of Utah, Salt Lake City, Utah 84112*

SYNOPSIS

Polyester fibers can be made into a satisfactory load-bearing unit which can be implanted to act as an artificial tendon. Initial basic studies have shown the amount of porosity in the fiber assembly which is essential to produce tissue ingrowth from both muscles and bone. Implant studies on dogs have shown the feasibility of this approach.

Fiber-reinforced composites can be made to simulate the mechanical properties of bone and can be implanted as plates to repair fractures. Implant studies on dogs have shown the feasibility of this approach, also.

INTRODUCTION

The following report is an indication that diverse uses of normal textile or industrial fibers are possible when applications are sought in the biomedical area. This research is far from complete, and this report is not meant to propose that a final use or a final product is yet available. This research is complete enough, however, to indicate the feasibility of these directions, and to suggest that more complete and detailed studies should be undertaken.

The majority of the following results have already been published or are in press. The excuse for the present repetition is that the majority of research workers in the fiber or polymer area are not apt to see the specialized journals where these results are published. Therefore, we feel it is appropriate to bring the results together in the context of this symposium in the nature of a review presentation.

IMPLANTABLE FLEXOR TENDONS FROM POLYESTER FIBERS [1]

Background

A brief discussion of the anatomy, function, and clinical problems encountered in repair of flexor tendons will be given in order to understand the problems in the development of a permanently implantable prosthesis.

Journal of Applied Polymer Science: Applied Polymer Symposium 31, 335–350 (1977)
© 1977 by John Wiley & Sons, Inc.

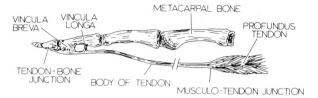

FIG. 1. Lateral view of flexor tendon.

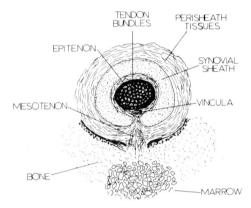

FIG. 2. Cross section of flexor tendon.

The body of a tendon is principally composed of dense collagen bundles, longitudinally oriented and parallel (Fig. 1). Fibroblasts surround these dense collagen bundles which are surrounded by a delicate framework, the endotenon, and are further grouped by the epitenon consisting of a thin, closely adherent, fibroblastic layer containing small blood vessels surrounding the tendon body (Fig. 2). The epitenon aids the gliding mechanism. A synovial sheath surrounds the whole tendon body, and between it and the epitenon is a scant amount of lubricating synovial fluid. Immediately surrounding the synovial sheath is the perisheath, consisting of highly vascularized tissues whose function is to supply the necessary fluids for maintenance of the enclosed structures and to serve as a basis for the repair mechanism in tendon injuries. The body of the tendon is connected at intervals to its synovial sheath and perisheath tissues by a meso-tenon, which is a delicate band of vascular connective tissue appearing as a cleft in the synovial sheath extending out to the tendon proper (Fig. 2). Certain areas of the mesotenon are highly specialized structures being long and mobile, and are designated as vincula. The vincula are very delicate and when injured do not reconstitute themselves. They act to supply the tendon body with a blood and a nerve supply.

Equally important as the tendon body is its attachment to the muscle and bone. At the musculotendinous junction, the muscular collagenous fibers blend indistinguishably into the collagen fibers of the tendon body. At the tendon–bone junction, the collagen bundles of the tendon body penetrate the cortex of the bone and extend well down into the medullary canal as Sharpey's fibers. They are diffused into the collagenous matrix of the osseous structure.

Successful surgical repair and restoration of function of a flexor tendon, particularly after injury within the fibrous flexor sheath, is difficult to achieve. The physiological processes necessary for healing invariably limit mobility of the tendon. The tendon itself is relatively avascular and possesses little capacity, if any, to provide a healing response. A damaged tendon primarily relies on infiltration of granulation tissue from the surrounding soft tissues of the perisheath structures to effect repair. While essential for adequate restoration of tendon continuity, this ingrowth provides the basis for rehabilitation problems following tendon surgery. As granulation tissue invades the damaged tendon-effecting repair, it forms dense fibrous connections or adhesions with the surrounding structures and tends to tether the tendon to its surroundings resulting in loss of excursion. Thus, while tendon continuity is successfully restored, loss of function results. In order to restore the function, these adhesions must be stretched or broken to permit motion. To facilitate stretching these adhesions, surgeons remove external constraints or bandages about 3 weeks following repair of flexor tendons and begin motion of the hand. Unfortunately, this early motion frequently disrupts the surgical repair despite elaborate suturing techniques. Studies indicate that 3 weeks following repair, a direct tendon anastomosis possesses only 12% of the tensile strength of a normal tendon and that 4 months is usually necessary for normal tensile strength to develop.

Our present research is aimed at developing a clinically acceptable prosthesis which directly attaches at the musculotendinous origin and bony insertion of the flexor digitorum profundus tendon of the hand to provide permanent function thus bypassing the adhesion problems encountered in tendon surgery. The project can be broken down into three basic areas of interest: (1) the musculotendinous junction; (2) the tendon–bone junction; (3) the body of the tendon.

This paper reports the initial findings in the first two areas. Our attachment concept for these two areas is essentially analogous, employing a fabric which permits tissue ingrowth through the inherent fabric porosity to provide a firm mechanical interlock with the biological system. The choice of polymeric fabric material to provide a scaffolding for tissue ingrowth over other porous materials such as calcium aluminate ceramics or sintered metals is based on three important criteria: (1) the fabric has a pliability which more closely approximates the physiological resiliency of the muscle and tendon tissue than does the more rigid porous material and is, therefore, more mechanically compatible with the biological system; (2) textile technology has reached the state-of-the-art in which the proposed prosthesis may be constructed in a continuous fiber process, uninterrupted from muscle origin to bone insertion, greatly enhancing long-term mechanical stability of the prosthesis; and (3) there is a nearly infinite selection of fabric structures such as weaves, knits, braids, velours, felts, etc., which can be fabricated to meet practically any biologically desirable properties. The following report summarizes our findings on four fabrics implanted in the muscle tissues of rabbits and bone tissues of dogs for periods of up to 8 weeks.

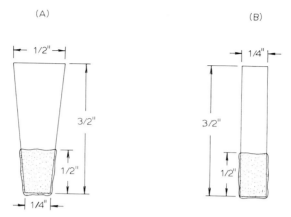

FIG. 3. Implant configuration of Dacron weaves and velours: (A) implantation at musculo–tendon junction; (B) implantation in bone.

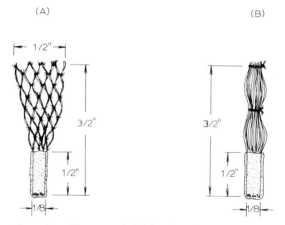

FIG. 4. Implant configuration of Dacron cord: (A) implantation at musculo–tendon junction; (B) implantation in bone.

Fabric Used

The materials used in the study were Dacron polyester fabrics, three of which were kindly supplied by the United States Catheter and Instrument Corporation (Glen Falls, N.Y.) and identified as: (1) U.S.C.I.C. #6079 fabric matrix Dacron felt; (2) U.S.C.I.C. #6108 one-sided Dacron velour; (3) U.S.C.I.C. #KG-1125 two-sided Dacron velour; (4) a proprietary fabric, identified as DN-4 braided Dacron cord supplied by the Eimco Corporation, Filter Media Division, Salt Lake City, Utah.

The design and preparation of the fabrics is as follows. (1) Using a template, each U.S.C.I.C. fabric was cut to the implant size and geometry shown in Figures 3A and 3B. The small tapered end of each implant was then dipped to a height of 0.5 in. in a silicone rubber solution consisting of 80 vol % Dow-Corning #92-009 dispersion coating and 20 vol % VM & P naptha. (2) The Eimco DN-4 braided Dacron cord muscle implants were hand-fabricated into "artificial" weaves by first teasing the cord into 16 individual fiber strands which were tied together two at a time in a regular fashion producing two planes of fabric, eight strands each, as shown in Figure 4A. The bone implants were again separated into 16 strands, teased and tied at the top and center as shown in Figure 4B. In both cases, the tying material was commercial-grade, cotton-coated Dacron polyester sewing thread. Prior to surgery, the implants were sterilized by steam-autoclaving at 250–270°C for 15 min.

Porosities of the implants were measured in terms of the quantity of water flow through a unit area of fabric per unit time (ml H_2O/cm^2)/min at a pressure head of 120 mm Hg using the technique described by Wesolowski in his investigations of fabrics for prosthetic arterial grafts. Results of porosity measurements on the implanted fabrics are shown in Table I. The Weaveknit material listed in Table I was used as a control fabric for comparison with the Wesolowski results.

Surgical Technique

A total of 71 rabbits were used in the muscle fabric evaluation studies. A single fabric was placed at the musculotendinous junction of both hind limbs of the rabbit, one limb then being used for mechanical testing, one for histology. Three rabbits were used for each fabric at each of the 2-, 4-, 6-, and 8-week intervals of the time study for a total of 12 implants carried out on each fabric.

A total of 12 dogs were used in the bone–fabric evaluation studies. All four fabric materials were placed in each hind limb of a single dog, again one limb being used for histology and one limb for mechanical testing. Three dogs were used for each of the 2-, 4-, 6-, and 8-week intervals of the time study for total of 12 implants carried out on each fabric.

In Vivo Mechanical Testing

Equipment

Equipment used in the mechanical testing studies consisted of a Statham Universal Transducing Cell, Model UC3-Gold Cell (Honeywell, Inc., Denver, Colo.), with a 0- to 50-lb Statham Model UL4 Load Cell Accessory, Statham

TABLE I
Fabric Porosity

Fabric	Porosity (ml H_2O/cm^2)/min 120 mm Hg
1. Weaveknit (Control)	4,000
2. U.S.C.I.C. # 6079 Felt	3,783
3. U.S.C.I.C. # 6108 Velour	4,145
4. U.S.C.I.C. #KG-1125 Velour	4,016
5. Braided Dacron Cord	32,889

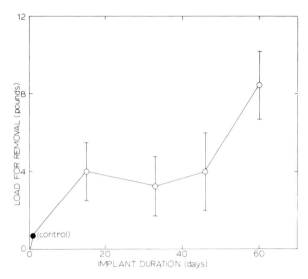

FIG. 5. Load to failure curve of single velour at musculo–tendon junction.

Analog Readout Meter, Model UR5, and a CEC Datagraph dual-channel recorder. The UC3 Transducer, UL4 readout meter, and the CEC recorder were operated in series during testing, after precalibration to produce a full-scale millivolt readout corresponding to 0–16 lb on one recorder channel and 0–40 lb on the second recorder channel.

The 0- to 16-lb scale was used primarily for the muscle implant load characterization while the 0- to 40-lb scale was used for the bone implant load characterizations. The serrated parts of the grip system were fabricated from the actual tensile-grip jaws of a large commercial testing machine and provided an excellent method of attachment of the testing device to the implants. No evidence of grip slippage or fabric failure in the gripped area appeared during the 2-month evaluation study.

Procedure

The following procedures were used in the *in vivo* mechanical evaluation of the fabric implants.

Muscle Procedure. (1) After sacrificing the rabbit, the position of the implant was determined, and the end tab exposed through a small incision. The silicone-coated portion of the implant was freed of all encapsulating tissue and attached to the grip mechanism. (2) The limb containing the implant was firmly pinned to the plate of the testing assembly by passing two B & S #20 gauge steel wires through the corresponding holes in the plate and twist-wrapping them around several segments of the limb. (3) The grip mechanism was attached to the testing device which was retracted by manual rotation of the threaded assembly, thus producing an extension and corresponding load on the implant, which was sensed by the cell and thus recorded.

Bone Procedure. After sacrificing the dog, the femur section containing the

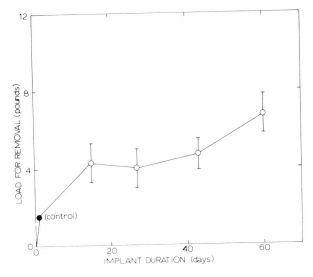

FIG. 6. Load to failure curve of double velour at musculo–tendon junction.

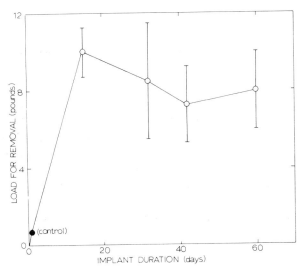

FIG. 7. Load to failure curve of felt at musculo–tendon junction.

four implants was removed distally and proximally to the implanted area with a bone saw. The removed section was then firmly fixed to the testing assembly plate as in (2) above and tested as in (3) above. Several points should be noted regarding the testing procedure: It was determined that surgical exposure of an implant end tab which was not silicone rubber-coated could produce wide variations in the amount of implant surface remaining exposed to the ingrown tissue, which would thus affect mechanical test results by introducing a non-uniform interface area. Thus, silicone rubber coating before implantation assured a uniform tissue ingrowth area upon surgical exposure of the end tab of the

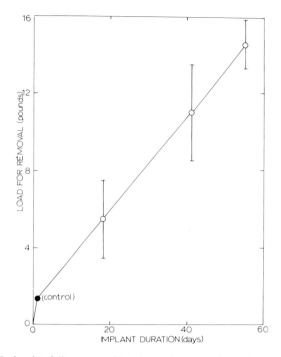

FIG. 8. Load to failure curve of braided cord at musculo-tendon junction.

implant, and correspondingly less variance in the data obtained. It should also
be noted that after sacrifice of the animal, 5 min was set as the maximum allowed
time limit for completing mechanical testing of the muscle implants, and 20 min
(5 min each) as the maximum allowed time limit for mechanical testing of all
four bone implants. This procedure was considerably easier than working with
an anaesthetized animal, and it was felt that this short time limit after sacrifice
produced very little test variance when compared to that produced in the living
animal and thus could safely be referred to as an "*in vivo*" mechanical evaluation.

In Vivo Mechanical Testing

Results of Muscle Implants

The graphs shown in Figures 5–8 indicate the average test results for the four
muscle implant fabrics during the 2-month study. Results are based on averaging
the load values obtained from three rabbits at each of the 2-, 4-, 6-, and 8-week
intervals.

Results of Bone Implants

Table II lists the test results for the bone implants, based on the individual load values obtained from three dogs at each of the 2-, 4-, 6-, and 8-week intervals. Those data with an asterisk (*) indicate external tensile failures in the fabrics before they could be removed from the bone.

Discussion

The significant outcome of the *in vivo* mechanical testing is the fact that in the initial 2-month study, implant strength appears to depend primarily on the internal fabric porosities, and not on their various surface morphologies. This is clearly seen in Figures 5 and 6, which represent the data for both the one- and two-sided U.S.C.I.C. Dacron velours.

The ratio of the total velour surface of the two-sided velour (U.S.C.I.C. #Kg-1125) compared to the one-sided velour (U.S.C.I.C. #6108) is about 2:1, respectively. Thus, if surface geometry is significant for ingrowth, the two-sided velour should show a measurable increase in implant strength over the 2-month period.

Comparison of Figures 5 and 6 shows no difference between the two fabrics. Thus, it is felt that the contribution of surface structure to implant strength can be considered negligible for implantations of less than 2 months duration. The main criteria for implant adhesion and corresponding tissue ingrowth appear to be fabric porosity, with the greater porosity yielding the greater adhesion.

TABLE II
In Vivo Mechanical Characterization of Bone Implants

		Load for Pull-out			
Dog.I.D. #	Time Implanted (Days)	#6079 Felt	Eimco Cord	#6108 Velour	# KG-1125 Velour
13	14	16.0*	18.8	8.8	8.8*
14	14	18.4	18.8	9.2	13.6*
21	14	8.0*	16.0	5.0	12.5*
18	28	22.5*	22.0*	22.0*	17.5*
15	42	22.0*	28.0	5.0*	8.0*
16	42	20.0*	26.0*	6.0*	15.0*
17	42	21.0*	23.0	5.8*	6.0*
10	56	9.0*	28.0	4.5*	7.0*
11	56	11.0*	25.0	**	**
12	56	14.0*	30.0*	**	**

* Fabric tensile failure outside bone.
** Implants severed during surgical exposure.

From the mechanical testing data for an implant geometry as used in this study, a fabric porosity of approximately 4000 (ml H_2O/cm^2)/min results in a fabric pull-out strength of between 7 and 8 lb in 8 weeks, whereas a fabric porosity of approximately 32,000 (ml H_2O/cm^2)/min results in a fabric pull-out strength between 14 and 16 lb in 8 weeks.

Studies on dogs have shown that animals can be successfully implanted with an artificial tendon device on the type described above with the braided cord and have full functional motion for extended periods of time.

IMPLANTABLE ORTHOPEDIC DEVICES FROM FIBER-REINFORCED COMPOSITES [2]

Background

Much recent research has indicated that, while currently utilized metallic implants perform adequately in many orthopedic applications, there is still much room for improvement. Thus, even the inert metals used today corrode in the sense that ions are given off to the surrounding tissues. These ions can cause adverse tissue reaction and in any case make the metals less than ideal implant materials. Also, although metals generally show adequate strength, fatigue failure of implants is still a clinical worry.

In addition to the above problems, the basic mechanical imcompatibility between the metal implants and bone can lead to problems. The elastic modulus of bone is approximately 17 GN/m^2 (2.5 $\times$ 10^6 psi), while the metals show moduli ranging from 117 GN/m^2 for titanium to 238 GN/m^2 for the chrome–cobalt alloys. This large difference in mechanical properties can cause relative motion between the implant and bone upon loading, as well as high stress concentrations at bone–implant junctions. In addition, in certain applications, such as fracture fixation plates, posthealing stress shielding results in Wolff's law remodeling and cortical thinning beneath the plate.

The fiber-reinforced polymer composite represents a class of materials with high structural strength, chemical inertness, and a design versatility which provides the potential solution to some of the problems encountered with the current metallic implants.

TABLE III
Candidate Materials

Ultrahigh Molecular Weight Polyethylene
Polypropylene
Polysulfone
Ethylene–Propylene–Acrylic Acid Terpolymer
Epoxy
Carbon or Graphite Fibre
Glass Fibre

TABLE IV
Tissue Culture Toxicity Test Results

Material	Trial Number 1 2 3 4 5 6
UHMW Polyethylene (Hercules 1900)	0 0 0 0 0 -
Ethylene-Propylene-Acrylic Acid Terpolymer (Exxon: Dexon) (with mold release)*	0 T 0 T 0 -
Ethylene-Propylene-Acrylic Acid Terpolymer (Exxon: Dexon) (without mold release)*	0 T 0 0 T -
Polypropylene	0 0 0 0 0 0
Polysulfone/Graphite Composite (Hercules 3004 AS)	T T T 0 0 T
Kevlar Fibre	0 0 0 0 0 0
Graphite Fibre (Type 3004 AS)	0 0 0 0 0 0
Kevlar/Epoxy Composite (3M-SP-308)	0 0 0 0 0 0
Glass/Epoxy Composite (3M-1002)	0 T 0 0 T -
Glass/Epoxy Composite (3M-SP-250)	0 0 0 0 0 -
Graphite/Epoxy Composite (3M-SP-288T3)	T T T 0 0 -
Negative Toxic U.S.P. Control	0 0 0 0 0 0
Positive Toxic Control (Poly(vinyl chloride))	Average 9.85 mm toxic zone

* The mold release used was Frekote 33, a nontransferable releasing interface.
Note: T = trace of reaction; less than 0.5 mm zone size; 0 = no reaction.

The following paragraphs discuss the development of a composite system specifically for orthopedic applications, from material selection to model implant system.

Material Selection

In the selection of any material for orthopedic applications, two major parameters must be considered prior to any design refinements: biocompatibility and mechanical strength. Thus, the process of choosing candidate polymer resins and fibrous reinforcements required some preliminary screening of materials for these two properties. The materials selected for this initial screening are shown in Table III.

The determination of the ultimate biocompatibility of a material is at best a difficult, long-term effort. For an initial screening test, however, a tissue culture toxicity test can serve the purpose with long-term animal studies taking place as the material's usefulness in biomedical applications becomes evident.

Table IV shows the results of the screening for candidate fibers, resins, and some composite systems. None of the candidate systems elicited more than a trace amount of reaction (less than a 0.5-mm zone around the disk or fiber), and all were considered satisfactory for additional testing.

The orthopedic implant is subject to two major types of hostile environment.

TABLE V
Strength Retention after Various Autoclaving Cycles (123°C) for Polypropylene, Polysulfone, and Graphite/Polysulfone

Autoclave Sequence		Total Autoclave Time Minutes	Percent Strength Retention		
			pp*	PS*	GPS*
A 1:	One 3-minute autoclave	3	95.7	101.3	–
A 2:	One 10-minute autoclave	10	95.7	100.9	95.4
A 3:	One 45-minute autoclave	45	96.2	106.1	98.9
B 1:	Two 3-minute autoclaves	6	98.9	100.4	–
B 2:	Three 3-minute autoclaves	9	93.6	105.3	–
B 3:	Four 3-minute autoclaves	12	98.7	–	–
B 4:	Ten 3-minute autoclaves	30	99.7	106.3	–
C 1:	Two 10-minute autoclaves	20	98.8	106.8	–
C 2:	Three 10-minute autoclaves	30	96.5	102.0	–
C 3:	Four 10-minute autoclaves	40	100.1	–	–
C 4:	Five 10-minute autoclaves	50	–	106.1	99.5

* Polypropylene, polysulfone, graphite/polysulfone, respectively.

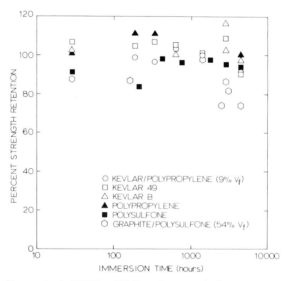

FIG. 9. Effects of immersion in PECF at 37°C on the strength of some candidate material systems.

The *in vivo* biological environment is known to be quite hostile. The possible attack mechanisms on a material *in vivo* stem from its composition as a "highly oxygenated saline solution" plus such potential reactive substances as enzymes and amino acids. In addition, the sterilization procedures used prior to implantation can be quite hostile.

There are several sterilization techniques. However, the technique of choice is the autoclave. As a result, we have conducted a preliminary study of the effects of multiple autoclaving on three of our candidate systems: polypropylene, polysulfone, and graphite/polysulfone. The test sequence included autoclaving periods of from 3 to 45 min with the shorter period tests being conducted for multiple autoclaves.

Table V presents the results of autoclaving three candidate materials at 123°C. No significant change in material mechanical strength is evident with up to ten 3 min autoclave cycles or autoclave times up to 50 min for the candidates.

Six of the candidate systems have been tested for strength retention after soaking in Homsy's pseudo-extracellular fluid (PECF) at 37°C. These tests are still in progress and the results are intermediate. However, after over 6 months, most of the systems have retained 95% of their original strength. The only exception is the graphite/polysulfone which has lost about 20% of its original strength. The results are shown in Figure 9.

The test results from the autoclave and PECF experiments provided the basic information necessary to select two material systems for an experimental implant design. The graphite/polysulfone (Hercules 3004 AS) and glass/epoxy (3M-SP-250) were selected for evaluation as fracture fixation plates. Both systems are available commercially, provide reasonable environmental stability, and represent a wide range of potential material properties.

THE FRACTURE FIXATION PLATE—A MODEL SYSTEM

Rationale

The fracture fixation plate is a nearly ideal system for demonstration of the fiber-reinforced polymer as an orthopedic implant. The stresses on the individual plate can be varied in several ways: (1) the plate size, (2) animal size, and (3) the bone chosen. In addition, the elastic properties of a bone plate have a significant effect on the ultimate performance of the implant. Too high a flexibility can result in nonunion of the fracture due to gross motion of the fracture ends, while a high stiffness can result in stress shielding of the bone after healing and subsequent cortical thinning. A high stiffness can also result in stress concentrations at the plate ends and stress fractures in these areas.

Design Optimization

The optimum design of a bone/plate system requires a trade-off among several parameters. Obviously, strength must be adequate to withstand the forces applied *in vivo* until healing occurs. Also, the plate flexibility must be low enough to provide adequate fixation and the elastic modulus of the system must be such as to permit the bone to assume as much load as possible subsequent to healing.

The parameters involved, then, are material geometry, material elastic

TABLE VI
Effects of Material and Geometric Properties on Bone/Bone Plate Interactions

Plate Type	Modulus (GN/m^2)	Strength (MN/m^2)	Thickness (cm)	Neutral** Axis (cm)	Flexural Strength ($N-m$)***
316L	193	690	.25	.23	13.7*
Titanium	117	1000	.30	.30	19.1
Composite:****					
Graphite/Polysulfone	69	690	.38	.36	21.1
Glass/Epoxy	35	750	.44	.47	30.7
Graphite/Poly(methyl methacrylate)[18]	17.2	241	.58	.53	13.5

* Based on fully plastic beam, ideal elastic–plastic behavior.
** Distance from surface of bone. Assumes bone modulus = $17.2 \ GN/m^2$, diameter = 1.5 cm, and cortical thickness = 0.38 cm.
*** Based on a beam with 1.27 cm width.
**** Typical properties. These can vary with ply orientations, volume fraction, etc.

properties, and material strength properties, as well as the properties of the bone being plated. A good way in which to understand the way in which these factors interact is to consider a bone/bone plate system in which the elastic, geometric, and strength properties of the bone plates vary.

Assume that the bone is a cylinder with diameter 1.5 cm. Further, assume that it has been determined, experimentally, that the flexural rigidity required of the plate to assure healing is 3.36×10^{-4} N-m^2 (1250 lb-in^2). With these parameters, it is now possible to determine the relative effects of various material properties upon the location of the neutral axis of the bone/bone plate system. In the unplated system the neutral axis is located at 0.75 cm from the bone surface, or at the center of the medullary canal. Table VI shows how the position of this neutral axis varies with the plate properties. Note that the plate with the elastic modulus matching that of bone permits the neutral axis to be in the medullary canal. This plate has properties similar to those which Woo demonstrated cause less cortical thinning in canine femurs than do metallic plates. Obviously, as the neutral axis moves into the cortex, the ultimate strains seen by the bone become less and remodeling becomes greater.

There are limits, of course, to how far the above process can be carried. However, the bone plate sizes in Table VI are reasonable and demonstrate the versatility of composite systems.

The design parameters discussed above have been based on the premise that the plate flexural stiffness which provides adequate fixation is known. In actuality, this parameter is essentially unknown. The next section discusses an experimental implant program undertaken to determine what it is.

TABLE VII

Mechanical Properties of Bone Plates Implanted on 20-25 kg Canine Femora

Plate	Modulus (GN/m^2)	Thickness (cm)	Flexural Rigidity $(\text{N-m}^2 \times 10^{-1})$	Neutral Axis* (cm)
Zimmer:				
Medium	193	.42	15.1	.072
Composite:				
Graphite/Polysulfone:				
#1	63.5	.25	1.05	.49
#2	42.2	.42	3.31	.44
#3	57.2	.43	4.81	.36
#6	53.3	.49	6.64	.33
Glass/Epoxy:				
GE-BP-1	45.1	.47	4.95	.39
GE-BP-2	39.6	.37	2.1	.49

* Distance from bone surface. Based on bone properties of $E = 17.2 \text{ GN/m}^2$, diameter = 1.5 cm, cortical thickness = 0.38 cm.

Implant Results

In order to determine the flexural rigidity required to assure the healing of a fractured bone, it was decided to implant plates of varying elastic and geometric properties onto transversely osteotomized canine femora.

Initially, four-hole, 8.3 cm × 1.27 cm plates were used. However, screw backout proved to be a problem with both graphite/polysulfone composite plates and 316L stainless steel control plates. As a result, it was decided to go to six-hole, 11.4 cm × 1.27 cm plates.

Due to the initial problems with screw backout, only preliminary results are available at this time. If we look at Table V, which gives the properties of the implanted plates, it appears that flexural stiffnesses of from 4.81×10^{-4} N-m^2 to 6.64×10^{-4} N-m^2 are sufficient to permit healing in dogs of 20-25 kg weight. These values are in the range of a light Zimmer plate compared to 15.1×10^{-4} N-m^2 for a medium plate, and as can be seen from Table VII the neutral axis shift is considerably less than for the metal plate.

Lighter plates have yet to be implanted, and when they are, bounds on permissible plate flexural rigidities for certain bone sizes will be obtainable.

SUMMARY

Preliminary investigation of the potential of fiber-reinforced polymers for orthopedic applications indicates that these materials offer a design versatility unobtainable with metals. Tissue culture toxicity tests were used to screen candidate fiber and resin systems for orthopedic composites. Subsequent immersion in simulated body fluids (Homsy's PECF) resulted in selection of two systems for further study: glass/epoxy and graphite/polysulfone.

In order to study the applicability of these composite materials to orthopedic applications, the fracture fixation plate was selected as a model system. The bone plate can serve to show the versatility of composites because it is a relatively high load system, yet also points up the problem of the mechanical incompatibility between bone and current metallic implants. This incompatibility can result in stress shielding and ultimate cortical thinning beneath the bone plate.

A series of composite plates having varying flexural rigidities and elastic moduli have been implanted on fractured canine femora. Initial results indicate that healing can still occur with flexural rigidities $\frac{1}{2}$ to $\frac{1}{3}$ of those encountered with metal plates while the potential for cortical thinning (as measured by a neutral axis shift in the bone/bone plate system) is considerably lessened.

This research would have been impossible without the inspiration and close guidance of Dr. H. K. Dunn and his co-workers in the Division of Orthopaedics, University of Utah, Medical Center. Support of the National Science Foundation and the Orthopaedic Research and Education Foundation is greatly appreciated.

REFERENCES

[1] H. K. Dunn, R. King, J. D. Andrade, Jr., and K. L. DeVries, *J. Biomed. Mater. Res. Symp. No. 4,* 109 (1973).
[2] G. B. McKenna, W. O. Statton, H. K. Dunn, *SAMPE,* in press; report to Orthopaedic Research and Education Foundation, 1975.

BIOACTIVE NYLON FIBERS

Y. CHARIT and R. BARBOY

Israel Fiber Institute, Jerusalem, Israel

SYNOPSIS

The modification of nylon and poly(vinyl alcohol) (PVA) fibers by the bactericides nitrovin, nitrofurylacrolein (NFA), nitrofufuralcarbazone, and glutaraldehyde to impart bactericidal properties to the fibers is described. It is shown that in a water–alcohol medium, the primary interaction between NFA and nylon is chemical; physical absorption is of secondary significance. In water as the medium, absorption and chemical reaction are of equal significance. After storage of 4 months under ambient conditions, bactericidal activity in treated nylon fibers was found in samples tested before and after autoclaving and before and after extensive washing and subsequent autoclaving. The pickup of the highly potent bactericide nitrovin on nylon is shown to be dependent on catalyst and treatment temperature. Nylon fibers were reacted with glutaraldehyde under various conditions and the reactions of the products with the different bactericides were investigated.

INTRODUCTION

Among the types of modified chemical fibers for medical purposes, such as bioactive, hemoactive, anastesionic, radioactive, radioopaque, therapeutic fibers, and others, one of importance is the antimicrobic fibers.

During the past 20–30 years, much has been written about imparting antiseptic properties to different textile materials by compounds of Cd, Cu, Ag, Hg, Sn, Zn, Pb, Zr, Gr; organic agents such as phenol derivatives and salicylic acid derivatives; compounds containing sulfur, quaternary ammonium salts, and others [1–7].

Since 1950, preparations of nitrofuran derivatives have become of special importance because of their high chemotherapeutic effectiveness against infections [8–14].

Much work in modifying fibers for biological activity has been carried out in the U.S.S.R. The Leningrad Textile Institute has reported work on antimicrobic synthetic fibers [15–24], and the Moscow Textile Institute reported in the main on cellulosic fibers [25–31].

Journal of Applied Polymer Science: Applied Polymer Symposium 31, 351–360 (1977)
© 1977 by John Wiley & Sons, Inc.

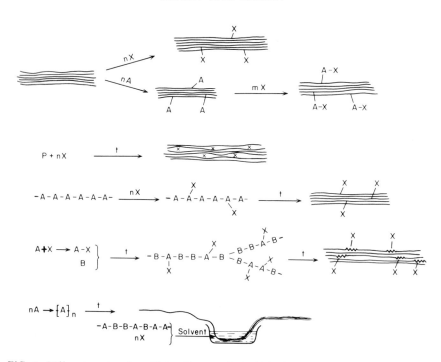

FIG. 1. Different routes of modifying fibers to obtain biological activity treatment: (X) active material; (A, B) monomers; (P) polymer.

RESULTS AND DISCUSSION

The choice of effective reagents is limited not only by the properties of the reagent, e.g., solubility, chemostability, toxicity, chemical functionality, melting temperature and thermal stability, the specific microbial activity, but also by the method of modifying the fiber. Some of the methods for modifying fibers are given schematically in Figure 1. Our work dealt with the modification of filaments of poly(vinyl alcohol) (PVA) and nylon 6 and 66 fibers. The bactericidal reagents employed in this study are listed in Table I.

Because fibers with bioactive properties are used mainly for medical applications, we carried out this work on filament yarns and not on staple. In the case of staple, it is possible that some staple fibers can separate from the yarn during use.

PVA

Multifilament poly(vinyl alcohol) yarn (Den 100) Solvron Type SX (Japan) was heat-set at 208–215°C for 10–15 min. They were then treated with formaldehyde and nitrofurylacrolein (NFA) and with NFA alone in the presence of sulfuric acid at 70°C for 3 hr in water–alcohol. These exhibited a bactericidal activity (Fig. 2). The extent of bactericidal activity is measured by the diameter of the inhibition zone around the fiber of the bacterial growth on Agar Muller Hinten of gram positive (*S. aureus*) and gram negative (*E. coli*).

TABLE I

Bactericidal Reagents

NAME	STRUCTURE	DECOMPOSITION TEMPERATURE °C
NITROFURYLACROLEIN	$O_2N-[furan]-CH=CH-C{\displaystyle <}^{O}_{H}$	114 – 115
NITROFURFURALSEMICARBAZONE (NITROFURAZON)	$O_2N-[furan]-CH=N-NH-CO-NH_2$	236 – 240
NITROFURFURALCARBAZONE	$O_2N-[furan]-CH=N-NH-CO-NH-NH_2$	~ 204
NITROVIN	$\left[O_2N-[furan]-CH=CH\right]_2=C=N-NH-\underset{\underset{NH}{\|}}{\overset{NH_2}{\|}}C \cdot HCL$	290 – 295
GLUTARALDEHYDE	$\overset{O}{\underset{H}{\|}}C-(CH_2)_3-C{\displaystyle <}^{O}_{H}$	

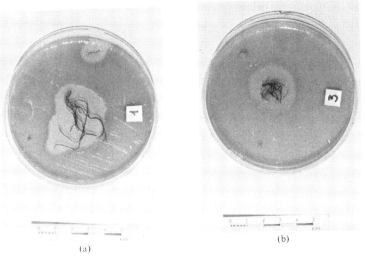

(a)　　　　　　　　　　(b)

FIG. 2. Bactericidal activity of PVA fibers modified with NFA: (a) sample 1 against *S. aureus* OXF.; (b) sample 3 against *E. coli*.

Interestingly, these fibers after autoclaving at 121°C for 20 min showed an increased bactericidal effect (Table II). The effect of autoclaving may be explained by partial hydrolysis of the PVA–NFA acetal linkage during the autoclaving process. The acetal was originally formed by the reaction of NFA with the 1,3-hydroxyl groups of PVA (Fig. 3).

TABLE II
Bactericidal Activity of NFA-Treated Poly(vinyl Alcohol) Fibers

Bacteria	Inhibition Zone Diameter, mm	
	before autocl.	after autocl.
Staph. aureus OXFORD	22	35
E. Coli K 12	--	30
Candida albicanse	--	20

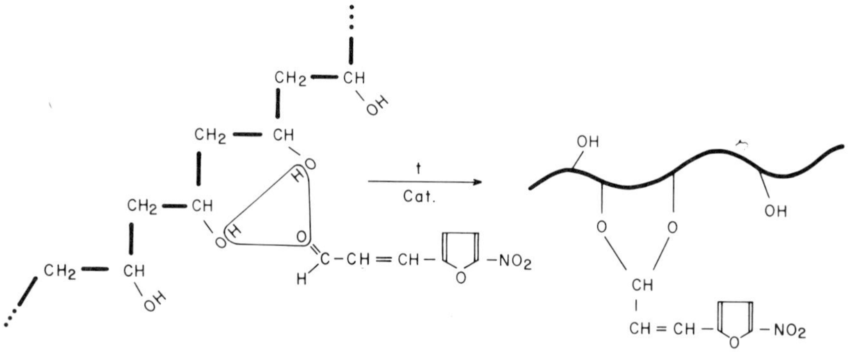

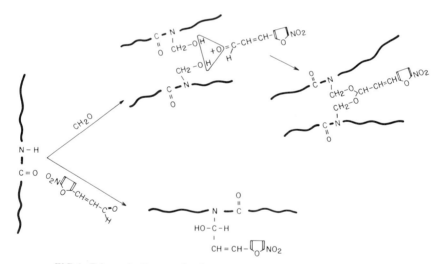

FIG. 3. Schematic diagram showing acetalation of PVA with NFA.

FIG 4. Schematic diagram showing alternate reactions of nylon with NFA.

Nylon

In order to impart bactericidal properties to multifilament nylon fibers (nylon 6 and 66), we employed the following bioactive compounds: NFA, nitrovin, nitrofurazone, nitrofurfural carbazone, and glutaraldehyde. Two methods for bonding the bioactive aldehydes to the nylon chain were investigated (Fig. 4). The first is the direct reaction of nylon with aldehyde groups present in some of the reagents (i.e., NFA, glutaraldehyde). Here, the amide N-H group of the

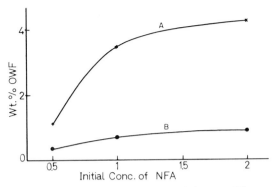

FIG. 5. Effect of reaction medium on NFA pickup: (A) water; (B) water and ethanol.

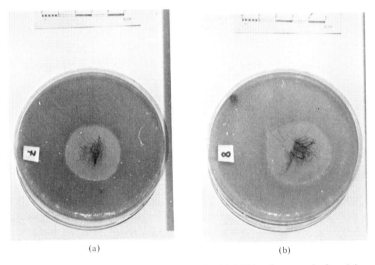

FIG. 6. Bactericidal activity of nylon fibers treated with NFA, after autoclaving: (a) sample 7 against *S. aureus* OXF; (b) sample 8 against *E. coli.*

nylon would add to the carbonyl group of the aldehyde. In the second method, the nylon is first treated with formaldehyde and the resulting *N*-methylol compound reacts in a subsequent step with NFA with the elimination of water to give an acetal.

The conditions for the treatment of nylon with NFA were: acetic acid (3%) as catalyst; temperature, 70°C in a water–alcohol medium or 88°C in water. In both cases, the reaction time was 2.5–3 hr. The ratio of fiber to solvent was 1:42; the NFA concentration varied from 0.5 to 2%. In Figure 5, the dependence of NFA concentration on the fiber is shown as a function of the initial NFA concentration for both water–alcohol and water cases. It can be seen that in water, the total concentration (absorbed + chemically bonded) of NFA on the fiber is higher than in water–alcohol medium.

The treated nylon fibers were washed in four stages as described below. (1) One part of fiber was washed with 50 parts of water containing 0.025 part

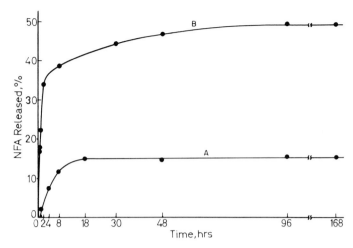

FIG. 7. Percent NFA of the total amount in the nylon fibers released into water vs. time: (A) nylon fibers treated with NFA in water–alcohol medium; (B) nylon fibers treated with NFA in water medium.

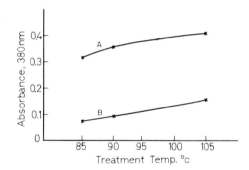

FIG. 8. Absorbance of nylon filaments after nitrovin treatment at various temperatures. Catalysts: (A) CH_3COOH; (B) H_2SO_4.

nonionic surface active agents (Triton X-100) and 0.01 parts sodium carbonate for 5 min at 38°C. The water was pressed out to give a 110% wet weight pickup. (2) One part fiber was washed with 50 parts water containing 0.025 parts Triton X-100 for 10 min at 38°C and pressed to 110% wet weight pickup. (3) One part fiber was washed with 50 parts water for 3 min at 38°C and pressed to 110% wet weight pickup. (4) One part fiber was washed with 50 parts water containing enough acetic acid to give pH = 5 for 3 min at 33°C and dried. The washings were carried out on treated nylon fibers that had stood for four months at ambient conditions after treatment.

The bactericidal activity of the treated fibers was determined on samples with different combinations of post treatments of washing and autoclaving at 121°C for 20 min. The results can be seen in Table III. Typical bactericidal activity is displayed in Figure 6.

Very definite bactericidal effects were obtained with initial NFA concentrations as low as 0.5%. This effect is maintained in the fiber even after the fiber is subjected to a sterilization treatment in an autoclave.

TABLE III
Bactericidal Activity of Nylon Filaments after NFA Treatment

Types of Bacteria	Post-Treatment	Diameter of Inhibition Zone, mm		
		0.5%*	1.0%*	2.0%*
STAPH. AUREUS	without washing	20	40	40
	after washing	20	30	20
	autoclave (without washing	17	37	33
	autoclave (after washing	+	10	35
ESCHER. COL.	without washing	±	25	30
	after washing	+	+	20
	autoclave (without washing)	±	16	20
	autoclave (after washing)	+	+	+

* Concentration of NFA in fiber treatment medium.
⁺ Inhibition 1–2 mm.
± Inhibition ≤1 mm.

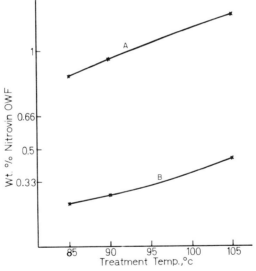

FIG. 9. Nitrovin pickup of nylon filaments at various treatment temperatures. Effect of catalysts: (A) CH_3COOH; (B) H_2SO_4.

Release properties of treated nylon yarns that had stood for 4 months under ambient conditions were evaluated. Different samples were placed in separate flasks each with water, and periodical stirring, and the amount of NFA released in the water was determined. The time of testing ranged from ½ hr to 7 days. The two curves in Figure 7 show the percent NFA of the total amount in the nylon samples released into the water as a function of time. Curve A represents results of release into water for nylon samples treated with NFA in water–alcohol medium, and curve B of release into water of nylon treated with NFA in water medium. It can be seen that the sample prepared in water showed greater release

TABLE IV
Biological Activity of Nitrofuran Derivatives

Nitrofuran derivative	MIC* (mg/ml)
NFA	10
NFA + Sulfodimidine	2.5
Nitrovin	0.075
Furaltadon	1.25
NFA + Aminohydantoin	1.25 - 2.5

* Minimal inhibitory concentration (*S. aureus* OXFORD).

of NFA than the sample treated in water–alcohol. Curve A shows that the sample lost about 15% of its NFA, and curve B about 50% of its NFA. This difference can be explained if it is assumed that most of the NFA released originated from that portion physically absorbed into the fiber. Thus, more NFA was physically absorbed with water treatment than with the water–alcohol treatment.

One of the shortcomings of NFA as for most nitrofuran derivatives is its low decomposition temperature. Consequently, reaction temperatures must be kept low, with the decomposition temperature being a limiting factor. Therefore, we recently undertook the study of the bonding of nylon with nitrovin [1,5-bis(5-nitro-2-furyl)-1,4-pentadien-3-one amidinohydrazone hydrochloride] whose decomposition temperature is 290–295°C. It also has the advantage of possessing high bactericidal activity (Table IV).

Due to its high decomposition temperature, nitrovin and nylon were interacted in solution using high-boiling solvents, such as DMF (bp 159°C) and N-methylpyrrolidone (bp 202°C), and by using melt-spinning techniques. Further work on melt-spinning, including the evaluation of bactericidal activity, is in progress.

In order to determine the amount of nitrovin on nylon fibers obtained from the solvent treatment process, a spectrophotometric study was carried out. The treated yarns were dissolved in formic acid and the solutions analyzed in the

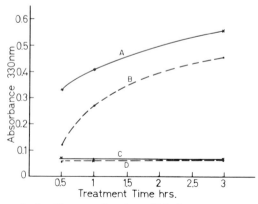

FIG. 10. Absorbance of nylon filaments after glutaraldehyde treatment. Treatment temperatures: (A, B) 100°C; (C, D) 50°C. Glutaraldehyde concentrations: (A, C) 1%; (B, D) 0.5%.

visible range (Fig. 8). The process involved heating nylon yarn at different temperatures in a solution of nitrovin and acid catalyst in $DMF + H_2O$ (1:1). Catalysts were acetic acid and sulfuric acid. Using the absorbance data, the total amount of nitrovin on the nylon yarns was calculated. From Figure 9, it can be seen that acetic acid is about three times as effective as sulfuric acid as a catalyst.

Nylon and Gluteraldehyde

It is known from the literature [32–34] that gluteraldehyde has bactericidal and bacteriostatic properties. We reacted gluteraldehyde with nylon in water using sodium bicarbonate as the catalyst. The effects of time, temperature, and initial concentration of gluteraldehyde on gluteraldehyde pickup were evaluated spectrophotometrically as before (Fig. 10).

Gluteraldehyde-treated nylon fibers were also treated in addition with various nitrofuran derivatives. Preliminary results indicate that the amount of nitrofural-carbazone in the fiber was significantly increased by this treatment.

This effect may have occurred because the groups of different compounds reacted with the free carbonyl functions of gluteraldehyde. Work is continuing along several of the lines discussed above.

The authors thank the Microbiological Laboratory of Tel Hashomer Hospital, Tel Aviv (Director of Department, Dr. G. Altmann) for the biological test results; and Abic, Ltd. and Teva, Ltd. for supplying the bactericidal compounds.

REFERENCES

[1] U.S. Pats. 2,791,518 (1957); 2,813,059 (1957); 2,689,809 (1957); 2,785,106 (1957); 2,374,754 (1945); 3,061,469 (1962); 3,105,773 (1963); 2,562,488 (1951).

[2] B. C. M. Dorset, *Text. Mfr., 80,* 421, 956 (1959).

[3] Brit. Pats. 885,848 (1960); 862,515 (1958).

[4] U.S. Pats. 2,294,339 (1941); 2,856,330 (1958).

[5] French Pat. 1,189,549 (1959); 1,268,607 (1961).

[6] M. Nopitsch and E. Mobis, *Melliand Textilber., 40* (6), 677 (1959).

[7] I. A. Somers, *Text. Rec., 80,* 67, 955 (1962).

[8] T. Takahashi, M. Saikachi, S. Yoshina, and Ch. Mizuno, *J. Pharm. Soc. Japan, 69,* 284 (1949).

[9] A. Sakura and J. Ogatta, *Arch. Antibiot. (Japan), 4,* 252 (1951).

[10] K. Miura, M. Yomato, and M. Ikeda, *Ann. Rep. Fac. Pharm. Kanazawa Univ., 2,* 66 (1952).

[11] S. Hillers and K. Venters, *Latvijas, PSR Zinatrin Akad. Vestis, 12,* 115 (1958).

[12] S. Millers, K. Medne, K. Venter, S. Germane, and A. Zile, *Dokl. Akad. Nauka S.S.R., 144,* 108 (1962).

[13] C. Y. Wang, C. W. Chin, K. Muraoka, P. D. Mickie, and G. T. Bryan, *Antimicrob. Ag. Chemother., 8,* 216 (1975).

[14] J. Labzari, V. Charavi, and N. Mazhari, *J. Pharm. Sci., 63* (9), 1485 (1974).

[15] S. Hillers, L. Volf, and A. Meos, U.S.S.R. Pat, 1,368,853 (1964).

[16] L. Volf, A. Meos, V. Kotetski, and S. Hillers, *Khim. Volokna, 6,* 16 (1963).

[17] V. Kotetski, Y. Charit, L. Volf, A. Meos, S. Hillers, and K. Venter, U.S.S.R. Pat. B 1959; Brit. Pat. 1,254,702 (1972); French Pat. 1,590,713 (1970).

[18] L. Aleksyuk, E. Sorokin, and L. Volf, *Khim. Volokna, 15* (3), 68 (1973).

[19] A. Tomtshin, Y. Charit, and A. Kutsenko, *Chem. Pharm. J., 7,* 10 (1973).

[20] A. Starkova, L. Volf, and T. Pavlova, *Khim. Volokna, 15* (1), 63 (1973).

[21] L. Volf and A. Meos, U.S.S.R. Pat. 168,849.

[22] Y. Charit and A. Kutsenko, *Vestn. Dermotol. Venerol., 11,* 39 (1969).

[23] L. Volf, V. Koletski, A. Meos, V. Chocklova, and V. Emets, *Khim. Volokna, 1,* 16 (1969).

[24] J. Schvarts and Y. Charit, U.S.S.R. Pat. 278,468 (1970).

[25] A. Virnik, J. Zdaneva, N. Plotkina, and E. Sharkova, *Text. Prom., 6,* 95 (1973).

[26] A. Virnik and Z. Rogovin, *Vestn. Dermotol., 1,* 50 (1968).

[27] Z. Rogovin and J. Vashkov, U.S.S.R. Pat. 176,363 (1965).

[28] T. Maltseva, A. Virnik, Z. Rogovin, and E. Sheglova, *Text. Prom., 4,* 15 (1965).

[29] A. Braitis, A. Virnik, and Z. Rogovin, *Med. Radiol., 8,* 25 (1967).

[30] B. Morin, J. Kriazev, and Z. Rogovin, *Vysokomol. Soedin., 7,* 1463 (1965).

[31] A. Virnik and Z. Rogovin, *Biol. Fiziol. Activ. Volokna Osn. Modifi. Cellul. IX Mendeleev Siezd. Izd. Nauka, 1965,* 181 (1965).

[32] A. D. Russell and T. J. Munton, *Microbiology, 11,* 147 (1974).

[33] P. M. Borich, *Advan. Appl. Microbiol., 10,* 291 (1968).

[34] T. J. Munton and A. D. Russell, *J. Appl. Bacteriol., 33,* 410 (1970).

HOLLOW FIBER MEMBRANE DEVELOPMENTS

ELIAS KLEIN

*Gulf South Research Institute, Lake Pontchartrain Laboratories,
New Orleans, Louisiana, 70186*

SYNOPSIS

New applications of hollow fiber membranes to problems in agricultural chemistry, in ion exchange, and in biology and medicine are reviewed. Fumigant and pheromone release fibers are discussed. Both cation- and anion-exchange fibers are evaluated for continuous ion exchange. Medical applications which are examined include the use of fibers as sorbent carriers, and as vascular supplies for tissue culture. Finally, some new developments in reverse osmosis and ultrafiltration fibers are summarized.

INTRODUCTION

New developments in membrane technologies have been greatly stimulated during the past decade by their potential application in reverse osmosis. Such developments have begun to be translated into hollow fiber geometries; bacterial filters, ultrafilters, and reverse osmosis barriers are now available in hollow fiber form. The translation from sheet membrane to hollow fiber membrane has generally been accomplished with little modification of the casting processes. Newer fibers include polysulfone substrates coated with ultrathin barriers [1-3], hollow fiber ultrafilters prepared from aromatic nylons [4], and tubular bacterial filters (Millipore Corp. product catalog).

In addition to the development of hollow fibers based on established precedents, new applications are under investigation based on excellent mass-transfer properties conferred by hollow fiber geometry. In this paper, a number of such new concepts will be reviewed, together with reports of new developments of the more conventional applications listed above.

The utilitarian properties of hollow fibers can be categorized according to several schemes. In Figure 1, we divide the potential application according to the function of the fiber wall. When the fiber wall serves as a selective barrier, one can consider separately those processes which result from purely hydraulic gradients, such as reverse osmosis and ultrafiltration, and those processes which operate as a result of a concentration gradient. The latter processes include classical dialysis, Donnan dialysis, and pervaporation. The fiber wall can also serve as a permselective container. When transport is primarily from the outside

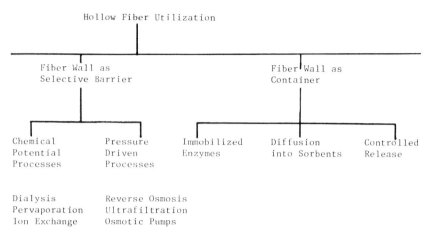

FIG. 1. Schematic diagram showing the potential application of hollow fibers according to the function of the fiber wall.

into the fiber bore, the fiber acts as a sorbent. When transport is principally from the bore to the outside, the fiber may function as a controlled release reservoir. For the application using enzymes immobilized in the bore or in the wall, substrate and product may be transported in both directions. In all such applications, the fiber generally functions as an extended surface having specific semipermeability properties.

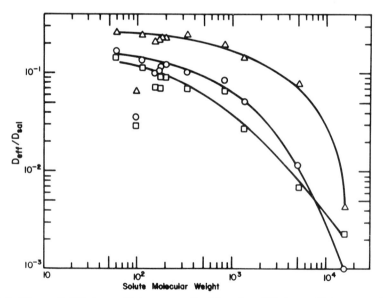

FIG. 2. Fiber wall diffusion coefficients normalized for solute diffusion coefficient at 37°C as a function of solute molecular weight: (O) Enka B2-AH; (△) Amicon; (□) Dow [9].

HOLLOW FIBERS IN BIOLOGY AND MEDICINE

The introduction of hollow fiber hemodialysis devices by Stewart et al. [5] has led to a series of devices useful in medicine and biology. In addition to the hydrolyzed cellulose acetate fibers described by the Dow Chemical Co., cellulose fibers prepared via the cuprammonium complex have been reported in recent years [6, 7]. The preparation of highly permeable polysulfone fibers for dialysis and ultrafiltration has been reported by Amicon Corporation [8] and by our laboratory [1, 2]. The geometries and transport properties of many hemodialyzer fibers are quite similar, reflecting in part the limitations on blood flow shear that can be tolerated, and the limits of ultrafiltration rate acceptable in a clinical device. Hemodialyzer fiber transport and mechanical properties have been examined [9]. Figure 2 shows the diffusivity of three such hollow fibers divided by the solute diffusion coefficient as a function of the molecular weight of the diffusing solute. The experiments were conducted at 37°C in 0.85% saline solution. The transport rates of low molecular weight solutes through the fiber walls are from 15 to 20% of the rates through an equivalent thickness of saline solution. As the solute molecular weight increases, the membrane interferes progressively with transport, exhibiting a sieving effect. The more open the fiber wall, the higher will be the molecular weight at which the membrane becomes impermeable to the solute. Conversely, membranes that exhibit a high relative diffusivity generally also have very high ultrafiltration fluxes.

Hemodialysis fibers with high convective flux have been examined for their applicability as hemodiafilters [10]. In this procedure, the metabolic wastes are not removed by dialysis; rather, the blood is diluted with saline, and the combined volume is ultrafiltered through a hollow fiber device. This device must be capable of retaining serum proteins and formed elements, but should permit passage of lower molecular weight metabolites and nutrients. Among the unresolved questions in this approach is the possibility of exacerbating depletion syndromes by the removal of essential nutrient and control biochemicals. Nevertheless, if

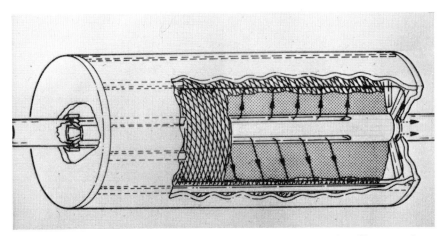

FIG. 3. Hemoperfusion cartridge using carbon-filled hydroxyethylcellulose fibers as sorbents [14].

it can be demonstrated that uremic symptoms can be alleviated by the removal of toxins of high molecular weight, this procedure may offer advantages over direct hemodialysis.

Improved performance in hemodialyzers will require identification of membranes having high diffusive solute fluxes without a concomitant increase in ultrafiltration rates. Currently available membranes which exhibit high permeability by large solutes generally also have excessive ultrafiltration rates. An alternative procedure would be to rely on sorption as a solute removal mechanism, thus avoiding the water fluxes associated with transmembrane pressures encountered in dialysis. Sorbents enclosed in highly permeable hollow fibers permit the attainment of rapid solute removal rates and do not permit any water transport to occur. Thus, one can project that extremely permeable fiber walls could be employed to hold particulate sorbents, or sorbents of molecular weight in excess of the fiber wall size cutoff.

The use of sorbents to remove metabolic wastes by hemoperfusion has been reported by Chang et al. [11], by Andrade and Kolff [12], and by Denti et al. [13]. Activated carbon isolated from contact with serum proteins and platelets has been used to remove serum components on a nonselective basis. Encapsulation procedures have relied principally on the use of carbon particles with permeable coatings, which are prepared by solvent deposition or interfacial polymerization. The primary concern in clinical use of such materials has been the possibility of embolism formation initiated by attrition of particles. These might be deposited in the circulatory extremities during treatments.

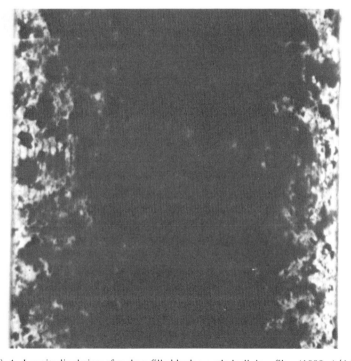

FIG. 4. Longitudinal view of carbon-filled hydroxyethylcellulose fiber (1000 μ) [14].

Two fiber forms of encapsulation have been reported recently. Davis et al. [14] have prepared a hydroxyethylcellulose fiber containing up to 70% activated carbon. The 1000 μ diameter fiber is wound in spool form and blood is allowed to perfuse the filaments (Fig. 3). Nosé et al. [15] reported on a hollow cellulosic fiber filled with activated carbon prepared by Enka-Glanzstoff. In the Nosé device, the fibers are formed into sheets within flexible bags to permit hemoperfusion both by diffusion and by ultrafiltration.

A longitudinal section of the hydroxyethylcellulose fiber is shown in Figure 4. In this example, the fiber contains only 25% (w/w) carbon, but the section

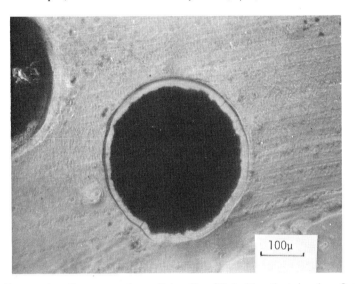

FIG. 5. Cross section of cuprammonium cellulose fiber filled with activated carbon. Outside diameter, 320 μ [45].

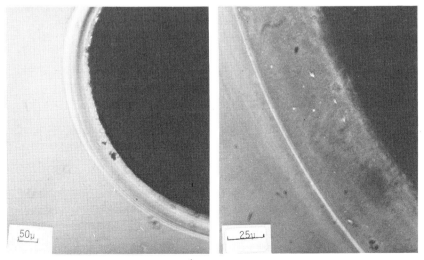

FIG. 6. Cross section of cellulose acetate fiber with activated carbon-filled bore [47].

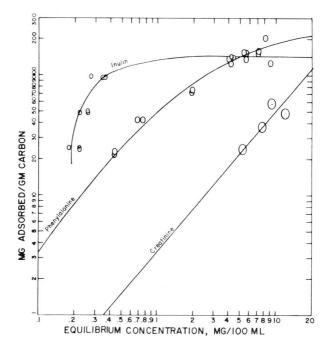

FIG. 7. Equilibrium isotherm of hydroxyethylcellulose fibers filled with activated carbon. Temperature, 37°C; solutes: creatinine, phenylalanine, inulin [47].

illustrates the proximity of the carbon to the surface of the fiber. Figure 5 shows a cross section of the Enka-Glanzstoff hollow fiber used by Nosé et al. The polymer annulus has a thickness of 60 μm; the inside diameter of 200 μm is filled with the carbon. The wall thickness and its resistance to solute transfer are the limiting kinetic factors to the removal rates of metabolites from serum both in the Davis fiber and in the Enka-Glanzstoff fiber. If the barrier thickness between the sorbent and the serum proteins is reduced, the danger of protein denaturation is heightened. If the barrier thickness is increased, the transfer rates decrease.

A cellulose acetate hollow fiber having an extremely high permeability and filled with activated carbon fiber is shown in Figure 6. The hydraulic permeabilities for these two carbon-filled hollow fibers are 5.0×10^{-4} and 700×10^{-4} ml/sec-atm, respectively.

The characterization of assemblies of filled fiber sorbents is a complex procedure. In Figure 7, we show the equilibrium isotherms of the Davis fibers for creatinine, phenylalanine, and inulin. When a reservoir containing these solutes (singly) is perfused through a cartridge containing the carbon-filled fibers, the reduction of solute concentration in the reservoir follows an approximately exponential pattern. As the solute concentration decreases, the equilibrium capacity of the carbon also decreases, as seen from the isotherms. Thus, the mass transfer from solution to fiber occurs with a continuously varying sorbent capacity, mediated by a diffusion process from solution, through the fiber polymer to the carbon.

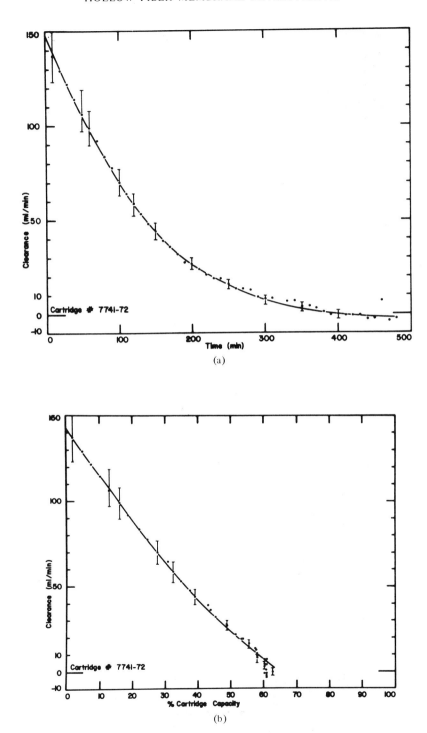

FIG. 8. Hemoperfusion cartridge "clearance" K as a function of (a) time of operation and (b) degree of saturation of activated carbon. Solute, 20 mg % creatinine; closed-loop recirculation, simulated hemoperfusion [46].

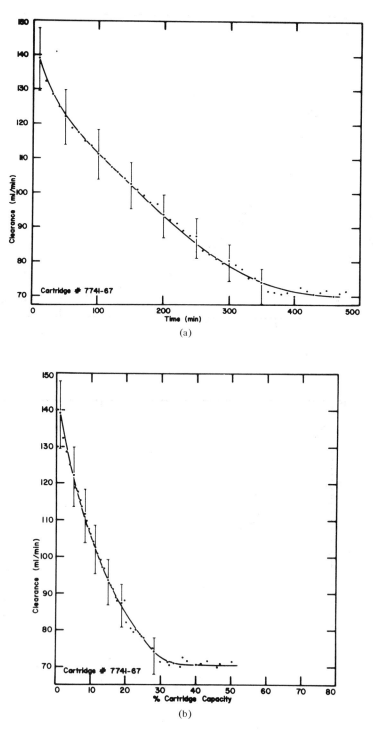

FIG. 9. Same plot as in Fig. 8 except that 15 mg % phenylalanine was used as solute instead of creatinine [46].

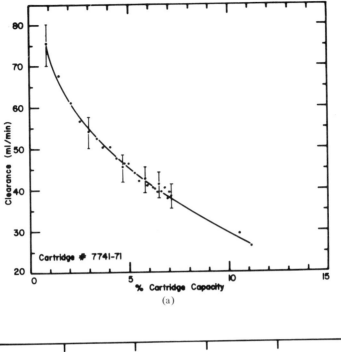

(a)

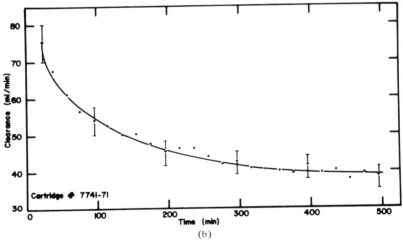

(b)

FIG. 10. Sample plot as in Fig. 8 except that 7.5 mg % inulin was used as solute [46].

Differentiation of the concentration versus time curve permits calculation of the "clearance" of the blood reservoir [16] by virtue of the definition:

$$K \text{ (clearance)} = \frac{1}{C} \frac{dM}{dt} = \frac{Q_B(C_{in} - C_{out})}{C_{in}}$$

where C is concentration, Q_B is blood flow rate, and M is solute mass. Plots of clearance versus time and clearance versus fraction of carbon capacity saturated are shown in Figures 8, 9, and 10 for the cartridges reported by Davis et al. [14].

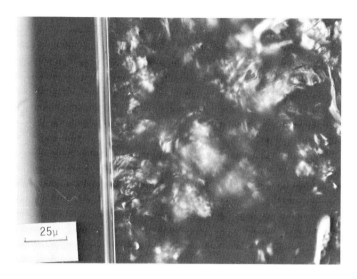

FIG. 11. Cellulose acetate hollow fiber filled with anion exchanger [46].

Clearance of inulin by this device varies from an initial value of 75 ml/min to 35 ml/min. A hemodialyzer of large surface area will have a clearance of approximately 10 ml/min, illustrating the very effective use to which such sorbents may be placed.

The use of hydroxyethylcellulose with an intimately mixed sorbent is restricted by the strongly alkaline solution which the spinning procedure requires. Thus, one would not expect to be able to utilize this procedure with enzymes or microsomal preparations. However, the fibers formed by extrusion of cellulose acetate and other solvent-soluble polymers permit a better isolation of fragile biologicals from the polymer solvent. Experiments in our laboratory with cellulose acetate indicate that a number of useful "fiber-fills" may be possible for use as medical hemoperfusion devices. For treatment of hepatic failures, the possibility of incorporating microsomes with their oxidative enzyme systems merits investigation. Acid/base imbalance may be treatable with ion-exchange materials encapsulated in hollow fibers. Figure 11 shows the encapsulation of a strong anion exchanger in the bore of a cellulose acetate hollow fiber. The fiber wall is permeable to both low and high molecular weight species, but not to serum proteins. To show the difference in charge densities, the section was stained with an indicator.

The use of hollow fibers as enzyme reactors has been studied in recent years [17, 18]. Lewis and Middleman [19] have analyzed the kinetics of substrate conversion in such fiber reactors. Davis [20] used cellulose fibers to study enzyme and membrane transport kinetics of alkaline phosphatase conversion of p-nitrophenylphosphate. A number of other applications are under study, but no large-scale production via such unit operations has been reported as yet. The fibers offer a form of enzyme immobilization by separating the proteins from their substrates and products. Such fibers are not suitable for substrates whose molecular weights approach those of the enzymes. However, for those applica-

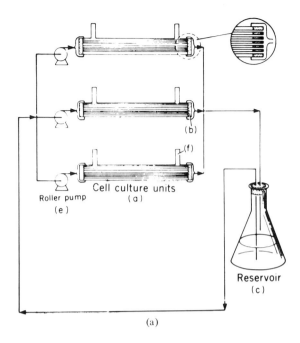

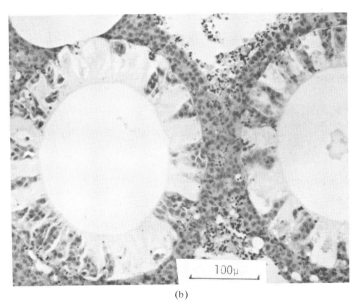

FIG. 12. Hollow fiber-supported tissue culture: (a) schematic for continuous perfusion circuit; (b) cross section of cell supported fibers after development of cell mass [21].

tions where the relative diffusion rates are suitable, enzyme reactor fibers have a unique advantage over other forms of immobilization: they can be refilled at relatively little cost.

Hollow fibers in tube-and-shell configurations also offer the possibility of providing an artificial vascular system. Knazek et al. [21] first reported the use of fibers for tissue culture supports (Fig. 12). More recently, Wolf and Munhelt [22] have shown the potential of such fibers for use in conjugating bilirubin with mouse liver cells. The cells are grown on the outside of the fibers, and the bilirubin-containing medium is circulated through the fiber bores. In similar experiments, Chick et al. [23] have shown the secretion of insulin to take place in β-cells grown on the surface of fibers, in response to glucose levels in the perfused medium in the bore. In our laboratory, experiments are under way to provide a biological sensor based on the response of cells growing on hollow fibers. By monitoring the metabolic rates of indicator cells grown on fiber surfaces, it should be possible to establish a form of "real time" toxicology. These experiments are in their early stages, but the basis for their development has already been established.

ION-EXCHANGE FIBERS

The bore-filled ion-exchange hollow fibers shown in Figure 8 require alternate loading and regeneration cycles for repetitive use. When the ion-exchange resin is in the wall of the hollow fiber itself, these cyclical steps can be converted to a continuous process based on simultaneous loading of the resin from one surface, and regeneration from the other surface.

Wallace [24, 25], Eiseman and Smith [26], and Smith [27] in a series of reports have described the use of ion-exchange membranes as dialytic barriers. The Donnan exclusion principle has been used in these experiments to yield a continuous ion-exchange process when dissimilar ionic species are allowed to flow on opposite sides of the membranes. Applications of this principle have been examined for the recovery of uranium, for the softening of water, and for the concentration of trace metals. Examination of the Donnan process indicates that it is an ion-exchange procedure in which resin loading and resin regeneration are carried out concurrently.

The continuous process offers substantial operational advantages over column resin exchangers: it is consistent with simple process control monitors; it reduces materially the required inventory of ion-exchange resin; and it can be used where column absorbers are not suitable. Fractional recovery of ion concentrations is possible because of the transport mechanism that governs the process. Selective chelating compounds can be used to reduce the activity of ions transported across the membrane, adding a new dimension to selective removal of ions of the same charge.

Cation-exchange hollow fibers were reported by Skiens [28], and both cation- and anion-exchange hollow fibers were reported by Klein [29] and by Rembaum et al. [30]. In our laboratory, ion-exchange hollow fibers have been prepared by posttreatment of hollow fibers prepared from aromatic polymers of controlled porosity, and by polymerization of ion-exchange resins in the pores of acrylic hollow fibers. Figure 13 shows a cross section of a cation-exchange fiber stained with Cu^{++} ions, and then examined by back-scattered x-rays in the scanning electron microscope (SEM). In the center panel is the x-ray fluorescence image

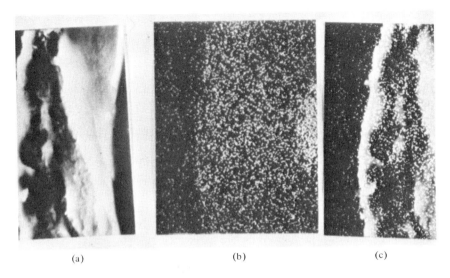

FIG. 13. Cationic-exchange fiber stained with Cu^{++}: (a) SEM scan of cross section; (b) x-ray fluorescence image of Cu from same section; (c) superposition of panels (a) and (b) [14].

derived from the metal stain. The right panel shows the SEM image of the cross section, and the left panel is a composite prepared by overlaying the two negatives.

Transport studies to describe the governing relations have been examined by Melsheimer et al. [31] and by Wendt et al. [32]. The latter report from our laboratory utilized transport experiments performed both with Nafion (DuPont) perfluorosulfonic acid copolymer hollow fibers, and with specially prepared polysulfone fibers. The transport relations are more complex than purely dialytic exchange, since ion transfer is dependent not only on concentration gradients across the membrane, but also on leakage flows of ion pairs. When such ion-pair fluxes can be ignored, i.e., when the fiber wall is perfectly selective, the transport relationships presented by Wendt et al. [32] can be used to describe ion-exchange processes.

The practical application of Donnan dialysis will require the development of stable, highly selective ion-exchange fibers. Stability of both fiber matrix and exchanger polymer is needed to permit extended, uninterrupted operation of devices based on such hollow fibers. Although the continuous ion-exchange process demands smaller volume excursions of the ion-exchange matrix than does cylindrical loading of beads, a high degree of chemical resistance is needed if permselectivity is to be maintained in the former process. In bead exchangers, loss of semipermeability leads to loss of capacity, but not necessarily to process failure. However, in the flow methods using fibers, the loss of exchange capacity can lead to uncharged pores which defeat the process.

The simplicity of Donnan dialysis is illustrated by its use in the removal of ions during concentration of drinking water samples for toxicological assays [33]. Both cation- and anion-exchange fibers can be used with acid and base counter-flows to exchange H^+ and OH^- for cations and anions, respectively; the dialysis is analogous to the use of mixed-bed desalting. However, with the fiber assemblies the dialysis is a continuous, rather than batch, process. In the cited study, it was necessary to remove calcium without introducing trace organics. By exchange of Ca^{++} for Na^+ across a Nafion fiber, the calcium was reduced to prevent precipitation and membrane fouling.

PERVAPORATION AND ORGANIC SEPARATIONS

Separations of ionic species as described in the preceding section have a parallel in organic separations. However, the governing relationships now are based not on electroneutrality, but on solubility phenomena. By judicious selections of polymer and organic compound combinations, a number of industrially significant membrane separations can be achieved. The use of hollow fibers in such applications permits the achievement of high surface-packing densities, and simple exchange of operating units for maintenance.

The pioneering studies by Binnings et al. [34] to develop organic separations based on selective permeation through membranes have led to continued interest. The literature for the past decade has dealt solely with separations based on sheet membranes, but recent work in our laboratory has shown promising separations using hollow fiber devices [35]. Remarkable separations of the methanol–hexane azeotrope were reported using a crosslinked derivative of polyphenylene oxide. Figure 14a shows the separation factor "α," defined as a function of feed side composition:

$$"\alpha"_{nm} = \frac{X_n^P X_m^F}{X_n^F X_m^P}$$

Here X_n^F, X_m^F, X_n^P, and X_m^P denote the weight fractions of component n and component m, respectively, in the feed solution and in the permeate.

The equilibrium composition of methanol–hexane consists of two phases, one of which contains 8% methanol at 25°C. As the methanol concentration in the mixture increases, the separation factor decreases (Fig. 14b). The separation factor reaches a plateau at a value of 446. The decrease in separation is attributed to the preferred distribution of the alcohol into the membrane and the resulting swelling of the polymer network. When approximately 2% alcohol is reached, the maximum swelling allowed by the crosslinked network has been achieved. At higher methanol concentrations no further effects are observed. The permeation rate of the methanol, however, continues to increase, as seen in the bottom of Figure 14b. The methanol transport rate is proportional to the activity gradient across the membrane, which increases with increasing volume fraction in the feed.

Separations of alcohol–hydrocarbon mixtures have also been achieved in our laboratories using asymmetric cellulose acetate membranes, but with much lower separation factors and fluxes. Detailed engineering analyses for each particular separation are required to assess the potential of this technology. The diffusion-controlled separations find their greatest potential in azeotrope separations,

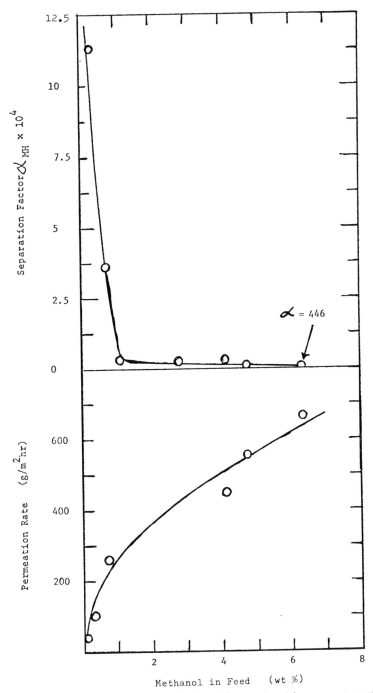

FIG. 14. Separation factor and flux density of methanol from methanol–hexane azeotrope through fiber permeator [35].

and in those tower distillations in which stage separation factors are low. The utility of hollow fibers would appear to lie in the high packing density per unit area that is permitted by this geometry. The vacuum receivers used in pervaporation provide long mean free paths for the permeates, so that downstream boundary layers are minimized.

PESTICIDES AND OTHER AGRICULTURAL CHEMICALS

The controlled release of pesticides and other chemicals in agricultural use offers advantages of improved environmental controls and efficient utilization of costly chemicals [36]. It is probable that continuous exposure of target insects to subacute levels of pesticides will effectively eradicate the insects with minimal loss of chemical to other sources [37]. Extended exposure of water hyacinth (*Eichhornia crassipes*) to 0.0025 times the single application dosage of 2,4-dichlorophenoxy acidic acid ester (2,4 D) was shown by Cardarelli [38] to be effective when a slow release procedure was used.

Hollow fibers permit the controlled release of chemicals and also provide adequate reservoir volumes for extended applications. By selection of wall permeability and fiber diameter, one can achieve a wide range of release rates. Thus, one might anticipate that this geometry would find numerous applications in the complex but growing technology of controlled release of agricultural chemicals. To date, very little research dealing with the utilization of fibers in this application has been reported.

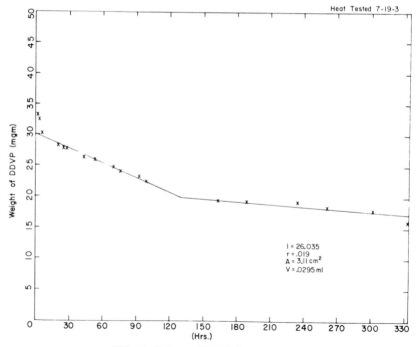

FIG. 15. Release of DDVP from vinyl fiber.

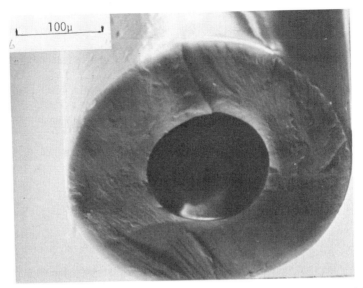

FIG. 16. Cross section of polysulfone fiber for high-pressure membrane support [1].

Pheromones have been identified as sex attractants that facilitate the repro-
ductive cycles of a large number of insects. A widely scattered release of minute
amounts of these chemicals into the atmosphere interferes with the directional
flight of the male insect in search of the female and thus reduces the rate of re-
production. Experiments with the release of elm bark beetle pheromones [39]
have been reported, using hollow fiber reservoirs prepared from polyethylene
and polyesters. In this application, the release rate is governed by the free
evaporation of the pheromone from its solution in the cylindrical fiber. One end
of the fiber assembly is closed and the other is open. As evaporation proceeds,
a cylinder of saturated vapor above the liquid phase becomes the diffusion-
limiting barrier. The kinetics are thus governed by the partial pressure of the
pheromone and by the geometry of the fiber assembly [40]. The walls of these
fibers are impermeable and do not contribute to the release rate.

The release of fumigants into closed atmospheres has been practiced for many
years, using a solution of the fumigant in a plasticized polymer as a reservoir.
As the fumigant is released, the activity in the plasticizer decreases and the release rate is reduced. Although such devices are capable of holding relatively
large quantities of fumigants, their release rates change over a wide range during
their useful life. Baker and Lonsdale [40] have reported on the kinetics of the
release of agents from cylindrical reservoirs having known permeabilities and
geometries. If the chemical activity inside the reservoir does not change, the
release rate is constant (after the initial lag or burst periods have passed).

In our laboratory, we have examined the use of a vinyl fiber for such applications, using the fumigant dimethyl 2,2-dichlorovinylphosphate (DDVP, Va-
pona) as the agent. By control of the wall permeability, it has been possible to
control the release rate to levels that are adequately low to provide a reservoir
which has a constant loss rate over a period of 3 weeks at room temperature.

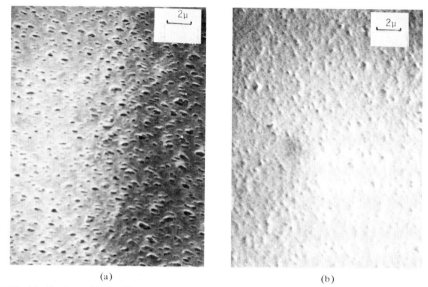

FIG. 17. Surface of polysulfone hollow fiber support: (a) before coating; (b) after deposition of NS-100 salt-rejecting membrane [1].

Figure 15 shows the loss of weight of a small test unit containing the DDVP; the permeability of the fiber was calculated to be 19 $\mu g/cm^2$-hr. The fibers can be assembled into refillable units for installation in storage areas, animal quarters, or primitive residences.

PRESSURE-MEDIATED PROCESSES

Hollow fibers for reverse osmosis have been available for some years, but new developments in this field continue to be reported. From our laboratory [1, 2] have come polysulfone fibers prepared to withstand 100 atm external pressure (Figs. 16 and 17). These are coated with an ultrathin film of a salt-rejecting membrane to form a two-layer structure (Fig. 18). Similar fibers have been reported by Davis et al. [3] using a different preparation for the polysulfone fibers.

Strathman [4] has reported preparation of both reverse osmosis fibers and ultrafilters based on aromatic nylons. These preparations offer potential use at temperatures up to 80°C. The effects of membrane fouling as a consequence of the ultrafiltration of polymer-containing solutions has been a subject of continuous study. The applications range from treatment of secondary sewage through the processing of serum proteins. For a number of applications, the fouling problems have been solved by membrane cleaning procedures. However, for other applications, the wetting phenomenon apparently overcomes the effectiveness of periodic cleaning cycles and results in reduced service life. It appears that a nonwetting ultrafilter, or one that is easily freed of surface-spread polymer, is needed for certain applications.

Hydraulic pressure in excess of 20 atm requires the use of pressure vessels to house fiber assemblies. Under certain circumstances, very large effective

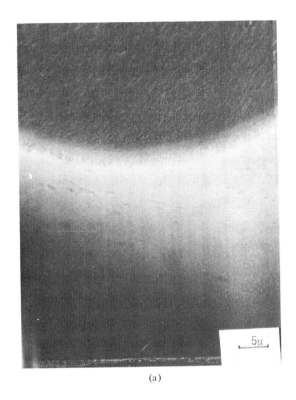

(a)

(b)

FIG. 18. Polysulfone hollow fiber coated with salt-rejecting furan resin: (a) cross section of fiber (2000×); (b) surface of furan resin coating (10,000×) [1].

driving pressures can be generated by the use of osmotic forces. These do not require pressure housing. The osmotic concentration of water to recover viruses was reported by Klein et al. [41] and Sweet et al. [42]. In a similar study, osmotic pumping [43] was used to find the rejection coefficient of reverse osmosis membranes for organic solutes. Strong saline solutions were used to "pump" water containing trace levels of the organics [33]. Osmotic pumping using hollow fibers has been proposed by Loeb [44] as a means of generating power. Wherever a large reservoir of strong electrolyte solution exists, the potential for transferring water into the saline is possible, whether to generate power or to dehydrate shipboard wastes prior to incineration.

The author expresses his appreciation to the staffs of the Chronic Uremia-Artificial Kidney

Program, NIAMDD; the Water Quality Division, Health Effects Research Laboratory, United States Environmental Protection Agency; and the National Science Foundation for financial support of various programs in his laboratory and reported here. The work from his laboratory is the result of the collaboration of his colleagues, Dr. I. Cabasso, and Messrs. F. Holland, P. May, and J. K. Smith.

REFERENCES

[1] I. Cabasso, J. K. Smith, and E. Klein, *ACS Coatings Plast. Prepr., 36,* 492 (1975).

[2] I. Cabasso, E. Klein, and J. K. Smith, "Polysulfone hollow fibers. I. Spinning and properties," *J. Appl. Polym. Sci.,* in press.

[3] R. B. Davis, A. E. Allegrezza, and M. J. Coplan, paper presented at the First Chemical Congress of North America, Mexico City, 1975.

[4] H. Strathman, paper presented at the International Symposium on Macromolecules, Jerusalem, 1975.

[5] R. D. Stewart, J. C. Cerny, and H. I. Mahon, *Univ. Mich. Med. Cent. J., 30,* 116 (1964).

[6] W. Bandel, paper presented at the 7th Work Session of Clinical Nephrology, Hamburg, Germany, 1974.

[7] K. Maede, K. Ohta, A. Saito, T. Shimoji, I. Amano, T. Manji, S. Kawaguchi, K. Kobayashi, Y. Fujisaki, and S. Eiga, *Trans. ASAIO, 20,* 344 (1974).

[8] R. A. Cross, W. H. Tyson, and D. S. Cleveland, *Trans. ASAIO, 18,* 279 (1968).

[9] E. Klein, F. Holland, A. Lebeouf, A. Donnaud, and J. K. Smith, "Transport and mechanical properties of hemodialysis hollow fibers," *J. Membrane Sci.,* in press.

[10] M. E. Silverstein, C. A. Ford, M. J. Lysaght, and L. W. Henderson, *Trans. ASAIO, 20,* 614 (1974).

[11] T. M. S. Chang, C. Coffey, C. Lister, E. Taroy, and A. Stark, *Trans. ASAIO, 19,* 87 (1973).

[12] J. D. Andrade and W. J. Kolff, in *Proceedings of Annual Contractors' Conference of NIAMDD* (DHEW Publication NIH-74-248), K. K. Krueger, Ed., NIH Public Health Service, 1973, p. 110.

[13] E. Denti, M. P. Luboz, and V. Tessore, *J. Biomed. Mater. Res., 9,* 143 (1975).

[14] T. A. Davis, D. R. Cowsar, S. D. Harrison, and A. C. Tanquary, *Trans. ASAIO, 20,* 353 (1974).

[15] Y. Nosé, P. S. Malchesky, F. Castino, I. Koshino, K. Scheucher, and R. Nokoff, paper presented at the Conference on Sorbents in Uremia and Hepatic Failure, Downstate Medical Center, Brooklyn, N.Y., November, 1975.

[16] E. F. Leonard, Ed., *Evaluation of Membranes for Hemodialyzers* (DHEW Publication No. NIH 74-605).

[17] L. R. Waterland, A. S. Michaels, and C. R. Robertson, *AIChE J., 20,* 50 (1974).

[18] P. R. Rony, *Biotechnol. Bioeng., 13,* 431 (1971).

[19] W. Lewis and S. Middleman, *AIChE J., 20,* 1012 (1974).

[20] J. C. Davis, *Biotechnol. Bioeng., 16,* 113 (1974).

[21] R. A. Knazek, P. M. Guillino, P. O. Kohler, and R. L. Dedich, *Science, 178,* 65 (1972).

[22] C. F. W. Wolf and B. E. Munhelt, *Trans. ASAIO, 21,* 16 (1975).

[23] W. L. Chick, A. A. Like, V. Lauris, P. M. Gallati, P. D. Richardson, G. Panol, T. W. Mix, and C. K. Colton, *Trans. ASAIO, 21,* 8 (1975).

[24] R. M. Wallace, *Ind. Eng. Chem. Proc. Res. Dev., 6,* 423 (1967).

[25] R. M. Wallace, U.S. Pat. 3,454,490 (1969).

[26] J. L. Eiseman and J. D. Smith, *Donnan Softening as a Pretreatment for Desalination Processes,* OSW Report No. 406, U.S. Govt. Printing Office, Washington, D.C., February, 1970.

[27] J. D. Smith, *Exchange Diffusion as a Pretreatment to Desalination Processes,* OSW R&D Report No. 655, U.S. Govt. Printing Office, Washington, D.C., May, 1971.

[28] W. E. Skiens, U.S. Pat. 3,186,941 (1965).

[29] E. Klein, P. May, J. K. Smith, and R. P. Wendt, *Polym. Lett., 13,* 45–48 (1975).

[30] A. Rembaum, S. P. S. Yen, E. Klein, and J. K. Smith, in *Electrolytes and their Application,* Reidl, The Hague.

[31] S. S. Melsheimer, H. M. Kelley, L. F. Landon, and R. M. Wallace, paper presented at the 74th National AIChE Meeting, New Orleans, 1973.
[32] R. P. Wendt, E. Klein, and S. Lynch, "Simplified flux equations for transport across ion exchange membranes," *Membrane Res.*, in press.
[33] I. Cabasso, E. Klein, and J. K. Smith, paper presented at the First Chemical Congress of North America, Mexico City, 1975.
[34] R. C. Binnings, R. J. Lee, J. F. Jennings, and E. C. Martin, *Ind. Eng. Chem.*, 53, 45 (1961).
[35] I. Cabasso and I. Leon, paper presented at the 80th National AIChE Meeting, Boston, 1975.
[36] W. C. Shaw, in *Proceedings of the 1974 Controlled Release Pesticide Symposium* (University of Akron, September, 1974).
[37] G. A. Janes, in *Proceedings of the 1974 Controlled Release Pesticide Symposium* (University of Akron, September,1974), p. 14.1.
[38] N. F. Cardarelli, in *Proceedings of the 1975 International Controlled Release Pesticide Symposium* (University of Akron, September, 1975).
[39] E. Ashare, T. W. Brooks, and D. W. Swenson, in *Proceedings of the 1975 International Controlled Release Pesticide Symposium* (University of Akron, September, 1975), p. 42.
[40] R. W. Baker and H. K. Lonsdale, in *Controlled Release of Biologically Active Agents*, A. C. Tanquary and R. E. Lacey, Eds., Plenum Press, New York, 1974.
[41] E. Klein, J. K. Smith, F. Morton, and B. H. Sweet, *Water Res.*, 5, 1067 (1971).
[42] B. H. Sweet, J. S. McHale, K. J. Harty, F. Morton, J. K. Smith, and E. Klein, *Water Res.*, 5, 823 (1971).
[43] E. Klein, J. Eichelberger, C. Eyer, and J. K. Smith, *Water Res.*, 9, 807 (1975).
[44] S. Loeb, U.S. Pat. 3,906,250 (1975).
[45] W. Bandel, private communication.
[46] E. Klein, F. Holland, I. Cabasso, A. Donnaud, and S. Verrett, "Evaluation of sorbent filled fibers for hemoperfusion," in preparation.
[47] T. Davis, private communication.

ION-EXCHANGE FIBERS: PREPARATION AND APPLICATIONS

R. MESSALEM, C. FORGACS and J. MICHAEL
Ben-Gurion University of the Negev, R & D Authority, Beer Sheva, Israel

O. KEDEM
Weizmann Institute, Rehovot, Israel

SYNOPSIS

The production of ion-exchange resins and ion-exchange membranes is a well-established technology. Of the many published and patented methods, one of these is the photochemical sulfochlorination of polyethylene sheets, followed by either hydrolysis or successive aminolysis and quaternization to obtain cation- and anion-exchange membranes, respectively. A new electrodialysis stack configuration was recently proposed by Kedem. This assumes the availability of woven spacers made of ion-exchange fibers, both cation- and anion-exchangers. Other possible applications are reported. Ion-exchange fibers are characterized by large exchange surface, low flow resistance, and ease of handling.

A new brand of ion-exchangers from low-density melt-spun polyethylene fibers was recently prepared. This method is based on the de Körösy-Shorr patent but utilizes the special geometry of the fibers and is tailored to achieve high mechanical strength and optimal radial distribution of ion-exchange sites. The fiber thus obtained has an average ion-exchange capacity of 1.6 meg/g dry weight, electrical conductivity of the order of 100 Ω·cm (better than the conductivity of ion-exchange material), and fairly high mechanical strength.

INTRODUCTION

The production of ion-exchange resins and ion-exchange membranes is a well-established technology [1–3]. Polymeric ion-exchange resins are fabricated, marketed, and utilized in bead form. They are readily produced by the method of emulsion polymerization. Diffusion (the rate-determining step in most ion-exchange processes) towards and from spherical solids exhibits the highest rate for a given size of resin particle. Methods for the incorporation of the resins were developed at a very early stage because it was not convenient to work with a heap of small beads. Today, for all practical purposes, ion-exchange columns are used exclusively. The theory and practice of these columns was developed to a large extent by utilization of knowledge applied in other unit operations (distillation, etc.).

Journal of Applied Polymer Science: Applied Polymer Symposium 31, 383–388 (1977)
© 1977 by John Wiley & Sons, Inc.

The use of ion-exchange columns poses certain problems and limitations. To enhance mass transport and thus obtain maximal utilization of the capacity of ion-exchange resins, the bead diameter should be as small as possible. Columns of fine particles, however, show very high hydraulic resistance and are readily "washed out" from the container. The monodispersity of the bead size is also of major importance as the inlet and outlet of the columns are designed to withhold particles above a certain diameter. The wear and tear of the ion-exchange material results in the mechanical deterioration of the beads, and thus small fragments are either lost or may even cause the plugging of the system.

Alternative concepts of ion-exchange material configuration have been suggested time and again. Ion-exchange belts, felts, screens, fibers, both solid [4–7] and hollow [8, 9], have been reportedly prepared and utilized. Until now, no commercial ion-exchange material (apart from beads and, to a smaller extent, granulates) has been available. (Ion-exchange membranes are not discussed here and are never used as ion-exchanger per se.)

The purpose of this paper is to report on the development of a new brand of ion-exchange fibers, their properties and certain applications. The development of these fibers was triggered by a recent suggestion of Kedem [10] concerning a new electrodialysis stack configuration. This configuration assumes the availability of woven spacers made of ion-exchange fibers, both cation- and anion-exchangers. The method for the fabrication of these fibers is based on a procedure developed by de Körösy and Shorr [11, 12] for the preparation of ion-exchange membranes. It consists of the photochemical sulfochlorination of polyethylene, followed by either hydrolysis or successive aminolysis and quaternization to obtain cation-exchanger material or anion-exchanger material, respectively (Fig. 1).

EXPERIMENTAL

Materials Used in Preparation of Ion-Exchange Fibers

Low-density polyethylene fibers were kindly supplied by Dr. G. Lopatin and Prof. M. Levin of the Israel Fiber Institute, Jerusalem. The fibers were melt-spun and cold-drawn at different ratios, up to 4:1.

Chlorine and sulfur dioxide of technical grade were used for sulfochlorination, a 10% solution of technical-grade NaOH for hydrolysis, 3-dimethylaminopropylamine of laboratory reagent grade (BDH, England) for amination, and methyl bromide (Bromine Compounds Ltd., Beer Sheva, Israel) for quaternization.

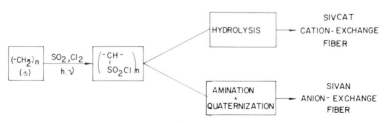

FIG. 1. Fiber preparation.

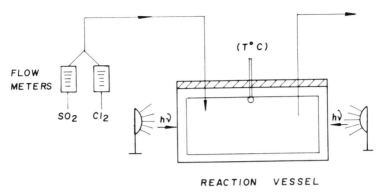

FIG. 2. Photochemical sulfochlorination system.

The Sulfochlorination Reaction

The central part of the sulfochlorination unit is the photochemical reactor. The experimental setup is shown in Figure 2. It is essentially a well-sealed, rectangular glass "aquarium" in which polyethylene fibers are loosely hung to allow for good illumination. The vessel is illuminated by two batteries of incandescent lamps. The gas mixture is supplied from high-pressure chlorine and SO_2 cylinders through pressure regulators. The gas flows are monitored by rotameters, bubbled through H_2SO_4 to dry, and then mixed. The reacting mixture is introduced through diffusers into the vessel. The outlet gas stream is either recirculated or scrubbed in an alkaline solution. It should be mentioned that HCl is formed during the reaction, so that the gas mixture cannot be recirculated indefinitely.

The photochemical reaction is carried out for 6 hr at 60°C. Ion-conductive cation-exchange fibers are then obtained by overnight hydrolysis in 10% NaOH solution at 60°C.

Anion-exchange fibers are prepared by overnight amination of the polyethylene sulfonyl chloride with $NH_2-(CH_2)_3-NH(CH_3)_2$, followed by quaternization with methyl bromide bubbling through nitromethane, for 12–14 hr.

Properties of Ion-Exchange Fibers

The following properties of the ion-exchange fibers have been monitored: S and Cl content, ion-exchange capacity, water content, and electrical conductivity.

The mechanical properties of the fibers have not yet been quantitatively determined. However, the end product seems to be superior to the raw material, and it can be easily processed on certain knitting machines.

S and Cl content were determined by standard analytical procedures. Ion-exchange capacity was determined by complete Na–Ca exchange, followed by EDTA titration of the exchanged Ca. The electrical conductivity of the fiber is determined by a Steyman's type cell [13], shown in Figure 3. The cell is made of a Plexiglas double-walled cylinder with two silver electrodes fixed on the water-cooled lid. One of the electrodes (6) can be moved up and down. A 10-cm

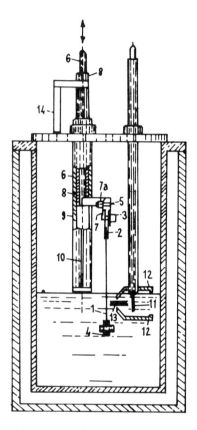

FIG. 3. Experimental setup for measuring the electrical conductivity of fibers and membranes.
See text for explanation of numbers.

piece of fiber is attached to the movable electrode, with a pair of metal clips (2).
The cell is filled with an electrolyte solution and kept at a constant temperature
(within 0.01°C) via a water thermostat. The sample is brought down into the
solution for complete equilibration, usually overnight. Then the electrode (6)
is moved upwards 1 cm at a time and the conductivity of the system is monitored
on a Wayne-Kerr WII high-precision conductivity bridge. The electrical conductivity of the fiber is calculated from both the slope of the resistance versus
fiber length and the measured diameter of the wet fiber. Representative characteristics of the fibers are shown in Table I.

RESULTS AND DISCUSSION

The above results show that the de Körösy-Shorr method can be successfully
adapted to the fabrication of ion-exchange fibers. The ion-exchange properties
of the product are compatible with those of ion-exchange beads. Electrochemical,
chemical, and mechanical properties of the material are also satisfactory.

The electrical conductivity of the fibers is a unique property which may find
future applications. Although conductivity of ion-exchange in columns has been

TABLE I
Conductive Ion-Exchange Fibers: Properties of Representative Samples

Sample Identification	SIVCAT-95	SIVAN-510
Material	PE	PE
Fixed groups	$-SO_3^-$	$-N^{\pm}$
Specific resistance (ohm x cm)	70-90	80-100
Exchange capacity (meq/gr dry fiber)	1.5-1.7	1.0-1.2
Water content (%)	10-12	10-12
Fiber diameter (μ)	200-230	200-230
Ionic form	Na^+	Cl^-
Sulfur content (%)	4.9	4.5
Chlorine content (%)	3.2	1.9

studied and conductivity of ion-exchange membranes is utilized in the electro-dialysis process, current transport *along* ion-exchange materials has not yet been utilized for practical purposes. Coupling of the effect of the electric field on the movement of ions in ion-exchange fibers with other displacement effects (like diffusion and movement of the fibers, and ion-exchange) may result in separation processes of novel characteristics.

The application of these fibers as conductive spacers in the process mentioned above [10] and in other possible applications (like separators in electrochemical energy storage systems) should also be mentioned.

As for ion-exchange itself, one can foresee the emerging use of ion-exchange fiber bundles and stacks instead of columns. Although mass transfer to and from spherical particles is higher than for cylindrical symmetry, this difference can be more than compensated for by the following phenomena: (1) As a fiber bundle or a net-stack has fixed position, there is no danger of "washing out" of resins; therefore, the fiber diameter can be much smaller than the smallest practical bead diameter (less than 20 μ as compared with 450–600 μ). (2) For similar reasons, the flow velocity through such a bundle or stack can be much higher; thus, the boundary layer thickness is reduced, and mass transfer rate will be further enhanced. (3) The smaller diameter will also result in better utilization of the ion-exchange sites in the resin; thus, the effective ion-exchange capacity will be much closer to that measured under equilibrium conditions. (4) As the mechanical wear and tear will be much smaller, the resin could be less crosslinked and more swollen. This again will lead to better utilization of the ion-exchange sites and further improved kinetics.

The development of these and other applications for the ion-exchange fibers described in this paper is presently being carried out in our laboratory.

REFERENCES

[1] F. C. Nachod and J. Schubert, Eds., *Ion Exchange Technology,* Academic Press, New York, 1956.

[2] Konrad Dorfner, Ed., *Ion Exchangers—Properties and Applications,* Ann Arbor Science Publishers, Ann Arbor, Mich., 1972.

[3] J. A. Marinsky, *Ion Exchange—A Series of Advances,* Dekker, New York, 1969.

[4] A. M. Maksimov and L. A. Volf, *Zh. Prikl. Khim.* (*Leningrad*), *46*(10), 2346 (1973).

[5] N. I. Prokhorenko, (*USSR*) *Nauch. Osn. Tekhnol. Ochist Vody, 1973,* 85.

[6] U. Mitsutaka, Japan. Pat. 7,421,394; *Chem. Abstr., 81,* 7898a;

[7] H. Asuo and Miyamoto (of Mitsubishi Rayon Co., Ltd.), Japan. Pat. 7,476,785.

[8] E. Klein, P. May, J. K. Smith, and R. Ward, paper presented at the 168th ACS National Meeting, Atlantic City, N.J., Sept. 8–13, 1974.

[9] R. Messalem, M.Sc. Thesis, Weizmann Institute, Rehovot (1969).

[10] O. Kedem, *Desalination, 16,* 105 (1975).

[11] F. de Körösy, J. Shorr, Isr. Pat. 14,720; Brit. Pat. 981,562; U.S. Pat. 3,388,080; Dutch Pat. 130,894.

[12] F. de Körösy and J. Shorr, *Dechema Monographs, 47,* 477 (1962).

[13] Cl. Steymans, *Ber. Bensenges. Phys. Chem., 71*(8), 818 (1967).

HOLLOW FIBER GLASS MEMBRANE

DOV BAHAT and JEAN FLICSTEIN
Department of Geology and Mineralogy,
Ben Gurion University of the Negev, Beer Sheva, Israel

SYNOPSIS

Glass hollow fiber membranes have a few advantages compared with polymer membranes. These are: high strength under compressive stresses, high stability under low pH and high-temperature conditions, and resistance to reactions of bacteria and organic solutions. The two disadvantages of glass membranes are fragility of glass, and complication in preparational operations.

Glass membrane is basically an ultrafilter and its performance is dependent on a large variety of parameters, including original glass composition, heat treatment and leaching processes, size and shape of pores, surface chemistry of pores, as well as chemical composition and dielectric properties of the solution. All these parameters are interdependent. Encouraging prospects of glass membrane performance would still be dependent on solving basic problems of mechanical properties of glass.

INTRODUCTION

It appears appropriate within the context of a seminar on membranes made of polymers to discuss the dependence of glass membrane performance on various surface properties of the glass, so that the reader may be able to elucidate similarities and differences between these two basic materials. The present study describes the basics of glass membrane preparation and then reviews various ideas on glass surface parameters in relation to membrane performance.

HOLLOW FIBER GLASS MEMBRANE

Glass Membrane

The operational principles of glass membrane preparation had been demonstrated about 40 years ago [1] and have been supported by theoretical considerations [2,3].

The glass membrane is prepared essentially from three oxides, SiO_2, B_2O_3 and Na_2O, within a certain chemical range [4]. Following melting, casting, and annealing, the glass is heat-treated above the annealing temperature range. Since it is unstable as a single phase, the alkali borosilicate glass separates into two phases, one rich in silica, and the other mainly sodium borate. The latter phase may be readily dissolved by acids (such as dilute HCl) and washed away by water, whereas the silica-rich phase remains undissolved and becomes a glass membrane full of interconnected pores. The morphology of the membrane pores is determined by a series of parameters. The first important factor is the position of the glass in the phase diagram in terms of chemical composition and temperature of heat treatment, and also the length of time at the particular temperature. This composition–temperature–time dependence determines the mechanism of phase separation and morphology of separated phases as well as the extent of chemical differentiation between the two phases, and size of various particles. Consequent leaching and sintering processes also affect the chemical composition, morphology, and size of membrane pores. In glass membranes, desalination is accomplished through pores of 50 Å (order of magnitude). This is essentially an ultrafiltration process.

Inherent Advantages and Disadvantages of Glass Membranes

Glass membranes have a few advantages compared with polymer membranes. Glass membranes can be theoretically operated at 500°C and above, which is more than an order of magnitude higher than the operational temperature of present polymer membranes. Glass hollow fibers have shown a compressive strength of 312×10^3 psi [5]. High compressive strength may be a great asset in reverse osmosis operations. Glass membranes enjoy from high chemical stability at low values of pH. Solubility of silica at 95°C is negligible in the presence of $8N$ HNO_3 [6]. Furthermore, glass membranes reject and show high resistance to reactions of bacteria and organic solutions [7,8].

The two disadvantages of glass membranes are fragility of glass and complication in the preparational process. Modern glass science and technology have not yet mastered the required control over glass brittleness, in spite of great strides in this direction. Glass melting still requires a high-temperature (1500°C) process.

Hollow Fiber Glass Membrane

Hollow fiber glass membranes have been successfully tried in reverse osmosis systems for water desalination operations [7–11]. The hollow fibers offer a high ratio of surface membrane per cell volume and enable the application of high pressures in the systems when penetration of solutions is done from outside the hollow fiber membrane through its wall inside. Under these conditions, the glass hollow fiber is brought under compression, a condition which enables the glass to stand the highest stresses.

Hollow fibers are produced either directly from the melt [10] or from a prefabricated tube by redrawing [7,11].

GLASS MEMBRANE PERFORMANCE

The dependence of the porous glass membrane performance on glass porosity is still to a large extent an open question, and a few correlations deserve a close examination. One may be tempted to classify the correlations into several defined parameters such as pore size and structure, physical properties of water in porous glass, chemical charges on glass surface, and chemical interactions of glass surface with various chemical components in the solution. It is, however, clear that such a correlation would be too artificial since all the above parameters are very closely interrelated.

The Role of Boron on the Glass Surface

Elmer et al. [12] indicated that if the porous glass is heated above 500°C, boron is known to migrate to the surface. According to Hair and Hertl [13], the presence of the boron on the surface enhances the reactivity of the silanol groups even more than the silicon. Chapman and Hair [14] have assigned the absorption of ammonia upon porous glass to boron atoms sites following the evacuation and heating to 700°C. On the basis of the above data, one can assume that the boron on the pore surface has an important role in the hyperfiltration of aqueous solutions. We are convinced that in order to elucidate it, one should study the performance of the membrane before being exposed to 500°C, after being exposed to heat treatments in the 500–700°C range, and after a partial sintering of the membrane at higher temperatures (700–900°C) which may homogenize the composition of the membrane material [10].

Hydrophobic Properties of the Glass Membrane

The glass membrane when formed has essentially a hydrophilic surface, covered with at least three types of hydroxy groups distinguishable by infrared absorption techniques. This surface may become hydrophobic following sintering and dehydration of Si-OH groups [15]. It may then appear to us that improvement in rejection of NaCl from about 18% to 80% by sintering the glass membrane [10] may be correlated with increasing the hydrophobic properties of the membrane. This should be further examined and established.

Belfort [16] and Belfort and Scherfig [17] compared the performance of two membrane series, one (Ag-39) with small average pore diameter 24 Å and one (Corning 7930) with average pore diameter 40 Å and found that the rate of salt transport in both membranes was similar, and the rate of water flux through the former membrane was an order of magnitude lower than in the latter. In the membrane with the smaller average pore size, approximately 80 vol % of the porous water (water within the porous glass) is restricted in motion, whereas only a small fraction of the porous volume is occupied by restricted water in large pores. We think that this important finding should be further examined and correlated with the chemical composition of the pore surface (AG-39 glass contains zirconia, and Corning 7930 does not), and its effects on the hydrophobic-hydrophilic properties of the membrane.

Adsorption of cations on to the membrane surface through donation of paired electrons to Lewis acid sites on the glass, as well as adsorption of other ions, may play an important role on the membrane performance. Dugger et al. [18] have shown that as the free energy of the single metal–surface bond becomes more negative, the tendency of the ion to adsorb increases. Glass surfaces which contain ions of high polarizability become hydrophobic [19], and it probably affects the membrane performance especially in terms of solution flux. In some cases, it may explain the increase in life time of the membrane, reported by Ballau et al. [9]. Adams [20] points out that 1 ppm copper in solution decreases the rate of water attack on glass by 20%, and 10 ppm decreases it by 50%.

Phase Separation

Perhaps the most fundamental problem in glass membrane preparation is the thermal treatment (or thermal treatments) which cause the phase separation. One may argue with Varshal [21] if indeed there are four immiscible liquids in the SiO_2–B_2O_3–Na_2O system or not, but the diversity of the metastable phase separation of this system versus composition and temperatures in terms of texture, size of particles, and composition of separated phases is quite established. Insufficient phase separation and subsequent partial leaching is responsible for a low chemical durability and may have been the source of inconsistencies in membrane performance [7,8] reported by various investigators on tests carried out over extended periods of time. It may well be that two cycles of heat treatment and leaching are required in order to establish a (relatively) required membrane structure and composition. Furthermore, few studies have been carried out on the effects of trace elements (such as As_2O_3) on surface tension, phase separation, and nucleation in glass [22] in relation to glass membrane.

Water Interaction with Glass Pores

We suggest that one approach to a systematic study of the role of hydroxy bonds on the membrane performance is their gradual removal from the membrane. Removing of the hydroxy groups by heating the membrane is not sufficient, since subsequent contact of the membrane with water would reversibly return the hydroxy group (quantitatively dependent on the extent of former heating). A permanent removal of the hydroxy groups may be carried out through an absorption of NH_3 on chlorinated glass membrane by $SOCl_2$ [23], or by replacing hydroxy groups first by fluorine atoms using ammonium fluoride solution, and then adsorption of ammonia on the glass surface [14]. An alternative method of removing hydroxy groups is by reaction with silanes [24]. The systematic study may be related both to total concentrations of hydroxy bonds and qualitatively to the various types of hydroxy bonds.

Glass membrane contracts on removing water by heating [15] due to an increase in negative pressure and surface free energy. When water is introduced to the membrane it expands. Polar absorbates cause greater extensions than nonpolar absorbates [25]. Various investigators have shown the far-reaching effects of water interaction with glass pores on properties of both liquids and glass. There is apparently a volume change of 0.3% in the solid [26] and structural deviations from normal in the liquid [27,28], at the contact of glass with water in the pores. Also, the ice obtained in the pores on freezing shows a particular x-ray pattern and a different nucleation behavior from normal [29,30]. Analogies to these have been mentioned in the polymer literature as well [31].

Belfort and Scherfig [17] have further correlated the effects of restricted water on membrane performance. They arrived at the conclusion that salt rejection by a porous glass membrane with very small average pore radii ($\sim$10–40 Å) is most probably a combination of exclusion due to a structural phenomenon of the ordered water and electrokinetic phenomena. For large average pore radii membranes ($\sim$100 Å) that have negligible bound water, rejection of salt is most likely due solely to electrokinetic phenomena. We feel that the above mechanism should be reviewed critically, especially in relation to changes in the extents of solubilities of water and ion–water complexes of the solutions in the restricted and unrestricted water layers. Finally, Vutukuri [32] and others have shown that in wet porous materials, energy balance between attraction and rejection forces of solid particles are affected to a large extent by the dielectric constant of the liquid. We should expect consequent effects also on glass membrane reverse osmosis performance, and a comparison of flux and rejection parameters of water ($\epsilon = 78.3$) to say methanol ($\epsilon = 33.0$) or n-hexane ($\epsilon = 2.0$) should be quite rewarding when pure or proper mixtures are applied under constant conditions.

Hydrocarbons can react with the hydroxy groups on the glass surface via hydrogen bonds [24]. This may be another starting point in the venture of understanding the dependence of membrane performance on various hydroxy bonds on the membrane surface. It may also provide some practical applications.

Shapes of Pores in Glass Membranes

Still another theoretical problem that deserves more attention is the shape or shapes of the pores in the glass membrane. Does it consist of alternate sections of cylindrical pores ($R_c = 28$ Å, $L = 50$ Å) and spherical cavities ($R_s = 2R_c$) as proposed by Basmadjian and Chu [33], or a bottle neck shape of a large body and a small mouth [34] is more reasonable geometry?

It appears to us that one should expect to have a wide variety of pore shapes in a porous glass. The preferred structure results from a spinodal liquid phase separation [2,3] and characterized by a high connectivity network of irregular channels in the glass. Alternatively, a more regular pattern of chains of spherical cavities formed by a "classic" nucleation and growth of separated drops is formed. It has, however, been shown that a transformation between above two systems in both directions is possible [35]. In addition, the improvements observed in membrane performance due to sintering [10] may also be related to changes in pore shapes (with or without related chemical changes on glass surface).

Belfort [16] reached the conclusion that the existence of large random cavities in the membrane is the prime reason for solute leakage and other consequent disadvantages. It seems possible to us that sintering of the membrane may be a partial remedy to this fault.

Prospects of Glass Membrane Performance

Using a 3.5% NaCl solution at 120 bar, product fluxes of about 1.0 $m^3/m^2/$day at 88% rejection were obtained with glass hollow fiber membranes [10]. Phillips et al. report even on a flux of 3.5 $m^3/m^2/$day. Such results are encouraging as they come close to performances of polymer membranes. With the rapid increase in knowledge of various aspects of phase separation and sintering in glass, better results may be expected.

Mechanical Properties

It seems at present that glass brittleness is the main limiting factor in the development of glass membranes. In cylindrical reverse osmosis cells commonly used today [7,10], a fracture in a single fiber may cause an immediate mixing of feed and product liquids. Future field operations will be dependent on techniques that will prevent the occurrence of such mishaps. Two complementary avenues of research may lead to such techniques: minimizing the prospects of fiber fracture and devising new systems in which a failure of a single fiber would not cause termination of the entire process.

Two parallel routes should be explored along the first avenue: (1) increasing glass strength by controlled sintering or chemical strengthening (ion exchange); and (2) proof-test techniques applied to space-shuttle windows [36] may be adopted. By this method, ensuring a minimum safe service life of specimen surviving the proof-test may be guaranteed.

The authors are most grateful to the National Council for Research and Development (Keren Idud Vizum) for financing the project.

REFERENCES

[1] H. P. Hood and M. E. Nordberg, U.S. Pat. 2,106,744 (Feb. 1, 1938).
[2] J. W. Cahn, *Acta Met.*, 9, 795 (1961).
[3] J. W. Cahn and R. J. Charles, *Phys. Chem. Glass*, 6, 181 (1965).
[4] K. Skatulla and K. Kuhne, *Silikattechnik*, 10, 105 (1959).
[5] J. A. Burgman, *Research Into Glass*, Vol. 2, P.P.G. Industries, 1970, pp. 215–218.
[6] T. H. Elmer and M. E. Nordberg, *J. Amer. Ceram. Soc.*, 41, 517 (1958).
[7] F. E. Littman et al., OSW report No. 379 (1968).
[8] F. E. Littman et al., OSW report No. 505 (1959).
[9] E. V. Ballou, T. Wydeven, and M. I. Leban, *Environ. Sci. Technol.*, 6, 1032 (1971).
[10] S. V. Phillips, D. S. Crozier, P. W. McMillan, and J. McC. Taylor, *Desalination*, 14, 209 (1973).
[11] D. Bahat and J. Flicstein, to be published.
[12] T. H. Elmer, I. D. Chapman, and M. E. Nordberg, *Amer. Ceram. Soc.*, 67, 2219 (1963).
[13] M. L. Hair and W. Hertl, *J. Phys. Chem.*, 77, 1965 (1973).
[14] I. D. Chapman and M. L. Hair, *Trans. Faraday Soc.*, 1965, 1507.

[15] T. H. Elmer, I. D. Chapman, and M. E. Nordberg, *J. Phys. Chem.*, *66*, 1517 (1962).

[16] G. Belfort, Ph.D. dissertation, Univ. of California, Irvine (1972).

[17] G. Belfort and J. Scherfig, *Int. Symp. Fresh Water Sea, 3*, 69 (1973).

[18] D. L. Dugger et al., *J. Phys. Chem.*, *68*, 757 (1964).

[19] L. R. Sanders, D. P. Enright, and W. A. Weyl, *J. Appl. Phys.*, *21*, 338 (1950).

[20] p. B. Adams, Dekker, New York, 1972, chap. 14, pp. 293–351.

[21] B. G. Varshal, *Struct. Glass, 8*, 84 (1973).

[22] D. Bahat, *J. Mater. Sci.*, *4*, 847 (1969).

[23] M. Folman, *Trans. Faraday Soc.*, *57*, 2000 (1961).

[24] L. R. Sayder and J. M. Ward, *J. Phys. Chem.*, *70*, 3941 (1966).

[25] W. B. Kipkie and A. McIntosh, *Can. J. Chem.*, *49*, 2352 (1971).

[26] C. Amberg and R. McIntosh, *Can. J. Chem.*, *30*, 1012 (1952).

[27] N. Sheppard and D. J. C. Yates, *Proc. Roy. Soc.* (*London*) *Ser. A, 238*, 69 (1956).

[28] G. Karagounis and O. Z. Peter, *Z. Elektrochem.*, *61*, 827 (1957).

[29] V. S. Brazhan, *Kolloid. Zh.*, *2*, 645 (1969).

[30] G. Karagounis, *Helv. Chem. Acta, 36*, 282, 1681 (1953); *ibid.*, *37*, 805 (1954).

[31] R. D. Schulz and S. L. Asummaa, *Recent Progress in Surface Science*, Vol. 3, Academic Press, New York, 1970, p. 291.

[32] V. S. Votukuri, *J. Rock Mech. Mining Sci. Geomech. Abstr.*, *11*, 27 (1974).

[33] D. Basmadjium and K. P. Chu, *Can. J. Chem.*, *42*, 946 (1964).

[34] L. S. Hersh, *J. Phys. Chem.*, *72*, 2195 (1968).

[35] T. S. Seward et al., *J. Amer. Ceram. Soc.*, *51*, 278 (1968).

[36] S. M. Wiederhorn et al., *J. Amer. Ceram. Soc.*, *57*, 319 (1974).

DIFFERENTIAL DYEING VIA LASER IRRADIATION

MARVIN S. ARONOFF

Israel Fiber Institute, Jerusalem, Israel

SYNOPSIS

Irradiation of nylon, polyester, cotton, wool, and polyester/cotton fabrics with a 10.6 μ laser beam altered dyeing response in the laser-treated zones. Increased or decreased pickup of different dye types is interpreted in terms of structural and chemical changes produced by laser treatment.

INTRODUCTION

Infrared laser radiation has been applied to the processing of textiles and other polymeric substances mainly as a cutting, drilling, and welding tool [1]. The present investigation was undertaken to determine if more subtle changes of potential commercial interest could be produced by laser treatment of polymerics under less drastic conditions.

Since infrared laser radiation concentrated to a small point delivers a high thermal dose, it is reasonable to assume that a polymer exposed to it will be altered structurally and chemically. Such changes should be reflected in the ability of the polymer to pick up different classes of dyes. Thus, dyeing response should give some clue to the changes produced. At the same time, alteration of dyeing response may itself be of some commercial significance.

The present paper reports a brief investigation into the effects of subcutting doses of laser radiation on dye pickup of some synthetic and natural fibers (nylon 66, polyester, cotton, cotton/polyester, and wool).

EXPERIMENTAL

Materials

Knit tubing was prepared from commercially available conventional bulked nylon 66 and polyester. Samples of cotton, 50/50 polyester/cotton, and wool were obtained from undyed bolts of commercially available fabric. All fabrics were scoured by standard procedures prior to laser treatment.

Journal of Applied Polymer Science: Applied Polymer Symposium 31, 397–405 (1977)
© 1977 by John Wiley & Sons, Inc.

TABLE I
Effects of Laser Radiation on Response to Dyes

	Color Intensity Trend in Irradiated Zones[a]		
	Disperse	Acid	Cationic
Nylon			
Energy Dose (j/cm^2)[b]			
2.6 – 3.6	+	–	0
4.3	++	–	0
5.3	+++	–	0
6.8	+++	+++	0
Polyester			
Energy Dose (j/cm^2)[b]			
2.6	+		+[c]
2.8 – 3.1	++		+
3.6 – 6.8	+++		+
Wool[b]			
Energy Dose (j/cm^2)			
2.5 – 4.9	+	0	+
Cotton[b]			
Energy Dose (j/cm^2)	Direct		
2.6 – 8.2 j/cm^2	–		

[a] General color trend for a given dye type on a given fabric: (+) denotes a distinct increase in color intensity relative to unirradiated background; (–) a distinct decrease or white zone; (0) no change.

[b] Power concentration (W/cm^2)/velocity (cm/sec) × spot size (cm). See Experimental section for details.

[c] No significant dye pickup in background.

Laser Treatment

A 40-W, CW, $CO_2/He/N_2$ laser (10.6 μ) operated between 25 and 35 W was used to irradiate the samples. The beam was concentrated by reflecting it from a planar mirror to a spherical mirror of 14 cm focal distance. The beam was focused (or defocused) on the fabric surface to give the desired spot size.

Fabrics were fastened flat against ruled paper backings, and these were taped to the side of a rotating drum whose velocity was controlled by a variable-speed motor-and-belt arrangement. As the drum rotated, the specimen passed through the laser beam creating a laser-exposed linear zone. The fabric was then moved up $\frac{1}{2}$–1 cm and the procedure repeated at the same or a different velocity until it was exposed to a series of different energy doses. Output power of the laser was monitored before and after each sweep by deflecting the primary beam into a Coherent Radiation model 201 thermopile by means of a 45° angle planar mirror.

Conditions used were as follows. Nylon: power concentration, 970–1620 W/cm^2; velocity, 36–70 cm/sec; spot sizes, 1.5 and 1.9 mm. Polyester: power concentration, 950–1025 W/cm^2; velocity, 28–70 cm/sec; spot size, 1.9 mm. Wool: power concentration, 875–940 W/cm^2; velocity, 38–70 cm/sec; spot size, 2.0 mm. Cotton, polyester/cotton (50/50): power concentration, 890–960 W/cm^2; velocity, 22–70 cm/sec; spot size, 2.0 mm.

Dyeing Procedures

All fabrics were scoured by standard procedures prior to dyeing. Specific dyes used according to class were as follows: (1) disperse—Cibacet Yellow 2 GC, Polanil Yellow 3 G, Polanil Marine Blue RE; (2) acid—Alizarin Fast Blue 2 BG, Benzyl Fast Red GRG; (3) direct—Chlorantine Fast Blue 2 BLL; and cationic—Astrazon Blue BG, Astrazon Red F 3 RL.

In general, all fabrics were dyed at 100°C for 75 min with any given dye. Fabric weight to volume ratios were 1:100–1:200. Otherwise, standard procedures were used. Fabrics were postscoured at 70–80°C for 30 min using standard procedures.

Nylon was dyed with disperse and acid dyes in one bath using the following formulation: fabric, 1.2 g; H_2O, 108 cc; Ultravon JU 5%/1.25 cc; Invadin BL 5%/0.23 cc; $(NH_4)_2SO_4$, 10%/0.35 cc; HOAc, 30%/0.44 cc; Cibacet Yellow 2 GC, 1%/3.6 cc; Alizarin Fast Blue 2 BG, 1%/3.6 cc.

Polyester was dyed with disperse and cationic dyes in one bath using the following formulation: fabric weight, 0.5 g; H_2O, 114 cc; Ultravon JU, 20%/1 cc; Na_2SO_4, 10%/0.5 cc; HOAc, 30%/0.2 cc; Cibacet Yellow 2 GC, 1%/1.5 cc; Astrazon Blue BG, 1%/1.5 cc.

RESULTS AND DISCUSSION

Nylon 66

Nylon 66 knit tubing exposed to doses of 10.6 μ laser radiation exhibited increased dye uptake or resist in the laser-treated zones (Fig. 1). Dyeing response was dependent on the type of dye used and the amount of energy delivered to the laser treated zone (Table I). Uptake of disperse dyes in the irradiated zones, as judged visually, increased with increasing amounts of energy delivered to the specimen (i.e., the dose). Cibacet Yellow disperse dye imparted a bright yellow color to the untreated background, but appeared more brown in the laser-treated zones.

Response to acid dyes was completely different (Table I). Here the irradiated zones picked up less dye than the background at lower energy doses, while at higher doses they picked up much more dye, judged by visual color intensity. Thus, with Alizarin Blue, the untreated background was blue in color, the zones irradiated at lower energies were almost white, while at higher energies there was a reversal with these zones appearing blue-black. The same pattern of behavior was found with Benzyl Fast Red acid dye.

Astrazon Blue cationic dye did not stain the untreated fabric to any great extent. Laser treatment produced virtually no change in this response. However, fabric which had been soaked with glacial acetic acid just prior to irradiation did pick up cationic dye in all of the laser-treated zones.

The differing responses of the laser-treated zones to different dye types suggest the different kinds of changes produced in nylon by laser radiation. Reduction of orientation and crystallinity changes, i.e., a more "open" fiber structure, can be expected from the thermal input of the laser. Consistent with this picture is

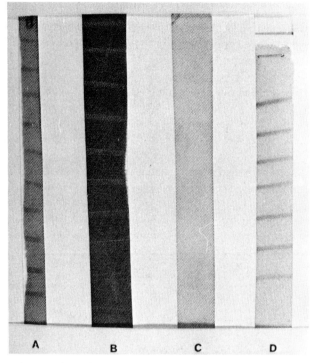

FIG. 1. Effect of laser radiation on dye pickup of nylon 66: (A) disperse dye; (B) acid dye; (C) cationic dye; (D) cationic dye, HOAc-treated before laser irradiation. Energy dose increases from top to bottom.

the observation that disperse dye pickup in the laser-treated zones is greater than that of the untreated zones and increases with exposure to greater energy doses. If the only result of laser treatment were to produce a "more open" fiber structure, a similar result could be expected from acid dyes; the effect of the laser treatment serves to "open" the structure, making more $-NH_2$ ends available for interaction. Resist to acid dye at lower energy doses shows that this cannot be the only effect produced by laser irradiation.

A reasonable explanation for resist to acid dye is thermal induced oxidation of the $-NH_2$ ends.

In contrast to this, zones exposed to higher energy doses picked up more acid dye than the unexposed zones indicating the availability of $-NH_2$ ends. Processes which become important only at higher energy doses (i.e., higher temperatures) may counterbalance destruction of $-NH_2$ ends by generating new ones. Hydrolysis of amide bonds and cyclization of adipoyl groups (an initial stage in gelation) are examples of such processes [2, 3]. Hydrolysis is probably unimportant in this case: higher energy doses did not increase pickup of cationic dye. Thus, there was no significant increase in $-CO_2H$ ends which would result from hydrolysis.

These results show that for nylon, response to acid dye can be controlled by the energy dose.

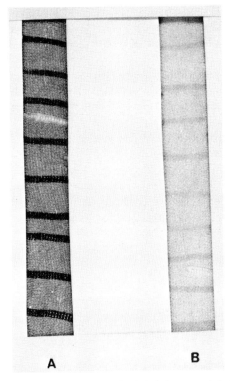

A   **B**

FIG. 2. Effect of laser radiation on dye pickup of poly(ethylene terephthalate): (A) disperse dye; (B) cationic dye. Energy dose increases from top to bottom.

Pickup of cationic dye in all zones in the case of fabric pretreated with acetic acid indicates generation of carboxyl groups. They may result from hydrolysis or transamidation driven by the thermal input of the laser.

Polyester

Laser treatment of knit tubing made of conventional polyester increased dye uptake in the laser treated zones (Fig. 2). With disperse dyes (Cibacet Yellow 2 GC, Polanil Yellow 3 G, Polanil Marine Blue RE), color intensity increased with the energy dose (Table I). This was especially apparent when zones exposed to the lower and higher doses were compared. Cibacet Yellow 2 GC dyed the untreated background bright yellow, while the laser-treated zones had a much deeper shade which increased with increasing dose and stood out clearly against the background. Polanil Marine Blue RE gave even more dramatic results: the untreated background yielded a light blue against which stood the dark blue stripes of the laser-treated zones, increasing to almost black with increasing energy dose (Table I: at the highest dose, the fabric was partially cut).

The response to disperse dyes is similar to that observed with nylon and can also be explained on the basis of reduced orientation and crystallinity changes. It is well known that heat-setting temperature markedly affects the response of polyester fiber to disperse dyes [4, 5]. As heat-set temperature increases to 160°C, dye uptake decreases; from 160–200°C it remains at a minimum, while above 200°C it increases and exceeds the dye capacity of the unheat-set fiber. This behavior has been ascribed to changes in orientation and crystalline structure of the polyester fiber.

Some localized melting is produced by laser radiation even at low energy doses. The response in the laser-treated zone should therefore be analogous to heat-setting response above 200°C.

Conventional polyesters lack sufficient functional groups to act as dye sites and are therefore dyed almost exclusively with disperse dyes. It is therefore of interest that some cationic dye (Astrazon Blue BG) was picked up by the laser-treated zones, while there was virtually no staining of the untreated background. This suggests that $-CO_2H$ groups may have been produced by laser irradiation. Such groups could result from hydrolytic, oxidative, or pyrolytic degradation [2, 6–8].

Cotton

Laser treatment of cotton produced resist to direct dye at all energy doses (Table 1). A pattern of white stripes (the laser-treated zones) whose presence became more pronounced with increasing energy dose was obtained on the normally dyed background (Fig. 3). The usual lack of affinity for acid, cationic,

FIG. 3. Resist to direct dye produced in cotton by laser irradiation. Energy dose increases from top to bottom.

and disperse dyes was unaltered. Resist to direct dye may result from pyrolytic cross-linking of the cellulose molecule as proposed by Basch and Lewin [9]. This would reduce both the number of available hydroxyl groups and accessibility to the lower ordered regions promoting resist to direct dyes.

Wool

The affinity of wool for acid dye was virtually unchanged by irradiation, while the affinities for both cationic and disperse dyes increased at all energy doses relative to untreated background (Table I).

At all dose levels, apparently no significant change in the number of sites available for acid dye was produced. This contrasts with the results found for nylon where resist was produced at the dose levels used here (Table I). Either destruction of dye sites ($-NH_2$ groups) is minimal or is counterbalanced by generation of new ones; those destroyed are an insignificant fraction of those available for acid dye pickup.

The color intensity patterns of the laser-treated zones was about the same with both disperse and cationic dye, increasing somewhat with increasing energy dose. Possibly, a more "open" structure was produced by laser treatment, increasing pickup of both cationic and disperse dyes.

Color and Shade Effects

Patterns differing from the background in color as well as shade can be obtained by taking advantage of altered dyeing behavior in the laser-treated zone. For nylon, this was done by dyeing in a single bath containing acid and disperse dyes of different colors. The background color resulted from a combination of the two colors; stripes from exposure to low energy doses had the disperse color (resist to acid dye), while those from high energy doses had the combination color in deeper shade.

For polyester, a similar procedure was used with cationic and disperse dyes of different colors. The background color resulted from the disperse dye, while the laser-treated zones gave the combination color.

A polyester/cotton (50/50) blend dyed with direct and disperse dyes of different colors in one bath gave good color differentiation between irradiated zones and background. Here the background had the combination color, while the laser-treated zones gave the disperse color. This was a result of the increased affinity of polyester for disperse dye combining with the resist of cotton to direct dye. These irradiation induced effects tended to limit dye pickup to the disperse dye in the treated zones. The effect was enhanced at high energy doses by localized melting of polyester which blocked the cotton.

Color and shade effects are summarized in Table II.

TABLE II

Color and Shade Effects Produced by Laser Treatment

Fiber Type	Dye Type[a]	Dye	Conc. (%)	Background Color	Laser Treated Zone (Stripe) Color	
					Low Energy Dose	High Energy Dose
Nylon 66	A	Alizarin Blue	3.0	Blue	White	Black - Blue
	A	Benzyl Fast Red	0.5	Orange-Red	White	Dark Orange-Red
	D	Cibacet Yellow	3.0	Yellow	Dark Yellow	Brown-Yellow
	A + D +	Alizarin Blue Cibacet Yellow	3.0 3.0 (0.3)	Dark Green (Light Green)	Yellow (Yellow)	Black (Middle Green)
Polyester	D	Polanil Marine Blue RE	3.0	Light Blue	Dark Blue	Black Blue
	D + C +	Cibacet Yellow Astrazon Blue BG	3.0	Yellow	Green	Green
Polyester-Cotton (50/50)	DR	Chlorantine Fast Blue 2 BLL	5.0	Blue	Light Blue	White
	D	Polanil Yell. 3G	5.0	Light Yellow	Medium Yellow	Dark Yellow
	DR + D +	Chlorantine Blue Polanil Yell.	5.0 5.0	Green	Green Yellow	Yellow

[a] A = acid dye, C = cationic dye, D = direct dye, DR = direct dye.

CONCLUSIONS

Irradiation of nylon, polyester, cotton, and wool fabrics with relatively low doses of infrared laser radiation altered their response to various dye types. These changes suggested that structural and chemical changes were produced in the polymeric substrates.

These phenomena might be used for a kind of fabric printing based on the altered receptivity to certain dye types of discrete zones in a given laser-treated fabric. Thus, patterns could be traced on a fabric by a laser under computer control. The fabric containing sensitized or desensitized zones could then be dyed to develop the pattern. Random dye effects in yarns might be produced by exposing a moving thread line to laser pulses of varying duration. Dyeing yarns or fabrics made from them will develop the pattern. (A recent patent discloses the use of laser radiation to produce random printed patterns by rapidly evaporating dye solutions from substrate fibers [10].) Further work is needed to gain an understanding of the interaction of laser radiation with polymers in order to apply it to such processes.

The author wishes to express his appreciation to Dr. Joseph Banbaji and Mr. Shlomo Hönig for their assistance and suggestions.

REFERENCES

[1] J. M. Murray, *J. Apparel Res. Found., 5,* 1 (1971).
[2] *Soc. Chem. Ind.* (London) *Monograph, 13,* (1961).
[3] W. Sweeny, *Kirk-Othmer Encyclopedia of Chemical Technology* (2nd ed.), Wiley, 1968, vol. 16, p. 11.
[4] F. Galil, *Tex. Res. J., 43,* 615 (1973).

[5] J. Andriessen and J. van Soest, *Textilveredlung, 3,* 618 (1968).

[6] R. J. P. Allan, R. L. Forman, and P. D. Ritchie, *J. Chem. Soc., 1956,* 3563.

[7] A. Mifune, S. Ishida, A. Kobayashi, and S. Sakajiri, *Kogyo Kagaku Zasshi, 65,* 992 (1962).

[8] R. B. Rashbrook and G. W. Taylor, *Chem. Ind. (London),* 1962, 215.

[9] A. Basch and M. Lewin, *J. Polym. Sci. Polym. Chem. Ed., 12,* 2053 (1974).

[10] T. Hiramatsu et al., Japan. Pat. 74 93, 681.

ASYMMETRIC CELLULOSE ACETATE FIBERS FOR THE DIALYTIC SEPARATION OF UREA AND SALT

GERALD TANNY*

Department of Plastics Research, The Weizmann Institute of Science, Rehovot, Israel

SYNOPSIS

A study of the dialytic separation of urea and salt was carried out utilizing both asymmetric cellulose acetate flat-sheet membrane and hollow fibers. The ratio of the urea/salt permeability, depending on the temperature of the membrane heat treatment, was between 19 and 43, which is comparable to published values for homogeneous Dow cellulose acetate hollow fiber. However, as a simple calculation implied, the absolute permeabilities could be significantly enhanced in an asymmetric fiber whose porous structure contained a large number of "finger-like" intrusions. An attempt was made to reversibly dry the asymmetric CA hollow fiber by prior immersion in solutions of glycerol and Triton X-100. These solutions failed to preserve the wet fiber permeability.

INTRODUCTION

One of the major problems facing current efforts to produce a wearable or at least highly portable artificial kidney is that of removing urea from the 1–2 liters of dialyzate which are recycled through the hemodialyzer. This dialyzate solution typically contains 0.13 M Na$^+$, 10^{-3} M K$^+$, 1.5×10^{-3} M Ca^{++}, and about 0.12 M Cl$^-$, so that part of the problem is to find a removal method which does not affect the required salt balance in the patient. A technique which removes all ions and the urea has been the subject of a patent [1].

One suggestion proposed which utilizes a hollow fiber membrane [2] is shown in Figure 1. The dialyzate is circulated through a cellulose acetate hollow fiber unit, which allows the urea but not the salt to dialyze across the fiber wall into the second loop. The latter contains an enzyme cartridge which breaks the urea down into ammonium carbonate, which is then taken up by a second cartridge containing ion-exchange resin. In the scheme proposed, dense hollow fibers with a 15 μ wall thickness were to be used. Since it seemed reasonable to believe that asymmetric hollow fibers or tubules of cellulose acetate could be produced with a larger urea flux and equivalent salt selectivity, the present study was undertaken. This contribution represents a preliminary report of the work completed to date.

* Present address: c/o Gelman Instrument Co., 600 South Wagner Road, Ann Arbor, Mich. 48106.

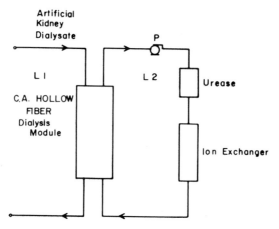

FIG. 1. Schematic arrangement for the use of a CA hollow fiber module in purifying the dialyzate of an artificial kidney: (L1) kidney dialyzate loop; (L2) dialyzate purification loop; (P) pump.

THEORY

For a dialytic process in which there is no flow of solvent (i.e., the osmotic force on the solvent has been balanced), the transmembrane flux of any species i from the sample solution to the dialyzate is given by

$$J_i = \omega_i \Delta \pi_i \tag{1}$$

where J_i is the flux of species i, $\Delta \pi_i$ is the osmotic force on species i, and ω_i is the diffusivity.

Equation (1) may also be conveniently written as

$$J_i = D_i \Delta C_i = D_i(C' - C_D) \tag{2}$$

where D_i possesses the units cm/min, C' is the sample concentration, and C_D the dialyzate concentration. For two circulating loops of constant volume, the mass balance is simply given by

$$J_i A = V(dC_D/dt) \tag{3}$$

where A is the specific membrane area and V the volume of the dialyzate compartment. Thus,

$$D_i A(C' - C_D) = V(dC_D/dt) \tag{4}$$

i.e.,

$$\ln\left[\frac{C_0' - (C_D)_0}{C_t' - (C_D)_t}\right] = \frac{D_i At}{V} \tag{5}$$

Plots of

$$\ln\left[\frac{C_0' - (C_D)_0}{C_t' - (C_D)_t}\right]$$

versus t can thus be used to obtain D_i.

EXPERIMENTAL

Membranes Tested

Thin-Film Composite Cellulose Triacetate (CTA)

This membrane consists of an 800 Å thick film of CTA resting on a porous substructure. It is produced by the ROGA Systems Division of Universal Oil Products (sample generously provided by Dr. R. Riley).

Asymmetric Cellulose Acetate (CA)

As-cast, nonheat-treated, asymmetric CA was obtained in tubular form (4 m × 28 mm i.d.) from Dr. B. Sachs of Israel Desalination Engineering (Tel-Baruch, Israel). Sections cut from the tube were subjected to heat treatment in water at 92, 95, and 97°C for up to 1 hr time in order to produce membranes with varying degrees of selectivity to salt and urea. Individual sections tested were first tested for salt rejection in order to establish that the membrane did not possess any gross leaks or flaws.

Membrane Testing

Membrane sections were tested in stirred Perspex cells which contained 43 cm^3 in each half-cell and exposed 10 cm^2 of surface area. The tests were run with a solution containing 100 mg % urea in 0.1 N NaCl versus 0.2 M sucrose. Samples (2 ml) for urea analysis (which also served for chloride analysis) were taken every ½ hr for 5 hr, with one sample taken after approximately 24 hr. The samples were replaced by 2 cc of 0.2 M sucrose solution and the changes in concentration due to sampling were taken into account when computing the concentration of urea and chloride in the compartment containing the sucrose. All membranes were equilibrated in 500 cc of a solution of 50 mg % urea in 0.1 N NaCl for 24 hr before they were tested.

Sodium chloride analyses were carried out by analyzing for chloride with an Aminco–Cotlove chloride titrimeter (American Instrument Co., Silver Springs, Md.). Urea analyses were performed colorimetrically with a Technicon auto analyzer, which had been calibrated with a series of urea standards immediately prior and subsequent to the urea samples.

Preparation of CA Hollow Fibers

The extrusion of CA hollow fibers was carried out on the simple apparatus shown in Figure 2. The extruder consists of a variable-speed syringe pump (A), a thermostated extruder head (B) containing the usual double needle arrangement for the introduction of either air pressure or nonsolvent to form the lumen of the hollow fiber, and a variable-speed rotating bucket (C) which serves as a windup and coagulation bath. The coagulant was at first delivered to the central needle by means of a peristaltic pump. However, the slight pulsations caused

FIG. 2. Hollow fiber extrusion apparatus: (A) variable-speed syringe pump; (B) double needle spinneret; (C) rotating coagulation bath.

nonhomogeneity in the fiber diameter and a syringe pump of appropriate capacity was substituted.

The solution typically used for extrusion was 25 g Eastman 398-3 cellulose acetate, 45 g acetone and 30 g formamide. The viscosity of the solution, which was measured on a Haake rotating viscometer prior to each extrusion, was 1700 ± 50 cps at 30°C.

Subsequent to extrusion the fibers were usually leached in water overnight, hand-wound onto glass spools, and heat-treated at 85°C for 1 hr. The ends of the fiber were then potted in either silicone rubber (GE, RTV-102) or polyester (Makhteshim, Israel).

Testing of CA Hollow Fibers

Potted fiber elements were first tested for leaks by circulating a dilute solution of Biebrich scarlet dye within the fiber. After the fiber was washed free of the dye, it was placed in a reservoir containing approximately 250 cc of 16 mM urea and 0.1 N NaCl (solution U), which was magnetically stirred. A known volume

TABLE I

Urea and Sodium Chloride Transport through Flat-Sheet Asymmetric CA Membranes

Membrane	Cure Temperature (°C)	Membrane Thickness (μ)	Urea Permeability (cm/min)x10^3	Sodium Chloride Permeability (cm/min)x10^4	D_{urea}/D_{NaCl}
C.A.	92	100	7.5	3.9	19
C.A.	95	100	7.4	3.3	22
C.A.	97	110	5.6	1.3	43
C.A.-Dow [a]	-	30	2-3	0.7	30-40

[a] Values from Sargent et al. [2].

of 0.2 M sucrose ($\sim$50–75 ml) was used as the dialyzate (solution S), and 3-ml samples of S and U were taken periodically for analysis. When the data were plotted, correction for the volume removed in sampling was made by treating the ratio V/A as a variable of time and correcting the left-hand side of eq. (5).

The S solution was circulated through the core of the hollow fibers by a peristaltic pump with a rate of 5–10 cc/min. Samples 5W and 6aW were single fibers, while 5-D and 7-W were five tubule bundles.

RESULTS AND DISCUSSION

Urea and NaCl Transport in Flat-Sheet Asymmetric CA

To establish an experimental basis of comparison between an asymmetric CA membrane and homogeneous Dow cellulose acetate hollow fibers, urea and salt transport were measured through sections of a tubular R.O. cellulose acetate membrane which were given heat treatments at several temperatures (Table I). As expected, the selectivity ratio of the membrane D_{urea}/D_{NaCl} increased at elevated curing temperatures, while the absolute urea flux is decreased. Nonetheless, the urea permeability of a 100 μ asymmetric membrane is significantly higher than that of a 30 μ homogeneous fiber. In principle, it should be possible to obtain even higher values, as the following calculation demonstrates.

Consider the cellulose acetate membrane to consist of two membranes in series, a dense selective layer A 0.1 μ thick and a porous sponge B $\sim$100 μ thick. The permeability is given by*

$$D_i = \overline{\mathcal{D}}K/\Delta x \qquad (6)$$

where $\overline{\mathcal{D}}$ is the membrane diffusion coefficient, K is the partition coefficient, and Δx the thickness. Utilizing $\overline{\mathcal{D}} = 1.3 \times 10^{-8}$ cm^2/sec and $K = 0.49$ for urea in CA [4], we obtain $D_i^A = 3.8 \times 10^{-2}$ cm/min for the dense layer. For the porous sponge, $\overline{\mathcal{D}} \sim 10^{-5}$ cm^2/sec, and $K = 1$, so that $D_i^B = 6 \times 10^{-2}$ cm/min.

$$\frac{1}{D_i} = \frac{1}{D_i^A} + \frac{1}{D_i^B} \qquad (7)$$

* Ref. [3], eqs. (3) and (4).

Thus, $D_i = 2.3 \times 10^{-2}$ cm/min, which is several times the value actually measured. Of course, the calculation has assumed zero tortuosity for the porous sponge, which cannot be correct. However, the exercise is instructive for two reasons: it shows where the probable maximum lies and it suggests that in changing casting variables one should move toward those conditions yielding what has been termed [5] "finger formation." These are large voids which can extend over almost the entire thickness of the membrane and are usually considered undesirable in membranes which must withstand pressure.

The effect of the acetyl content in CA was determined by measurements of urea and salt transport through an ultrathin film composite of cellulose triacetate (CTA) on a CN–CA support. The flux of both species was beneath the detectable limits of our analytical technique even over 24-hr time periods. Since these membranes display 99.8% rejection in reverse osmosis, this was expected for the salt flux. However, the urea flux is surprisingly low if one considers that the diffusive water permeability [6] of the CTA is only a factor of 2.5 smaller than that of cellulose acetate (39.8% acetyl). The strong dependence on the acetyl content suggests that urea permeability might be enhanced at lower acetyl contents, without substantially changing the ratio of D_{urea}/D_{NaCl}; further work in this direction is presently under way.

Performance of CA Tubules

The most successful fibers were spun with a core coagulant of water at room temperature and an external leach bath at 0–5°C in the rotating bucket. Biebrich scarlet dye penetration tests showed that after heat treatment the selective layer was on the outer surface of the fiber.

Values of the urea and salt permeability of potted tubules were also obtained from plots of $\ln [C_0'/C_t' - (C_D)_t]$ versus t, as shown for example in Figure 3.

The results of a number of preliminary extrusion experiments are given in Table II. We may note that in one instance (sample 6a-W), the urea dialyzance can reach values as large as 1.05×10^{-2} cm/min and maintain a ratio of

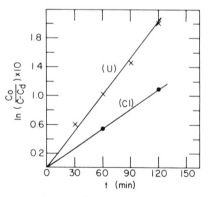

FIG. 3. Measurement of urea (U) and sodium chloride (Cl) flux for sample CA 7-W (Table II).

TABLE II
Urea and Sodium Chloride Transport through CA Hollow Tubules

Sample #	Tubule Diameter (cm)	Area (cm^2)	D (cm/min) Urea	D (cm/min) NaCl
5W	0.061	7.3	3.48×10^{-3}	1.8×10^{-3}
5-D	0.05	32.6	3.83×10^{-4}	---
6 a-W	0.06	5.7	1.05×10^{-2}	5.75×10^{-4}
7-W	0.07	26.4	3.26×10^{-3}	1.72×10^{-3}

$D_{\text{urea}}/D_{\text{NaCl}} = 18$. While results of this nature are not totally reproducible due to a combination of factors in our present equipment, they are encouraging.

The potting of wet cellulose acetate presents a major problem because of the lack of adhesion between most potting agents and the wet cellulose acetate, and the need to keep the fibers wet while the potting agent is cured. For tests on single tubules, it was possible to dry the ends of the tubule while preventing the middle from dehydrating. However, for larger modules this is impractical, because tubes with dried ends tend to crack at the boundary between wet and dry CA, thus rendering the tubules extremely difficult to handle. For this reason, an attempt was made to dry the tubules by first immersing them in a bath of 66:4:30 (w/w) water/Triton X-100/glycerol for 15–30 min followed by air-drying, a procedure used to produce dry CA reverse osmosis membranes [7]. The results (sample 5-D, Table II) indicate that the tubule's porous structure underwent severe collapse, thereby lowering the urea permeability by two orders of magnitude. Since the handleability of the tubule is considerably improved when dry, further work is under way to find a successful drying technique.

Note added in proof: Thanks in part to the advice of Dr. E. Klein of the Gulf South Research Institute, the problem of potting the fiber bundles without leaks was solved and values of D for urea and NaCl approached those of Table I for CA 92. The fibers were not dried but were exchanged with glycerine which was forced (with the aid of a fine needle) down the fiber core prior to potting.

This work was carried out with the financial support of the David Rose Foundation. The author gratefully acknowledges the encouragement and advice of Prof. O. Kedem and Dr. T. Bohak, and the technical assistance for various portions of the work afforded by E. Kazes and Mr. Y. Bodell.

REFERENCES

[1] L. B. Morantz and M. A. Greenbaum, Ger. Offen. 1,960,504; *Chem. Abstr., 73,* P63135a.
[2] J. A. Sargent, F. A. Gotch, R. S. Olson, M. C. Evans, L. Sturkey, J. W. Herr, M. Keen, and M. Seid, Contract No. AK-2-70-2302, 2nd Annual Progress Report, National Institute of Arthritis, Metabolism and digestive Diseases, N.I.H., Bethesda, Md.
[3] O. Kedem and A. Katchalsky, *J. Gen. Phys., 45,* 143 (1961).
[4] J. E. Anderson, S. J. Hoffman, and C. R. Peters, *J. Phys. Chem., 76,* 4006 (1972).
[5] R. Matz, *Desalination, 10,* 1 (1972).
[6] R. Riley, Universal Oil Products, private communication.
[7] K. D. Vos and F. O. Burris, R & D Progress Report No. 348, Office of Saline Water, U.S. Dept. of Interior, Washington, D.C.

Author Index

Aronoff, M. S., 397

Backer, S., 63
Bahat, D., 389
Banbaji, J., 117
Barboy, R., 351
Bhat, G. R., 55
Bittencourt, E., 201
Brookstein, D., 63

Charit, Y., 351

Deshpande, A. B., 257

Economy, J., 23

Flicstein, J., 389
Forgacs, C., 383

Grosberg, P., 83
Gutman, B., 133
Guttmann, H., 163

Hearle, J. W. S., 137
Hersh, S. P., 37, 55
Ho, K. H., 83
Hoffman, A. S., 313

Katz, I., 193
Kedem, O., 383
King, R. N., 335
Klein, E., 361

Lewin, M., 163, 193
Liepens, R., 201, 257
Lopatin, G., 127

McKenna, G. B., 335
Messalem, R., 383
Michael, J., 383
Mocherla, K. K. R., 183

Pearce, E. M., 257
Perel, J., 133

Ross, J. H., 293

Sello, S. B., 229
Shabtai, D., 163
Shiloh, M., 105
Shimshoni, M., 133
Stannett, V., 201
Statton, W. O., 183, 335
Stenzenberger, H. D., 91
Surles, J. R., 201

Tanny, G., 407

Van Krevelen, D. W., 269
Veldsman, D. P., 251

Walsh, W. K., 201
Wilfong, R. E., 1

Yoon, H. S., 257

Zimmerman, J., 1

Published Applied Polymer Symposia

1965   No. 1   High Speed Testing, Vol. V
                Co-Chaired by A. G. H. Dietz and Frederick R. Eirich
1966   No. 2   Thermoanalysis of Fibers and Fiber-Forming Polymers
                Edited by Robert F. Schwenker, Jr.
        No. 3   Structural Adhesives Bonding
                Edited by Michael J. Bodnar
1967   No. 4   Weatherability of Plastic Materials
                Edited by Musa R. Kamal
        No. 5   High Speed Testing, Vol. VI: The Rheology of Solids
                Co-Chaired by Rodney D. Andrews, Jr. and Frederick R. Eirich
        No. 6   Fiber Spinning and Drawing
                Edited by Myron J. Coplan
1968   No. 7   Polymer Modification of Rubbers and Plastics
                Edited by Henno Keskkula
1969   No. 8   International Symposium on Polymer Modification
                Edited by K. A. Boni and F. A. Sliemers
        No. 9   High Temperature Resistant Fibers from Organic Polymers
                Edited by J. Preston
        No. 10 Analysis and Characterization of Coatings and Plastics
                Edited by Claude A. Lucchesi
        No. 11 New Polymeric Material
                Edited by Paul F. Bruins
        No. 12 High Speed Testing, Vol. III: The Rheology of Solids
                Co-Chaired by Rodney D. Andrews, Jr. and Frederick R. Eirich
1970   No. 13 Membranes from Cellulose and Cellulose Derivatives
                Edited by Albin F. Turbak
        No. 14 Silicone Technology
                Edited by Paul F. Bruins
        No. 15 Polyblends and Composites
                Edited by Paul F. Bruins
1971   No. 16 Scanning Electron Microscopy of Polymers and Coatings
                Edited by L. H. Princen
        No. 17 Mechanical Performance and Design in Polymers
                Edited by O. Delatycki
        No. 18 Proceedings of the Fourth International
                Wool Textile Research Conference
                Edited by Ludwig Rebenfeld
1972   No. 19 Processing for Adhesives Bonded Structures
                Edited by Michael J. Bodnar
1973   No. 20 United States—Japan Seminar on Polymer Processing and Rheology
                Edited by D. C. Bogue, M. Yamamoto, and J. L. White
        No. 21 High-Temperature and Flame-Resistant Fibers
                Edited by J. Preston and J. Economy
        No. 22 Polymeric Materials for Unusual Service Conditions
                Edited by Morton A. Golub and John A. Parker
1974   No. 23 Scanning Electron Microscopy of Polymers and Coatings. II
                Edited by L. H. Princen
        No. 24 Technological Aspects of Mechanical Behavior of Polymers
                (Witco Award Symposium)
                Edited by R. F. Boyer

                All of the above symposia may be individually purchased from
                the Subscription Department, John Wiley & Sons, Inc.